STUDENT'S SOLUTIONS AND STUDY GUIDE

TO ACCOMPANY
TOPICS IN CONTEMPORARY MATHEMATICS

SIXTH EDITION

Ignacio Bello

Hillsborough Community College

Jack R. Britton

University of South Florida

HOUGHTON MIFFLIN COMPANY   BOSTON   NEW YORK

Editor in Chief: Charles Hartford
Associate Editor: Elaine Page
Manufacturing Coordinator: Florence Cadran
Marketing Manager: Charles Cavaliere

Copyright © 1997 by Houghton Mifflin Company. All rights reserved.

No part of this work may be reproduced or transmitted in any form or by any means, electronic or mechanical, including photocopying and recording, or by any information storage or retrieval system without the prior written permission of Houghton Mifflin Company unless such copying is expressly permitted by federal copyright law. Address inquiries to College Permissions, Houghton Mifflin Company, 222 Berkeley Street, Boston, MA 02116-3764.

Printed in the U.S.A.

ISBN: 0-669-41786-6

123456789-CRS-00 99 98 97 96

PREFACE

This **Student's Solutions and Study Guide** has been especially prepared to accompany *Topics in Contemporary Mathematics,* Sixth Edition by Ignacio Bello and Jack Britton. It has been created and designed with the student in mind and should be used as an extra tool to aid understanding of the material in the text. The Guide is composed of the following parts:

Part I provides study tips and solutions to the odd numbered exercises appearing in each section of the text. The study tips clearly outline the ideas that should become your learning objectives as you study each section. They also include helpful hints for understanding the more complicated mathematical ideas you are studying. Solutions to **all** Practice Tests given at the end of each chapter in the textbook are provided.
Many times students need additional practice with these tests, so we have included extra Practice Tests in this Guide.

Part II includes **two** additional Practice Tests per chapter: a multiple choice and a fill-in-the-blank, free response version. The questions in these tests are intended for complete mastery and for that reason **do not** follow the same order as the Practice Tests in the book. You can use these tests to review further so you can be thoroughly prepared before actual exams are administered by your instructor.

Part II contains the answers to all additional Practice Tests.

Good luck with your studies and your course!

CONTENTS

Part I	Study Tips; Odd-Numbered Problem Solutions; Solutions of Textbook Practice Tests	1
Part II	Additional Practice Tests and Answers	333
Part III	Answers to All Additional Practice Tests	430

PART 1 STUDY TIPS; ODD-NUMBERED PROBLEM SOLUTIONS; SOLUTIONS OF TEXTBOOK PRACTICE TESTS

Chapter 1 Sets and Problem Solving

Exercise 1.1 1
Exercise 1.2 4
Exercise 1.3 7
Exercise 1.4 10
Exercise 1.5 15
Exercise 1.6 19
Practice Test 1 21

Chapter 2 Logic

Exercise 2.1 27
Exercise 2.2 29
Exercise 2.3 34
Exercise 2.4 38
Exercise 2.5 41
Exercise 2.6 47
Exercise 2.7 52
Practice Test 2 55

Chapter 3 Numeration Systems

Exercise 3.1 60
Exercise 3.2 64
Exercise 3.3 66
Exercise 3.4 69
Practice Test 3 71

Chapter 4 Number Theory and the Real Numbers

Exercise 4.1 74
Exercise 4.2 79
Exercise 4.3 81
Exercise 4.4 85
Exercise 4.5 88
Exercise 4.6 90
Exercise 4.7 92
Exercise 4.8 94
Practice Test 4 98

Chapter 5 — Equations, Inequalities and Problem Solving

Exercise 5.1 103
Exercise 5.2 110
Exercise 5.3 115
Exercise 5.4 118
Exercise 5.5 123
Exercise 5.6 129
Practice Test 5 132

Chapter 6 — Functions and Graphs

Exercise 6.1 139
Exercise 6.2 146
Exercise 6.3 151
Exercise 6.4 156
Exercise 6.5 162
Exercise 6.6 169
Practice Test 6 178

Chapter 7 — Geometry

Exercise 7.1 185
Exercise 7.2 188
Exercise 7.3 193
Exercise 7.4 195
Exercise 7.5 200
Exercise 7.6 204
Exercise 7.7 206
Practice Test 7 208

Chapter 8 — Mathematical Systems and Matrices

Exercise 8.1 213
Exercise 8.2 222
Exercise 8.3 227
Exercise 8.4 231
Exercise 8.5 235
Practice Test 8 239

Chapter 9   **Counting Techniques**

    Exercise 9.1 248
    Exercise 9.2 249
    Exercise 9.3 251
    Exercise 9.4 253
    Practice Test 9 255

Chapter 10   **Probability**

    Exercise 10.1 258
    Exercise 10.2 261
    Exercise 10.3 265
    Exercise 10.4 269
    Exercise 10.5 272
    Exercise 10.6 276
    Practice Test 10 278

Chapter 11   **Statistics**

    Exercise 11.1 282
    Exercise 11.2 288
    Exercise 11.3 291
    Exercise 11.4 295
    Exercise 11.5 299
    Exercise 11.6 304
    Exercise 11.7 308
    Practice Test 11 311

Chapter 12   **Your Money and Your Math**

    Exercise 12.1 316
    Exercise 12.2 318
    Exercise 12.3 320
    Exercise 12.4 323
    Practice Test 12 325

Appendix   **The Metric System**

    Exercise A.1 327
    Exercise A.2 329

CHAPTER 1
SETS AND PROBLEM SOLVING

EXERCISE 1.1

STUDY TIPS Word problems, sometimes called "story problems" or "statement problems" are at the heart of math anxiety. So says Sheila Tobias, author of *Overcoming Math Anxiety*. Her solution? Figure out some way to help people conquer their fear and disability in solving word problems. Let us do this together and do it now!

We start this book with a discussion of how to solve problems. This section is very important because the procedures and techniques discussed here will be used in the rest of the book. Why should you do this? Read the Getting Started again to remind you. The Exercises at the end of this section are designed to train and help you solve problems. Here are the answers to the problems in Exercise 1.1.

1. **Step 1.** Understand the problem.
   Step 2. Devise a plan.
   Step 3. Carry out the plan.
   Step 4. Look Back.

3. What does the problem ask for? What is the unknown? After all, if you don't know the question, how can you find the answer?

5. Go to the "Occasional Plan" row and look at the last column. It says that the transaction fee is 20 cents. Since you made 15 transactions, the answer is 15 × 20 cents or **$3.00**

7. If you are planning to make 15 transactions per month, the Light Use plan will cost 15 × 15 cents or $2.25 while the Occasional Plan is $3.00, thus the **Light Use Plan** is **less expensive**.

9. To answer this question find the cost of each plan when you make 20, 40 and 60 calls. For 60 calls the cost for the Light Use Plan is

   Monthly   Number of   Cost per
   Fee   paid calls   call
   $1.95 + (60 - 10)0.15 = $1.95 + $7.50 = $9.45

   For 62 calls it will be 30 cents more or $9.75. After that the Standard Use Plan ($9.95 for 100 free calls) is less expensive.

11. To get the 2nd term (**2**), you add 1 to the 1st term.
    To get the 3rd term (**4**), you add 2 to the 2nd term.
    To get the 4th term (**7**), you add 3 to the 3rd term.
    To get the 5th term (**11**), you add 4 to the 4th term.
    The 6th and 7th terms are 11 + 5 = **16** and 16 + 6 = **22.**

13. Note that the odd numbered terms are always 1's and the even numbered terms are multiples of 5. Thus, the seventh and 9th terms are 1's and the eighth term is the next multiple of 5 after 15, that is 20. Hence, the next three terms are **1, 20 and 1**.

15. Going clockwise, the shaded region is moved 1 place, 2 places, 3 places and so on. The next three moves will move the shaded region 4, 5 and 6 places. The answer is shown.

17. The numbers in the denominator are obtained by doubling. Thus, the next three terms are $\frac{1}{16}$, $\frac{1}{32}$ and $\frac{1}{64}$. Note that each term is half the preceding term.

19. The odd numbered terms are 1, 2, 3, 4, 5, ... and the even numbered terms are 5, 6, 7, 8, 9, ... The next three terms are 7, 4, 8 as shown.

    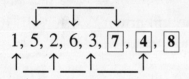

21. (a)
    ```
          •
         • •
        • • •
       • • • •
      • • • • •
           15
    ```
    (b) The rows are constructed by adding one more dot than on the preceding row. The next triangular numbers after 10 are 10 + 5 = **15**, 15 + 6 = **21** and 21 + 7 = **28**.
    (c) Following the pattern after the 7th triangular number which is 28, the 10th triangular number is 28 + 8 + 9 + **10** = **55**.

SECTION 1.1 Problem Solving 3

23. (a)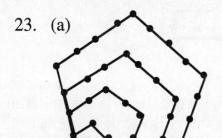

(b) At each step, increase the length of the bottom and left lower side of the pentagon by one unit. The number of dots on each side is increased by one unit.

(c) The 6th pentagonal number is **51**.

25. (a) Here is a summary of the information shown in the figure:

    Sides   4 5 6 7
    Diagonals   1 2 3 4

    The number of diagonals is **three** less than the number of sides. Thus, 10 - **3** = 7 diagonals can be drawn from one vertex of a decagon.

27. (a) It is always **4**.
    (b) If you pick any number and follow the instructions you eventually get to a number less than or equal to 10. For any of these numbers the pattern leads to the number 4.

29. (a) $(1 + 2 + 3 + 4)^2 = 1^3 + 2^3 + 3^3 + 4^3$
         $(1 + 2 + 3 + 4 + 5)^2 = 1^3 + 2^3 + 3^3 + 4^3 + 5^3$
         $(1 + 2 + 3 + 4 + 5 + 6)^2 = 1^3 + 2^3 + 3^3 + 4^3 + 5^3 + 6^3$

    (b) The square of the sum of the first n counting numbers equals the sum of the cubes of these numbers.

31. The number of units of length of the pendulum is always the square of the number of seconds in the time of the swing.

33. (a) **12, 15, 18**
    (b) We can see from the table that each unit increase in size corresponds to a $\frac{1}{3}$ of an inch increase in length. Thus, a 2 unit increase in size (from 6 to 8) corresponds to a $\frac{2}{3}$ increase in length, from 9 to $9\frac{2}{3}$ **in.**

39. 1 + 1 + 2 + 3 + 5 = 12. The sixth term is 3 + 5 = 8, so the seventh term is 5 + 8 = **13**, which is 1 more than the sum of the first five terms.

41. The fourteenth term is 377 (check this!), so the sum of the first twelve terms is one less or **376**.

CHAPTER 1 SETS AND PROBLEM SOLVING

EXERCISE 1.2

STUDY TIPS Most mathematics textbooks start with a discussion about sets, so this section can help you later. The word "set" is an undefined term but at the same time it is listed in the Guinness Book of Records as the "word with the most meanings" (194 in all!) In this section you learn:
(1) How to determine if a set is well defined (Hint: Stay away from subjective words like "good", "bad" and "beautiful")
(2) How to describe a set (There are three ways: writing the set in words, listing the elements separated by commas between braces and using set builder notation.)
   Special caution: the empty set is denoted by $\emptyset$ or $\{\ \}$, but not $\{\emptyset\}$
(3) Determine if two sets are equal. (When two sets are equal, they have exactly the same elements, but these elements do not have to be written in the same order.)
(4) Find the subsets of a set.
Definitions 1.4 and 1.5 mean the same thing but Definition 1.5 is used to convince you that the empty set is a subset of every set A (there is no element in the empty set that is not in A.) To find the subsets of a set start by finding the subsets with no elements, then the subsets with one element and so on. How do you know when you are finished? If you have 1 element, you need $2^1 = 2$ subsets, if you have 2 elements you need $2^2 = 4$ subsets and so on.

1. People do not agree on the meaning of "grouchy", so this set is not well-defined.

3. Well-defined.   5. Well-defined.

7. People do not agree on what is "good", so this set is not well-defined.

9. (a) *Incorrect.* The letter D is not an element of A.
   (b) *Correct.* Desi is an element of A.
   (c) *Incorrect.* Jane is an element of A, not the other way around.
   (d) *Correct.* The letter D is not an element of A.
   (e) *Incorrect.* Jane is an element of A

11. x is an element of the set X. Fill the blank with $\in$.

13. A is not an element of the set X. Fill the blank with $\notin$.

15. The set consisting of the first and last letters of the English alphabet.

17. The set consisting of the names of the first Biblical man and woman.

19. The set of counting numbers from 1 to 7. Note that the numbers do not have to be in any specific order.

21. The set of odd counting numbers from 1 to 51.

23. The set of counting numbers starting with 1 and then adding 3 successively until the number 25 is obtained.

25. {Dioxin, Xylene} is the set that was found in everybody's tissue.

27. {1, 2, 3, 4, 5, 6, 7} 29. {0, 1, 2, 3, 4, 5, 6, 7}

31. The word "between" is to be taken literally. The 3 and the 8 are not elements of this set. The answer is {4, 5, 6, 7}.

33. There are no counting numbers between 6 and 7, so this set is empty. The answer may be written as ∅ or as { }.

35. {4, 5, 6, . . .}

37. {WangB, Gull} 39. {ENSCO, WDigitl, TexAir}

41. {WhrEnf, Ech Bg} 43. {TexAir}

45. The set A = {4, 8, 12, 16, ...} and the set B = {2, 4, 6, 8, ...}, so these sets are not equal.

47. Both A and B are empty, so these sets are equal.

49. (a) Every element of A is an element of B, and every element of B is an element of A. Thus, A = B.
    (b) ≠ is correct because 0 is an element of C but not of A.
    (c) ≠ is correct because 0 is an element of C but not of B.

51. ∅, {a}, {b}, {a, b} The first three of these are proper subsets of the given set.

53. ∅, {1}, {2}, {3}, {4}, {1, 2}, {1, 3}, {1, 4}, {2, 3}, {2, 4}, {3, 4}, {1, 2, 3}, {1, 2, 4}, {1, 3, 4}, {2, 3, 4}, {1, 2, 3, 4} All but the last one of these are proper subsets of the given set.

55. ∅, {1}, {2}, {1, 2}   The first three of these are proper subsets of the given set.

57. Since there are **4** elements in this set, there are 2^4 or 16 subsets.

59. There are **10** elements in A, so A has 2^{10} or 1024 subsets.

61. Note that $32 = 2^5$, so the answer is **5**.

63. Since $64 = 2^6$, the answer is **6**.

65. Yes. Since ∅ has no elements, there is no element of ∅ that is not in ∅. Furthermore, every set is a subset of itself.

67. B ⊆ A because every counting number that is divisible by 4 is also divisible by 2. (A is not a subset of B because a number can be divisible by 2 and not by 4. For instance, 6 is divisible by 2, but not by 4.)

69. (a) There are 5 toppings, so you have 5 choices.
    (b) There are 10 subsets with 2 elements in each. So you have 10 choices. (Try writing these out.)
    (c) You can choose which two toppings you don't want. So the answer is 10 as in (b).

71. Since $256 = 2^8$, you would need **8** different condiments.

77. (a) If g ∈ S, then Gepetto shaves himself, which contradicts the statement that Gepetto shaves all those men and only those men of the village who do not shave themselves. Hence, g ∉ S.
    (b) If g ∈ D, then Gepetto does not shave himself, and so by the same statement, he does shave himself. Thus, there is again a contradiction and g ∉ D.

79. The word "non-self-descriptive" cannot be classified in either way without having a contradiction. If it is an element of S, then it is a self-descriptive word, which contradicts the definition, "non-self-descriptive is a non-self-descriptive word". On the other hand if non-self-descriptive is a non-self-descriptive word, then it is an element of S, which is again a contradiction.

SECTION 1.3 Set Operations

EXERCISE 1.3

STUDY TIPS You have to learn how to do three operations with sets: form the **intersection** of sets A and B (make sure all the elements are in A **and** also in B), form the **union** of two sets (the elements in the union can be in A **or** in B **or** in both), find the **difference** of sets A and B (the elements have to be in A but not in B.) In addition, you have to learn how to find the complement of a set. Hint: To find the complement A' of a set, A you have to know the universal set $\mathcal{U}$. The complement consists of all the elements that are in $\mathcal{U}$ and **not** in A, that is, the difference of $\mathcal{U}$ and A.

1. (a) $A \cap B$, the set of all elements in both A and B, is $\{1, 3, 4\}$.
   (b) $A \cap C$, the set of all elements in both A and C, is $\{1\}$.
   (c) $B \cap C$, the set of all elements in both B and C, is $\{1, 6\}$.

3. (a) $A \cap (B \cup C)$, the set of all elements in both A and $(B \cup C)$, is $\{1, 3, 4\}$.
   (b) $A \cup (B \cap C)$, the set of all elements in A or in $(B \cap C)$, is $\{1, 2, 3, 4, 5, 6\}$.

5. $A \cup (B \cup C)$, the set of all elements in A or in $(B \cup C)$, is $\{1, 2, 3, 4, 5, 6, 7\}$.

7. $A \cap (B \cap C)$, the set of all elements in A and in $(B \cap C)$, is $\{1\}$.

9. (a) $A \cap B$, the set of all elements in both A and B, is $\{c\}$.
   (b) $A \cap C$, the set of all elements in both A and C, is $\varnothing$. Note that the set $\{a, b\}$ is an element of A, but a and b, separately, are not elements of A.

11. (a) Correct. The set $\{b\}$ is a subset of the set $A \cap B$.
    (b) Incorrect. The set $\{b\}$ is not an element of the set $A \cap B$.

13. (a) Correct. The set $\{a, b, c\}$ is a subset of the set $A \cup B$.
    (b) Correct. The set $\{a, b, c\}$ is an element of the set A, so it is an element of the set $A \cup B$.

15. (a) A', the set of elements in $\mathcal{U}$ but not in A, is $\{b, d, f\}$.
    (b) B', the set of elements in $\mathcal{U}$ but not in B, is $\{a, c\}$.

8 CHAPTER 1 SETS AND PROBLEM SOLVING

17. (a) $(A \cup B)'$, the set of elements in $\mathcal{U}$ but not in $(A \cup B)$, is $\varnothing$. Note that $(A \cup B)$ includes all the elements in $\mathcal{U}$.
    (b) $A' \cup B'$, the set of elements in A' or in B' is $\{a, b, c, d, f\}$.

19. (a) $(A \cap B) \cup C'$, the set of elements in $(A \cap B)$ or in C', is $\{c, e\}$. Note that $(A \cap B)$ is $\{e\}$.
    (b) $C \cup (A \cap B)'$, the set of elements in C or not in $(A \cap B)$ is $\{a, b, c, d, f\}$.

21. (a) $A' \cap B$, the set of elements not in A but in B, is $\{b, d, f\}$.
    (b) $A \cap B'$, the set of elements in A but not in B, is $\{a, c\}$.

23. (a) The elements in C' are c and e and those in $(A \cap B)'$ are a, b, c, d, and f. Thus, $C' \cup (A \cap B)' = \{a, b, c, d, e, f\}$
    (b) $C' = \{c, e\}$, $A \cup B = \mathcal{U}$, so $(A \cup B)' = \varnothing$. Thus, $C' \cup (A \cup B)' = \{c, e\}$.

25. (a) This is the set $\mathcal{U}$ with the elements in A taken out. The answer is $\{b, d, f\}$.
    (b) This is the set $\mathcal{U}$ with the elements in B taken out. The answer is $\{a, c\}$.

27. (a) This is the set in $\mathcal{U}$ and not in B. The answer is $\{2, 3\}$
    (b) This is the set $\mathcal{U}$ with the elements in B taken out. The answer is the same as in Part (a), $\{2, 3\}$.

29. $\varnothing'$, the set of all elements in $\mathcal{U}$ that are not in the empty set, is $\mathcal{U}$.

31. The set of all elements that are in both A and the empty set is $\varnothing$.

33. This is the set of all elements that are in both A and $\mathcal{U}$, so the answer is **A**.

35. This is the set of elements that are both in A and not in A, so the answer is $\varnothing$.

37. This is a double negative; the elements that are not in "not A" are, of course, in A. Thus, the answer is **A**.

39. To include A, you must have the elements 1, 2, 3. Then, to include B, you must have the element 4, and to include C, you must have the element 5. Thus, the smallest set that can be used for $\mathcal{U}$ is $\{1, 2, 3, 4, 5\}$.

SECTION 1.3   Set Operations   9

41. {Beauty, Consideration, Kindliness, Friendliness, Helpfulness, Loyalty}

43. This is the set of traits that are in both M_w and M_m, so the answer is {Intelligence, Cheerfulness, Congeniality}.

45. {Intelligence, Cheerfulness}

47. {Is aware of others, Follows up on action}

49. {Follows up on action}

51. (a) M', the set of employees who are not male, is **F**.
    (b) F', the set of employees who are not female, is **M**.

53. (a) Male employees who work in the data processing department.
    (b) Female employees who are under 21.

55. These employees would have to be in sets D and S, both. Thus, the answer is $D \cap S$.

57. These employees would have to be in sets M and D, both. Thus, the answer is $M \cap D$.

59. Male employees or employees who are 21 or over.

61. (a) $F \cap S = \{04, 08\}$ is the set of full-time employees who do shop work.
    (b) $P \cap (O \cup I) = \{02, 05, 07\}$ is the set of part-time employees who do outdoor field work or indoor office work.

63. A and B have no elements in common.

65. All the elements of A are elements of B, and all the elements of B are elements of A.

67. (a) This is the set of characteristics that are in both columns of the table.
    (b) The same answer as in Part (a).
    (c) This is the set of characteristics that occur in either column of the table: {Tall, Short, Long neck, Short Neck, Skin-covered horns, Native to Africa}
    (d) G' = {Short, Short neck}   (e) O' = {Tall, Long neck}

10 CHAPTER 1 SETS AND PROBLEM SOLVING

69. This is the second item in the 35 and older column: 685,000.

71. This is the set of 12 - 17 year old females: $F \cap A$.

73. There are no persons that are both male and female. This set is empty.

EXERCISE 1.4

STUDY TIPS In this section you will learn to make pictures associated with the operations of intersections and unions. These pictures can be used to verify the equality of certain sets. Think of sets A and B as sets of points inside two circles. To make a picture (called a Venn Diagram) of an intersection, draw vertical lines inside of A and horizontal lines inside of B. Where the lines **intersect** is the intersection. If there is no intersection, the sets are called **disjoint**. To form the **union**, draw vertical lines inside of A and continue drawing vertical lines inside B, the result is the **union** of A and B. What about the complement of A? As before, you need the universal set $\mathcal{U}$, represented by a rectangle outside the circles A and B. The complement of A consists, as before, of all points in $\mathcal{U}$ **not** in A.

1. **Step 1**: Draw a diagram with two circles and label the regions w, x, y, z, as shown.

   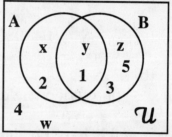

   Step 2: Select the element that is in both A and B, the number 1. Write 1 in Region y.

   Step 3: Select the element that is in A but not in B, the number 2. Write 2 in Region x.

   Step 4: Select the elements that are in B but not in A, the numbers 3 and 5. Write 3 and 5 in Region z.

   Step 5: Select the element that is not in A or B, the number 4, and write 4 in Region w. This completes the diagram.

3. Draw a Venn diagram labeled as in the figure Then look at the regions corresponding to the various sets. A: Regions 1, 3; B' (regions outside of B) Regions 1, 4; A ∩ B' (regions in both A and B'): Region 1. Shade Region 1 for the desired diagram.

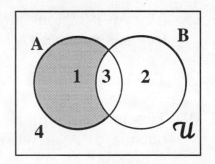

5. Draw a diagram as in Problem 3. Then find the regions corresponding to the various sets. (A ∪ B) (regions in A or B): Regions 1, 2, 3; (A ∩ B) (regions in both A and B): Region 3; (A ∪ B) - (A ∩ B): Take the region in (A ∩ B) away from the regions in (A ∪ B), leaving Regions 1, 2. Shade Regions 1, 2 for the desired diagram.

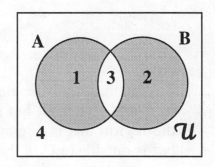

7. Draw a diagram as in Problem 3.
The region corresponding to A' is the entire region outside of circle A. The region corresponding to B' is the entire region outside of circle B. Thus, the region corresponding to A' ∩ B' is the region outside both circles, the shaded region.

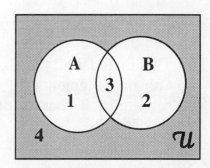

9. A: Regions 1, 3, 5, 7; (B ∪ C) (regions in B or C): Regions 2, 3, 4, 5, 6, 7; A - (B ∪ C): Take the regions common to A and (B ∪ C) away from the regions in A, leaving the answer, Region 1.

11. (A ∩ B ∩ C) (regions common to A, B, C): Region 7; (A ∩ B) (regions in both A and B): Regions 3, 7; (A ∩ B ∩ C) - (A ∩ B): Take the regions common to (A ∩ B ∩ C) and (A ∩ B) away from those in (A ∩ B ∩ C). This leaves no regions, so we get the empty set, ∅.

13. (A ∪ B') (regions in A or outside of B): Regions 1, 3, 4, 5, 7, 8; C: Regions 4, 5, 7; (A ∪ B') ∩ C (regions in both (A ∪ B') and C): Regions 4, 5, 7.

12 CHAPTER 1 SETS AND PROBLEM SOLVING

15. (A ∩ B') (regions in A and not in B): Regions 1, 5. (A ∩ B') ∪ C (regions in either (A ∩ B') or C: Regions 1, 4, 5, 6, 7.

17. This consists of the region that is outside all three of A, B, C: Region 8

19. This includes all the elements that are outside of A or outside of B. Thus, everything except the region common to the two circles is shaded.

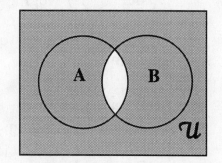

21. This includes all the elements that are in A and also in B, but not in C. Thus, the region that is common to A and B but is outside of C is shaded.

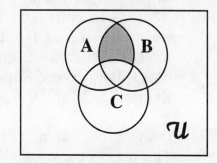

23. This includes all the elements of B that are not elements of A. Thus, the region in B that is outside of A is shaded.

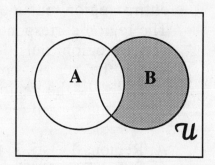

25. This includes all the elements that are in both B and C, but not in A. Thus, the region common to B and C, but outside of A is shaded.

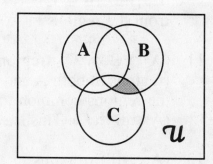

27. Because the intersection of A and B is given equal to B, all of B is contained in A, as shown in the diagram.

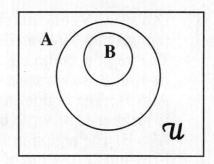

29. Note that A ∩ B contains all the elements common to A and B, so that A ∩ (A ∩ B) is the same as (A ∩ B). Because A ∩ (A ∩ B) = A, all the elements of A must be in B. Thus, the circle A was drawn inside the circle B.

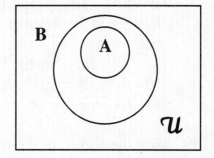

31. (a) A: Regions 1, 3, 5, 7; (B ∪ C): Regions 2, 3, 4, 5, 6, 7. Thus, for A ∪ (B ∪ C), we have Regions 1, 2, 3, 4, 5, 6, 7. Similarly, we have the following correspondences. A ∪ B: Regions 1, 2, 3, 5, 6, 7; C: Regions 4, 5, 6, 7. So for (A ∪ B) ∪ C, we have Regions 1, 2, 3, 4, 5, 6, 7. This verifies the given equality.
    (b) A: Regions 1, 3, 5, 7; (B ∩ C): Regions 6, 7. Thus, for A ∩ (B ∩ C), we have Region 7. Also, (A ∩ B): Regions 3, 7; C: Regions 4, 5, 6, 7. Thus, for (A ∩ B) ∩ C, we have Region 7. This verifies the given equality.

33. (a) A: Regions 1, 3, 5, 7: A': Regions 2, 4, 6, 8. Thus, for A ∪ A', we have Regions 1, 2, 3, 4, 5, 6, 7, 8, the same set of regions that represents 𝒰. This verifies the given equality.
    (b) From Part a, we see that A and A' have no elements in common. Therefore, A ∩ A' = ∅.
    (c) A - B: Regions 1, 5; A ∩ B': Regions 1, 5. This verifies the equality.

35. A ∩ B (The regions that are in both A and B): Regions 3, 7.

37. Draw a Venn diagram with the regions numbered as shown. A ∩ B: The region common to A and B is Region 3, so write a, b in Region 3. A ∩ B': The region in A and not in B is Region 1, so write c, e in Region 1. A' ∩ B: The region in B but not in A is region 2, so write g, h in Region 2 (A ∪ B)': The region that is outside the union of A and B is Region 4. Write d, f in Region 4. Now, you can read off each of the required sets:

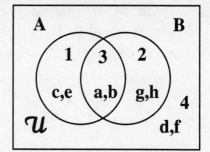

   (a) A = {a, b, c, e}, B = {a, b, g, h}, 𝒰 = {a, b, c, d, e, f, g, h}
   (b) A ∪ B = {a, b, c, e, g, h}
   (c) (A ∩ B)' = {c, d, e, f, g, h}

39. See the figure in the textbook answer section.

41. Arizona, California, Florida, Texas

43. The set of elements common to A and B.

45. The set of elements in 𝒰 and not in either A or C.

47. **False**.   49. **False**.

51. An AB⁺ person has all three antigens and thus may receive blood from any donor.

53. **No**, because the B⁻ person does not have the A antigen.

55. **No**, because the O⁻ person does not have the Rh antigen.

57. Look at the diagram. It gives the answer, 16 or 2^4.

59. (a) This requires everything common to A, B, and D, that is not in C. Thus, the answer is **Region 11.**
    (b) This requires everything that is outside of the union of A, B, and C. Thus, the answer is **Regions 8 and 16.**

SECTION 1.5 The Number of Elements in a Set

EXERCISE 1.5

STUDY TIPS To find the number of elements in set A, denoted by n(A), and called the **cardinal number of A,** simply count the elements in set A. To find the number of elements in the union of A and B, you need the formula $n(A \cup B) = n(A) + n(B) - n(A \cap B)$ If A has 15 elements and B has 20 (as in Exercise 1), you can not simply add 20 and 15, you have to use the formula and subtract the number of elements in the intersection so that these are not counted twice.

Hint: When doing survey problems like Example 3, try to start from the bottom and diagram the students belonging to each of the given categories.

1. Use Equation (1): $\mathbf{n(A \cup B) = n(A) + n(B) - n(A \cap B)}$. Then, since n(A) = 15, n(B) = 20, and n(A ∩ B) = 5, it follows that
$n(A \cup B) = 15 + 20 - 5 = \mathbf{30.}$

3. Use the same equation as in Problem 1. Thus, n(A) = 15, n(A ∩ B) = 5, and n(A ∪ B) = 30, which we substitute into Equation (1) to get
30 = 15 + n(B) - 5, that is, 30 = 10 + n(B). Thus, n(B) = **20.**

5. Let T be the set of people who subscribe to Time and N be the set who subscribe to Newsweek. Since 100 people were surveyed and 10 subscribe to neither magazine, 90 subscribe to one or both. Hence, we can use Equation (1) in the form
$n(T \cup N) = n(T) + n(N) - n(T \cap N)$
With n(T ∪ N) = 90, n(T) = 75, and n(N) = 55, we get the equation 90 = 75 + 55 - n(T ∩ N), that is, 90 = 130 - n(T ∩ N). Therefore, n(T ∩ N) = 40, which says that **40** people subscribe to both.

7. To do this problem, first make a Venn diagram. Let C be the set of students taking Chemistry, E be the set taking English, and M the set taking Mathematics. Draw the diagram with three circles as shown. Then start at the end of the given listing to fill in the proper numbers in the various regions. Because 25 are taking all three courses, write **25** in the region common to all three circles. Since 35 are taking Math and Chem, and we have accounted for 25 of these, write **10** in the remainder of the

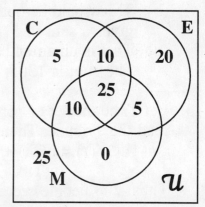

region common to M and C. Likewise, 35 are taking English and Chem, and we have accounted for 25 of these, write **10** in the remainder of the region common to E and C. Similarly, because 30 are taking English and Math, and we have accounted for 25 of these, write **5** in the remainder of the region common to E and M. Then, because 50 are taking Chem, and we have accounted for 10 + 25 + 10 = 45 of these, write **5** in the region of C that is outside of E and M. There are 40 in Math, and we have accounted for 5 + 10 + 25 = 40, so write **0** in the region of M that is outside of E and C. Since there are 60 taking English, and we have accounted for 5 + 10 + 25 = 40 of these, write **20** in the region of E that is outside of C and M. Up to this point, we have accounted for 20 + 5 + 5 + 10 + 10 + 25 = 75 students. Because 100 students were surveyed, write **25** in the region outside of the three circles. This completes the diagram and we can now read off the answers.

(a) **No** students were taking Math and neither Chem nor English. (See the region of M that is outside of C and E.)

(b) There were **10** students taking both Math and Chem, but not English. (See the region that is inside both M and C, but is outside of E.)

(c) There were **10** students taking both English and Chem but not Math. (See the region inside both E and C, but outside of M.)

9. You can get the answers directly from the table.
    (a) Add the numbers in Row P to get 1 + 14 + 7 = **22**.
    (b) Add the numbers in the SK column to get **36**.
    (c) This is the number in Row J, Column SK, **6**.

11. (a) This is the item in the Materials Row of Column 1-2, 120, meaning **$120,000.**
    (b) This would be obtained by adding the items in Column 1-2, which gives a sum of 510, so the answer is **$510,000.**
    (c) This is obtained by adding the items in the first row of the table. The sum is 1305, so the answer is **$1,305,000.**

13. Let Ti be the set of persons who read the Times, and Tr be the set who read the Tribune. Then we can use Equation (1) in the form
    $$n(Ti \cup Tr) = n(Ti) + n(Tr) - n(Ti \cap Tr).$$ This gives
    $$n(Ti \cup Tr) = 130 + 120 - 50 = 200.$$
    Thus, **200** people were surveyed.

SECTION 1.5 The Number of Elements in a Set 17

15. Let O stand for onions, M for mustard, and C for catsup. Draw a Venn diagram as shown. The numbers are obtained as follows: Start at the end of the list. Since 5 had all three condiments, write **5** in the region common to the three circles. Because 25 had onions and catsup, the region common to O and C must have a total of 25. We have already put 5 into this region, so **20** goes into the remainder of the region. Next, we see that 20 had mustard and catsup, so the region common to M and C must have a total of 20. Since 5 have already been put in this region, **15** must be put in the remainder of the region. Since 15 had onion and mustard, the region common to O and M must have a total of 15. Because 5 have already been put in this region, **10** must be in the remainder of the region. Now we see that 50 had catsup, so circle C must contain a total of 50.

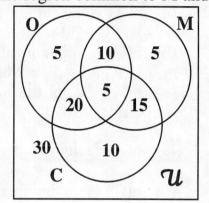

The diagram shows that we already have put 5 + 15 + 20 = 40 in C, so **10** must be put in the remainder of the region. Next, we see that 35 had mustard. Thus, circle M must have a total of 35. Since we have entered 5 + 15 + 10 = 30 in this circle, **5** must go into the remainder of this circle. Since 40 had onions, and the diagram shows that 5 + 10 + 20 = 35 have already been put in circle O, an additional **5** must go in this circle. Now, the diagram shows that 70 persons have been put into the three circles. Since 100 were surveyed, we must put **30** into the rectangle outside of the three circles. This completes the diagram and we can read off the answers to the questions.

(a) **5** (See the region inside O and outside M and C.)
(b) **30** (See the region in the rectangle outside of the three circles.)
(c) 5 + 5 + 10 = **20** (See the non overlapping regions of the three circles.)

17. Let G be the set of people who liked ground coffee, and I be the set of people who liked instant coffee. Then we can use Equation (1) in the form $n(G \cup I) = n(G) + n(I) - n(G \cap I)$ to find how many of those surveyed liked coffee (either ground or instant or both). We were given $n(G) = 200$, $n(I) = 270$, and $n(G \cap I) = 70$, so that
$$n(G \cup I) = 200 + 270 - 70 = 400.$$
Since 50 people did not like coffee, 450 people were surveyed, and the company had to pay out **$450**.

18 CHAPTER 1 SETS AND PROBLEM SOLVING

19. The Venn diagram for the data reported in Problem 18 is shown at the right. The sum of the numbers in the diagram is **28**, which is the correct number of persons interviewed to give this data.

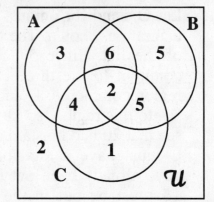

21. (a) n(A) is the sum of the numbers in A, 70 + 50 = **120**.
    (b) n(C) is the sum of the numbers in C, 30 + 50 = **80**.
    (c) n(A ∩ C) is the number in the overlapping portion of the two circles, **50**.

23. (a) n(A') is the number of students in $\mathcal{U}$ and not in A, which is
        200 - (70 + 50) = **80**.
    (b) n(C') is the number of students in $\mathcal{U}$ and not in C, which is
        200 - (50 + 30) = **120**.
    (c) n(A' ∩ C') is the number of students outside of both A and C, that is, 200 - (70 + 50 + 30) = **50**.

25. (a) This is the sum of the numbers in E and T that are not in the region common to these two circles, 35 + 20 + 12 + 6 = **73**.
    (b) This is the sum of the numbers in E and T that are not in M, 35 + 8 + 12 = **55**.
    (c) This is the sum of all the numbers in the diagram except the number in the region common to all three circles. Thus, we get 35 + 8 + 12 + 20 + 6 + 10 = **91**.
    (d) This is the sum of the numbers in all the regions common to at least two of the circles, 8 + 6 + 20 + 4 = **38**.
    (e) Since the sum of all the numbers in the diagram is 95 and 100 persons were interviewed, **5** persons must have had none of these three types of investment.

27. **False.** A counterexample is A = {1, 2} and B = {m, n}.

29. **False.** A counterexample is A = {1, 2} and B = {1, 2, 3}

31. See diagram in the answer section of the textbook. Thus, with the added information, the statistics in the cartoon are possible.

33. If there are 4 elements and each element can be either included or not included (two choices for each element), then there is a total of $2 \times 2 \times 2 \times 2 = 2^4 = 16$ different subsets.

SECTION 1.6   Infinite Sets   19

EXERCISE 1.6

STUDY TIPS How do you know if a set is infinite? The only elementary way is to show that the set can be put into one-to-one correspondence with one of its proper subsets. Exercise 1 shows a one-to-one correspondence between the set of natural numbers and one of its proper subsets, the set of odd numbers. Thus, the set of natural numbers is infinite. Mathematicians have even tried to do arithmetic with infinite numbers. To see how different this is examine questions 24-27 and the Using Your Knowledge.

1. One such correspondence is
   ```
   1  2  3 ...  n ...
   ↕  ↕  ↕      ↕
   1  3  5 ... 2n - 1 ...
   ```
   This shows that the sets N and O are equivalent.

3. One such correspondence is
   ```
   2    4    6  ...    2n    ...
   ↕    ↕    ↕         ↕
   102  104  106 ... 100 + 2n ...
   ```
   Thus, the two sets E and G are equivalent.

5. One such correspondence is
   ```
   202  204  206 ... 200 + 2n ...
   ↕    ↕    ↕        ↕
   302  304  306 ... 300 + 2n ...
   ```
   Thus, the two sets G and T are equivalent.

7. We can set up the correspondence
   ```
   2   4   8   12
   ↕   ↕   ↕   ↕
   6   12  24  36
   ```
   Thus, the sets P and Q are equivalent.

9. The correspondence
   ```
   -1  -2  -3 ... -n ...
   ↕   ↕   ↕       ↕
   1   2   3  ...  n ...
   ```
   shows that I⁻ and N are equivalent.

11. n(A) = **26**   13. n(C) = **50**

15. The set E can be put into one-to-one correspondence with the set N of counting numbers. So n(E) = $\aleph_0$.

17. The set {100, 200, 300, . . . } can be put into one-to-one correspondence with a subset of itself, {200, 300, 400, . . .}. This shows that the set is infinite.

19. The set $\{\frac{1}{3}, \frac{2}{3}, \frac{3}{3}, \ldots\}$ can be put into one-to-one correspondence with a subset of itself, $\{\frac{2}{3}, \frac{3}{3}, \frac{4}{3}, \ldots\}$. Thus, the set is infinite.

21. Sets B and D are equal and equivalent.

23. Set A is neither equivalent nor equal to any of the other sets.

25. Fill in the blank with $\aleph_0$. This is justified by considering the combination of two sets such as {1, 3, 5, . . . } and {2, 4, 6, . . .} into the set N = {1, 2, 3, 4, 5, 6, . . .} . Since all three of these sets have $\aleph_0$ as their cardinality, it follows that $\aleph_0 + \aleph_0 = \aleph_0$.

27. Fill in the blank with $\aleph_0$. This is justified by considering the multiplication table for the set N of counting numbers. The set of products in this table can be put into one-to-one correspondence with the set N itself. This shows that $\aleph_0 \cdot \aleph_0 = \aleph_0$.

29. (a) The next two points are $\frac{7}{9}$ and $\frac{8}{9}$.

    (b) $\frac{1}{3} + \frac{2}{9} + \frac{4}{27} + \frac{8}{81} + \cdots$ The sum gets closer and closer to 1.

31. To room **223**.   33.   Rooms 1, 3, 5, . . ., 2n + 1, . . .

35. They would move to room **666**.

PRACTICE TEST 1

STUDY TIPS There are two additional practice tests, with answers, at the end of this manual Find out if your actual test will be multiple choice or fill in the blank, and how long you will have to take it. Then take the corresponding practice test using the time limit set by your instructor. Find your weaknesses and remedy them before your actual test! We will have more test taking tips later.

1. Look at the difference between successive terms as shown

	1		2		7		19		41		76		**127**
Differences		1		5		12		22		35			
Differences			4		7		10		13				
Difference				3		3		3					

   The third differences are constant, so the next number can be constructed by addition. Add the last diagonal from bottom to top. We obtain the next number in the pattern, $3 + 13 + 35 + 76 = 127$. Now, we can use the 127 to continue the last three rows as shown. The next term now is $3 + 16 + 51 + 127 = 197$

	1		2		7		19		41		76		127		**197**
Differences		1		5		12		22		35		51			
Differences			4		7		10		13		16				
Difference				3		3		3		3					

   If you do this one more time, you will find the next term to be **289**. Once you get four terms you can show that all the following terms can be obtained from the formula: $a_{n+1} = 3 + 3a_n - 3a_{n-1} + a_{n-2}$

2. The counting numbers between 2 and 10 are 3, 4, 5, 6, 7, 8, and 9, so the set is {3, 4, 5, 6, 7, 8, 9}.

3. (a) This is the set of vowels in the English alphabet. In set-builder notation, it is {x | x is a vowel in the English alphabet}.
   (b) This is the set of even counting numbers less than 10. In set builder notation, it is {x | x is an even counting number less than 10}.

4. The proper subsets are ∅, {$}, {¢}, {%}, {$, ¢}, {$, %}, {¢, %}.

5. (a) Both blanks take the symbol ∈, because A ∪ B is the set of all

5. (a) Both blanks take the symbol $\in$, because $A \cup B$ is the set of all elements that are in A or B.
   (b) The first blank takes the symbol $\in$. The second blank takes the symbol $\notin$, because $A \cap B'$ is the set of elements that are in A and not in B.
   (c) A' is the complement of A, that is, the set of elements in $\mathcal{U}$, but not in A. Thus, the first blank takes the symbol $\in$, and the second blank takes the symbol $\notin$.
   (d) Recall that A - B is the set of elements in A with the elements that are also in B removed, that is, it is the set of elements that are in A and not in B. Thus, both blanks take the symbol $\in$.

6. (a) A', the complement of A, = {King}.
   (b) In this problem, $A \cup B = \mathcal{U}$, so the complement of $A \cup B$ is the empty set, $\emptyset$.
   (c) $A \cap B$ is the set of elements that are in both A and B, so the answer is {Queen}.
   (d) The complement of $A \cap B$ is {Ace, King, Jack}. Taking this set away from $\mathcal{U}$ gives $\mathcal{U} - \{A \cap B\}' = \{Queen\}$.

7. (a) In Problem 6 (c), we found that $A \cap B = \{Queen\}$. The union of this set with the set C gives $(A \cap B) \cup C = \{Ace, Queen, Jack\}$.
   (b) In Problem 6(a), we found that A' = {King}, so that $A' \cup C = \{Ace, King, Jack\}$, which has only the element King in common with B. Thus, $(A' \cup C) \cap B = \{King\}$.

8. The two diagrams for this problem are shown here.

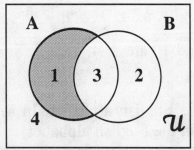 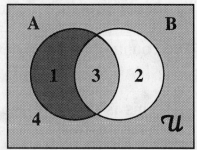

Refer to the diagram on the left. Since A - B is the set A with the elements common to A and B removed, this corresponds to Region 1, that is, circle A with Region 3 removed. Shade Region 1. Now, refer to the diagram on the right. $A \cap B'$ is the set of elements common to A and the complement of B. To show this, shade circle A one way and the region outside of B another way. The cross-hatched region corresponds to $A \cap B'$. The diagrams show that $A - B = A \cap B'$.

9. In the diagram at the right, first shade the region outside the circle C. This region corresponds to C'. Then cross hatch that portion of the shaded region that lies inside both A and B. The cross-hatched region corresponds to the set A ∩ B ∩ C'.

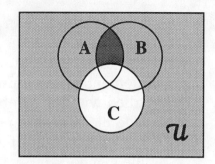

10. (a) A ∪ B is the set of elements in A or in B, so the corresponding regions are 1, 2, 4, 5, 6, 7. C' is the set of elements in 𝒰 but not in C, so the corresponding regions are 1, 2, 5, 8. The regions common to (A ∪ B) and C' are 1, 2, 5.

    (b) A' is the complement of A, so the corresponding regions are those in 𝒰 that are not in A, Regions 2, 3, 6, 8. Similarly, the regions corresponding to B' are those in 𝒰 that are not in B, Regions 1, 3, 4, 8. The regions common to B' and C, Regions 3, 4, correspond to B' ∩ C. Thus, the regions corresponding to A' ∪ (B' ∩ C) are those in A' or in (B' ∩ C), Regions 2, 3, 4, 6, 8.

11. A ∩ B corresponds to the regions inside both circles A and B. Regions 5, 7; (A ∩ B) ∪ C corresponds to the regions in (A ∩ B) or in C, Regions 3, 4, 5, 6, 7. Similarly, (A ∪ C) corresponds to the regions in A or C, Regions 1, 3, 4, 5, 6, 7; (B ∪ C) corresponds to the regions in B or C, Regions 2, 3, 4, 5, 6, 7; (A ∪ C) ∩ (B ∪ C) corresponds to the regions common to (A ∪ C) and (B ∪ C), Regions 3, 4, 5, 6, 7. This verifies the equation (A ∩ B) ∪ C = (A ∪ C) ∩ (B ∪ C).

12. (A ∩ B)' is the complement of A ∩ B, so the corresponding regions are those in 𝒰 and not in A ∩ B This gives us Regions 1, 2, 3, 4, 6, 8 (all except 5 and 7). A' is the complement of A, so it corresponds to Regions 2, 3, 6, 8; B' is the complement of B, so it corresponds to Regions 1, 3, 4, 8. Since A' ∪ B' must correspond to the regions in A' or in B', we get Regions 1, 2, 3, 4, 6, 8, as before. This verifies the equation (A ∩ B)' = A' ∪ B'.

13. The (b) part is correct. (A ∪ C) ∩ B.

14. None of these. B ∩ C is represented by regions 6, 7.

24   CHAPTER 1   SETS AND PROBLEM SOLVING

15. (a) The element 2 is in B but not in A, so the blank is filled with the symbol $\notin$.
    (b) The element 4 is in both B and C, so is in (B $\cap$ C). Thus, 4 is an element of the union of A and (B $\cap$ C). Therefore, the blank is filled with the symbol $\in$.
    (c) The element 4 is not in A. Consequently, 4 is not an element of the intersection of A with any other set. The blank is filled with the symbol $\notin$.

16. (a) A $\cup$ C = {1, 3, 4, 5, 7, 8}, so n(A $\cup$ C) = 6.
    (b) B $\cap$ C = {4, 8}, so n(B $\cap$ C) = 2 $\neq$ 3.

17. (a) A $\cup$ C = {1, 3, 4, 5, 7, 8}, so that (A $\cup$ C) $\cap$ B = {4, 8}. Thus, the blank is filled with =.
    (b) A $\cap$ C = {1,5}, so that (A $\cap$ C) $\cup$ B = {1, 2, 4, 5, 6, 8}. Thus, the blank is filled with $\neq$.

18. You can use the equation n(A $\cup$ B) = n(A) + n(B) - n(A $\cap$ B) for both parts of this problem.
    (a) n(A) = 25, n(B) = 35, and n(A $\cap$ B) = 0. Therefore
    $$n(A \cup B) = 25 + 35 - 0 = \mathbf{60}$$
    (b) n(A) and n(B) are as in Part a, but n(A $\cap$ B) = 5. Therefore,
    $$n(A \cup B) = 25 + 35 - 5 = \mathbf{55.}$$

19. (a) You can use the same equation as in Problem 18, but with n(A) = 15, n(B) = 25, and n(A $\cup$ B) = 35. With these values, you get 35 = 15 + 25 - n(A $\cap$ B), or 35 = 40 - n(A $\cap$ B), which means that n(A $\cap$ B) = **5**.
    (b) n(A' $\cap$ B') = 8 means there are 8 elements outside of both A and B. Since n(A $\cup$ B) = 35, n($\mathcal{U}$) = 35 + 8 = **43**.

20. The shaded region is the region inside of B and outside of A, so it may be described as B - A or as B $\cap$ A'. (A' $\cap$ B is also correct.)

21. First make a Venn diagram showing the various sets. Then start at the end of the list and work back through it. Since no students are taking all three of the courses, put a 0 in the region common to the circles F, G, and S. Because 70 are taking no language, write **70** in the rectangle outside of the three circles. As 15 are taking German and Spanish, write **15** in the region common to G and S, but outside of F. 30 are taking French and 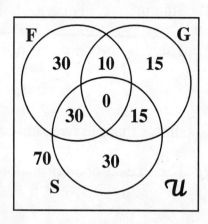 Spanish, so write **30** in the region common to F and S, but outside of G. 10 are taking French and German, so write **10** in the region common to F and G but outside of S. 75 are taking Spanish and of these, you have accounted for 0 + 30 + 15 = 45. (See the diagram.) Hence, you must write **30** in the region inside S but outside both F and G. 40 are taking German, and you have accounted for 0 + 10 + 15 = 25 of these, so write **15** in the region inside G, but outside F and S. 70 are taking French, and you have accounted for 0 + 10 + 30 = 40 of these, so write **30** in the region inside F, but outside G and S. This completes the diagram and you can get the answers from it.
    (a) To get the number of students taking two languages, add the numbers common to two of the circles: 10 + 15 + 30 = **55.**
    (b) You can read this directly as the number **30**, which is inside S and outside the other two circles.
    (c) This number is the sum of all the numbers inside S, but outside F: 15 + 30 = **45.**

22. (a) This is the sum of all the numbers in the diagram:
        10 + 5 + 15 + 2 + 4 + 3 + 8 = **47.**
    (b) This is the sum of the numbers that are in A, but not in B:
        10 + 2 = **12.**
    (c) This is the sum of the numbers that are in B or C, but not in A:
        15 + 3 + 8 = **26.**
    (d) This is the sum of the numbers that are in both B and C, but not in A, so the answer is **3**.
    (e) This is the sum of all the numbers in the diagram that are not in either B or C, so the answer is **10**.

23. Statement (b) is correct. $D' \cup C' = (D \cap C)'$

24. Statement (c) is correct. $D \cup (C \cap E) = (D \cup C) \cap (D \cup E)$

26 CHAPTER 1 SETS AND PROBLEM SOLVING

25. Statement (c) is correct. $D \cup (C \cup E) = (D \cup C) \cup E$

26. The one-to-one correspondence
$$\begin{array}{cccc} 1 & 3 & \ldots & 2n-1 & \ldots \\ \updownarrow & \updownarrow & & \updownarrow \\ 2 & 4 & \ldots & 2n & \ldots \end{array}$$
shows that the two sets have the same cardinal number.

27. The one-to-one correspondence
$$\begin{array}{cccc} 1 & 2 & \ldots & n & \ldots \\ \updownarrow & \updownarrow & & \updownarrow \\ 4 & 16 & \ldots & (2n)^2 & \ldots \end{array}$$
shows that the two sets are equivalent.

28. $n\{1, 4, \ldots, n^2, \ldots, 144\} = \mathbf{12}$

29. Since you can set up a one-to-one correspondence between the given set and the set $\{1, 2, \ldots, n, \ldots\}$, the cardinality of the given set is the same as that of the set of counting numbers, $\aleph_0$.

30. The one-to-one correspondence
$$\begin{array}{cccc} \frac{1}{2} & \frac{1}{3} & \frac{1}{4} & \ldots & \frac{1}{n} & \ldots \\ \updownarrow & \updownarrow & \updownarrow & & \updownarrow \\ \frac{1}{2} & \frac{1}{4} & \frac{1}{6} & \ldots & \frac{1}{2n-2} & \ldots \end{array}$$
between the given set and a subset of itself shows that the given set is infinite.

CHAPTER 2
LOGIC

EXERCISE 2.1

STUDY TIPS In order to communicate, you need statements: (Declarative sentences that can be classified as true or false, but not both simultaneously.) Thus, questions, commands, and exclamations are not statements. Can you imagine how boring it would be if we used **simple** statements only? To make more complicated (compound) statements we need to learn how to connect simple statements. The connectives we use are the conjunction **and** ($\wedge$), the disjunction **or** ($\vee$) and the negation **not** (~). Simple statements can be negated by inserting the word not or by writing "It is not the case" in front of the statement. To negate conjunctions and disjunctions, you need De Morgan's Laws. Here they are:

$\sim(p \wedge q)$ is $\sim p \vee \sim q$. In words, not (p and q) is (not p or not q).

$\sim(p \vee q)$ is $\sim p \wedge \sim q$. In words, not (p or q) is (not p and not q).

(Take our word for this right now and memorize these two rules if necessary. We will prove these laws using truth tables later.)
You also need to know that:
    The negation of   **All** p's are q's   is   **Some** p's **are not** q's.
    The negation of   **No** p's are q's   is   **Some** p's **are** q's.
Knowing these facts or memorizing them will make this section easier.

1. This sentence cannot be classified as either true or false. It is **not** a statement.

3. This sentence is either true or false, so it **is** a statement. It is a **compound** statement with two components: *Jane is taking an English course. She has four themes to write.*

5. This sentence is a question and so is **not** a statement.

7. This sentence is either true or false, so it **is** a statement. It is a compound statement with the two components: *Students at Ohio State University are required to take a course in history. Students at Ohio State University are required to take a course in economics.*

28   CHAPTER 2   LOGIC

9. This statement is a conjunction and is symbolized by $a \wedge f$.

11. This statement is a disjunction and is symbolized by $d \vee f$.

13. This statement is a conjunction and is symbolized by $b \wedge p$.

15. This statement is a disjunction and is symbolized by $a \vee m$.

17. The conjunction of p and q: $p \wedge q$.

19. The conjunction of (not p) and (not q): $\sim p \wedge \sim q$.

21. Dagwood loves Blondie, or Blondie does not love Dagwood.

23. Dagwood loves Blondie but Blondie does not love Dagwood.

25. It is not the case that Dagwood and Blondie love each other.

27. Bill's store is not making a good profit.

29. My dog is not a spaniel.

31. I like to work overtime.

33. These two are negations of each other because if either is true, the other is false.

35. Some men are not mortal.

37. All basketball players are 6 feet tall.

39. Since this statement is a disjunction, you must negate both components: He is not bald and he does not have a 10-inch forehead.

41. This statement is negated by replacing the "Some" by "No": No circles are round.

43. Nobody up there loves me.

45. Somebody does not like to go on a trip.

47. Some persons occupying your covered auto are not insured.

49. All expenses are subject to the 2% limit.

51. Statement d is consistent with the given information.

53. (d ∧ p) ∨ r   55. r ∧ (t ∨ g)

57. (a) The diagram is neither a rectangle nor a square.
    (b) The diagram is a square or not a rectangle.
    (c) The diagram is a square and a rectangle.

EXERCISE 2.2

STUDY TIPS Sometimes we do not know under what conditions compound statements are true or false. Here is the information you need:
The conjunction p ∧ q is **true** only when both p and q are **true**.
The disjunction p ∨ q is **false** only when both p and q are **false.**
If a statement p is **true** its negation ~ p is **false**.
If a statement p is **false** its negation ~ p is **true.**

With this information, you can determine the truth or falsity of compound statements using truth tables. Make sure you are consistent when making a truth table. Look at the solution to Problem 5. Columns 1 and 2 (for p and q) will be the same for every truth table we construct. Aside from determining under what conditions a compound statement is true, truth table can be used to find equivalent statements, which by definition, means statements with the same truth tables. Remember De Morgan's Laws? Now you can prove these Laws by using the idea of equivalent statements.

1. To write the disjunction, use the word "or": Today is Friday **or** Monday.

3. Here, you only have to insert the word "not": Today is **not** Friday.

5. To write the disjunction, use the word "or": He is a gentleman **or** a scholar.

7. To write the conjunction, use the word "and": He is a gentleman **and** a scholar.

9. This statement is the conjunction of the given statements: **g ∧ s**.

11. (a) p ∧ q   (b) p ∨ q   (c) The statement in (a) is false, the one in (b) is true.

13. This is the conjunction of the negation of statement q and the statement p: **~q ∧ p**.

15. This is the disjunction of the two given statements: **p ∨ q**.

17. **p ∨ q**. This is the disjunction of p and q. Since p is true, the disjunction is **true**

19. **~p ∧ ~q**. This is the conjunction of (not p) and (not q). Since (not p) is false, the conjunction is **false.**

21. **~q ∧ ~p**. This is the conjunction of (not q) and (not p). Since (not p) is false, the conjunction is **false.**

23. **g ∨ ~j**. This is the disjunction of g and ~j. Since g and ~j are both false, the disjunction is **false.**

25. **(g ∨ j) ∧ ~(g ∧ j)**. This is the conjunction of (g ∨ j) and ~(g ∧ j). Since j is true, the disjunction (g ∨ j) is true. Since g is false, the conjunction (g ∧ j) is false, so that ~(g ∧ j) is true. Therefore, the statement (g ∨ j) ∧ ~(g ∧ j) is **true.**

27.

1	2	4	3
p	q	p ∨	~q
T	T	T	F
T	F	T	T
F	T	F	F
F	F	T	T

To construct the table, first fill in Columns 1 and 2: Under p, 2 T's, 2 F's and under q, TFTF. This gives all the combinations of truth values for p and q. Next fill in Column 3, the negation of q: FTFT. Then fill in Column 4, the disjunction of p and ~q. This combines Columns 1 and 3. Since the disjunction is false only when both components are false, write F in the third line and T's in the other lines. This completes the table and shows that **TTFT** are the truth values for p ∨ ~q.

SECTION 2.2   Truth Tables: A Problem Solving Tool

29.
1	2	3	4
p	q	~p	∧ q
T	T	F	F
T	F	F	F
F	T	T	T
F	F	T	F

First fill in Columns 1 and 2 just as in Problem 27. Then fill in Column 3, the negation of p, FFTT. Then do Column 4, the conjunction of ~p and q. This combines Columns 3 and 2. Since the conjunction is true only when both components are true, write T in the third line and F's in the other lines. This completes the table and shows that **FFTF** are the truth values for ~p ∧ q.

31.
1	2	5	4	3
p	q	~	(p ∨	~q)
T	T	F	T	F
T	F	F	T	T
F	T	T	F	F
F	F	F	T	T

First fill in Columns 1 and 2 just as in Problem 27. Then do Column 3, the negation of q: FTFT. Next, do Column 4, the disjunction of p and ~q (Columns 1 and 3). Since the disjunction is false only when both components are false, write F in the third row, where p and q are both F's and write T's in the remaining rows. Column 5 is the negation of Column 4: **FFTF** are the truth values for ~(p ∨ ~q).

33.
1	2	6	3	5	4
p	q	~(	~p	∧	~q)
T	T	T	F	F	F
T	F	T	F	F	T
F	T	T	T	F	F
F	F	F	T	T	T

First fill in Columns 1 and 2 as in Problem 27. Then do Column 3, the negation of p: FFTT; and Column 4, the negation of q: FTFT. Next, do Column 5, the conjunction of ~p and ~q. Since the conjunction is true only when both components are true, write T in the bottom row, where ~p and ~q are both T's, and write F's in the remaining rows. Column 6 calls for the negation of Column 5, so write T where Column 5 has F, and F where Column 5 has T. The final result gives **TTTF** for the truth values of ~(~p ∧ ~q).

35.
1	2	3	6	4	5
p	q	(p ∧ q)	∨	(~p	∧ q)
T	T	T	T	F	F
T	F	F	F	F	F
F	T	F	T	T	T
F	F	F	F	T	F

Fill in Columns 1 and 2 as in Problem 27. Then do Column 3, the conjunction of p and q. The conjunction is true only when both components are true, so write T in the first row and F in the other rows. Next, do Column 4, which calls for the negation of p: FFTT; and then do Column 5, the conjunction of ~p and q. This takes a T in the third row, where both Columns 4 and 2 have T's, and takes F's in the other rows. Finally, Column 6, the disjunction of Columns 3 and 5, takes T in the rows where Column 3 or 5 is T, and F where both Columns 3 and 5 are F's. The required truth values are **TFTF**.

32 CHAPTER 2 LOGIC

37.

1	2	3	5	4
p	q	r	p ∧	(q ∨ r)
T	T	T	T	T
T	T	F	T	T
T	F	T	T	T
T	F	F	F	F
F	T	T	F	T
F	T	F	F	T
F	F	T	F	T
F	F	F	F	F

First, write four T's and then four F's in Column 1; then two T's, two F's, two T's, two F's in Column 2; in Column 3, write alternately T and F. This gives all the possible combinations of T's and F's for p, q, and r. Next, do Column 4, the disjunction of Columns 2 and 3. The disjunction is false only when both components are false, so write F in the 4th and 8th rows, where Columns 2 and 3 are both F's, and write T in the other rows. Column 5 calls for the conjunction of Columns 1 and 4, so write T in the first three rows, where Columns 1 and 4 are both T's, and write F in the remaining rows. This completes the table.

39.

1	2	3	4	7	6	5
p	q	r	(p ∨ q)	∨	(r ∧	∼q)
T	T	T	T	T	F	F
T	T	F	T	T	F	F
T	F	T	T	T	T	T
T	F	F	T	T	F	T
F	T	T	T	T	F	F
F	T	F	T	T	F	F
F	F	T	F	T	T	T
F	F	F	F	F	F	T

First write four T's and then four F's in Column 1; then, two T's, two F's, two T's, two F's in Column 2; in Column 3, write T and F alternately. This gives all the possible combinations of T's and F's for p, q, and r. Next, do Column 4, by combining Columns 1 and 2 to form the disjunction of p and q. Thus, write F in the last two rows, where p and q are both F's, and write T's in the remaining rows. Then do Column 5, the negation of q: FFTTFFTT. Column 6 is the conjunction of r and ∼q, Columns 3 and 5, so write T in the third and seventh rows, where r and ∼q are both T, and write F in the other rows. Finally, Column 7 is the disjunction of Columns 4 and 6, so write F in the last row, where Columns 4 and 6 are both F, and write T in the remaining rows. This completes the table.

41. (a) Since this is the conjunction of the two statements, it will be true when p and q are both true.
(b) This will be false if at least one of p and q is false.
(c) This is the disjunction of the two statements, so it will be true if at least one of p and q is true.
(d) This will be false only if both p and q are false.

SECTION 2.2 Truth Tables: A Problem Solving Tool

43.

1	2	3	5	4	6	8	7
p	q	r	p ∨	(q ∧ r)	(p ∨ q)	∧	(p ∨ r)
T	T	T	T	T	T	T	T
T	T	F	T	F	T	T	T
T	F	T	T	F	T	T	T
T	F	F	T	F	T	T	T
F	T	T	T	T	T	T	T
F	T	F	F	F	T	F	F
F	F	T	F	F	F	F	T
F	F	F	F	F	F	F	F

Columns 5 and 8 of the above truth table show that the two statements have the same truth values, so they are equivalent.

45.

1	2	4	3	5	7	6
p	q	~(p ∨ q)		~p	∧	~q
T	T	F	T	F	F	F
T	F	F	T	F	F	T
F	T	F	T	T	F	F
F	F	T	F	T	T	T

Columns 4 and 7 of the table show that the two statements have the same truth values, so they are equivalent.

47.

1	2	3	5	4	6
p	q	(p ∧ q)	∨	~p	q ∨ ~p
T	T	T	T	F	T
T	F	F	F	F	F
F	T	F	T	T	T
F	F	F	T	T	T

Columns 5 and 6 of the table show that the two statements have the same truth values, so they are equivalent.

49. (a) p ∧ q is true only when p and q are both true. Thus, the truth values are TFFF as in the table. p ∧ ~q is true only when p and ~q are both true, that is, when p is true and q is false. The truth values are FTFF as in the table. ~p ∧ q is true only when p is false and q is true, so that the truth values are FFTF as in the table. ~p ∧ ~q is true only when p and q are both false, so the truth values are **FFFT** as in the table.

(b) The disjunction of two statements is true whenever at least one of the statements is true. Since (p ∧ q) is true only in the first row and (~p ∧ ~q) is true only in the last row, the disjunction of these two has the truth values **TFFT** as stated.

(c) A statement with the truth values FTTF is (p ∧ ~q) ∨ (~p ∧ q). Similarly, (p ∧ ~q) ∨ (~p ∧ q) ∨ (~p ∧ ~q) has truth values FTTT. A much simpler statement with truth values FTTT is ~(p ∧ q), the negation of p ∧ q, as can be seen from the table.

51. **None** of the applicants is eligible. Statement (d) is false for Joe, (t) is false for Mary and (m) is false for Ellen.

53. 7 is greater than or equal to 5.

55. 0 is less than or equal to 3.

57. $\frac{1}{2}$ is greater than $\frac{1}{8}$.

59. I will not go fishing or the sun is not shining. This would be true if either or both of the statements: "I will not go fishing." and "The sun is not shining." is (are) true.

61. $[(e \wedge g) \vee a] \wedge h \wedge (c \vee n \vee \sim t)$.

63. **Mr. Baker** is the carpenter.

EXERCISE 2.3

STUDY TIPS Remember that the statement **p → q** is **false** only when p is true and q is false. (See the photo of the chiropractor's sign stating "*If you stop here, your pain will too*" and Problem 52 if you are not convinced.) In this statement, p is called the antecedent and q the consequent. Hint: Statements such as "All math students make A's" can be translated into the corresponding conditional "If you are a math student then you make A's." Example 4 gives a very important equivalence to remember: **p → q ⇔ ~p ∨ q**. We will use these facts in Section 2.6. Finally, keep in mind that the biconditional
p ↔ q is **true** only when p and q have the same truth values.

1.

1	2	3	5	4	6
p	q	~q	→	~p	p → q
T	T	F	T	F	T
T	F	T	F	F	F
F	T	F	T	T	T
F	F	T	T	T	T

Columns 5 and 6 are identical, so
~q → ~p is equivalent to p → q.

SECTION 2.3   The Conditional and the Biconditional

3.
1	2	3	4	5
p	q	~p → q		p ∨ q
T	T	F T		T
T	F	F T		T
F	T	T T		T
F	F	T F		F

Since Columns 4 and 5 are identical, **~p → q and p ∨ q are equivalent.**

5. This is a statement of the form p → q, with p being "2 + 2 = 4," and q being "8 = 5." Since p is true and q is false, the conditional has the truth value **F**.

7. This is a statement of the form p → q, with p being "2 + 2 = 22," and q being "4 = 26." Since p is false, the conditional has the truth value **T**.

9. Since the antecedent, 2 + 2 = 22, is false, x may be **any number**.

11. Since the consequent, 2 + 2 = 32, is false, x may be any number for which the antecedent is false, that is, **any number except 4**.

13. It is false whenever the antecedent, "You got the time," is true, and the consequent, "We got the beer," is false.

15.
1	2	3	4	6	5
p	q	r	(p → q)	↔	(p ∨ r)
T	T	T	T	T	T
T	T	F	T	T	T
T	F	T	F	F	T
T	F	F	F	F	T
F	T	T	T	T	T
F	T	F	T	F	F
F	F	T	T	T	T
F	F	F	T	F	F

The first three columns are filled out as usual to get all the possible combinations of p, q and r. The conditional, Column 4, has F's in the rows where the antecedent p is true and the consequent q is false. The other rows are all T's. Column 5 has T's in the rows where either p or r is true and has F's in the remaining rows. Column 6 has T's in the rows where Columns 4 and 5 both have T's and has F's in the remaining rows.

36 CHAPTER 2 LOGIC

17.

1	2	3	5	4
p	q	r	p →	(q ∧ r)
T	T	T	T	T
T	T	F	F	F
T	F	T	F	F
T	F	F	F	F
F	T	T	T	T
F	T	F	T	F
F	F	T	T	F
F	F	F	T	F

The first three columns are filled out as usual to get all the possible combinations of p, q and r. Column 4 is the conjunction of Columns 2 and 3, so has T only in the rows where q and r are both T. Column 5, the conditional, has F in the rows where the antecedent p is true and the consequent (q ∧ r) is false (second, third and fourth rows). The remaining entries are T's.

19. The final columns in the tables in Problems 17 and 18 are identical, so the two statements are **equivalent**.

21. The antecedent is "I will buy it," and the consequent is "It is a poodle." Thus, the symbolic form is **p → q**.

23. The antecedent is "It is not a poodle," and the consequent is "I will not buy it." Hence, the symbolic form is **~q → ~p**.

25. The antecedent is "It is a poodle," and the consequent is "I will not buy it. Thus, the symbolic form is **q → ~p**.

27. Here, the antecedent is "You are out of beer," and the consequent is "You are out of Schlitz." Hence, the symbolic form is **~b → ~s**.

29. **~a ∨ b** The temperature is not above 80º, or I would go to the beach.

31. **~a ∨ g** Eva does not have a day off, or she would go to the beach.

33. Take p to be "You got the time," and q to be "We got the beer." Then ~p ∨ q becomes "**You do not have the time or we got the beer.**"

35. If it is a dog, then it is a mammal.

37. If it is a man, then it is created equal.

39. If it is a rectangle with perpendicular diagonals, then it is a square.

SECTION 2.3   The Conditional and the Biconditional   37

41.
```
     1 2   4  3      6  5
     p q  ~(p → q)   p ∧ ~q
     T T   F  T      F  F
     T F   T  F      T  T
     F T   F  T      F  F
     F F   F  T      F  T
```
Since Columns 4 and 6 are identical, ~(p → q) and p ∧ ~q **are equivalent**.

43. Johnny does not play quarterback and his team does not lose.

45. I kiss you once, but I do not kiss you again.

47. Evel Knievel is careless, but he will not lose his life.

49. If Johnny plays quarterback, then his team wins.

51. If Joe had not had an accident, then he could get car insurance.

53. The antecedent "You eat the spinach and the liver" is false, so the conditional is true. She has **not** broken her original promise.

55. Let p be: "Mary is in Tampa," and q be: "Mary is in Florida." Then the given statement is p → q, which is equivalent to ~p ∨ q. Statement (d) is symbolized by ~q → ~p which is equivalent to ~(~q) ∨ ~p, that is, q ∨ ~p. Since q ∨ ~p is equivalent to ~p ∨ q, statement **(d)** is logically equivalent to the given statement.

57. Let p be: "You studied hard," and q be: "You passed the course." The given statement is p → q. The negation of p → q (See Problem 17) is p ∧ ~q, which is the symbolic form of statement **(d)**.

59. Let p be: "You stop here," and q be: "Your pain will stop." Then the given statement is p → q, which is false only if p is true and q is false. Thus, if p is false, then q can be true or false. Hence, the statement: "If you did not stop, your pain will not stop either," **does not** follow logically from the given statement.

61. The student has to take the placement examination only if the student has satisfied the Freshman requirements (perhaps by advanced courses in High School) and is being admitted to Sophomore standing, but is entering college for the first time.

63. **r → a**

65. **No.** It only says that an adjustment will be made if a report is made in 10 days.

EXERCISE 2.4

STUDY TIPS There are many ways of expressing an idea so here we concentrate on finding statements that mean the same as the conditional $p \to q$. (Table 2.16 has these equivalent forms. Memorize them.) In addition, the conditional is equivalent to its **contrapositive** $\sim q \to \sim p$ and to $\sim p \vee q$. You have to learn how to write the inverse, converse and contrapositive of the conditional $p \to q$ ($q \to p$, $\sim p \to \sim q$ and $\sim q \to \sim p$, respectively.) Finally, in this section we learn to discover when one statement **implies** another statement. In general, to show that the statement p **implies** the statement q, simply show that $p \to q$ is a **tautology** (a statement that is always true). If the truth table for p and q is given, make sure that there is no case in which p is true and q is false since in this case the statement $p \to q$ is false.

1. The *contrapositive* is: If n is divisible by 2, then n is an even number.

3. This statement can be written in the symbolic form **q → p.**

5. This statement can be written in the symbolic form **q → p.**

7. If one is a mathematics major, then one takes calculus.

9. If the measure gets a two-thirds vote, then it carries.

11. If we have a stable economy, then we have low unemployment.

13. If birds are of a feather, then they run together.

15.

p	q	p → q	Converse q → p	Inverse ~p → ~q
T	T	T	T	T
T	F	F	T	T
F	T	T	F	F
F	F	T	T	T

The converse, $q \to p$, is true except when q is true and p is false (third row). The inverse, $\sim p \to \sim q$, is true except when p is true and q is false (third row). Thus, the **converse** and the **inverse** have the same truth values, and hence are equivalent.

17. $p \leftrightarrow s$

SECTION 2.4 Variations of the Conditional

19. (a) *Converse*: If you are not strong, then you do not eat your spinach.
    Inverse: If you eat your spinach, then you are strong.
    Contrapositive: If you are strong, then you eat your spinach.
    (b) *Converse*: If you are strong, then you eat your spinach.
    Inverse: If you do not eat your spinach, then you are not strong.
    Contrapositive: If you are not strong, then you do not eat your spinach.
    (c) *Converse*: If you eat your spinach, then you are strong.
    Inverse: If you are not strong, then you do not eat your spinach.
    Contrapositive: If you do not eat your spinach, then you are not strong.

21. If the square of an integer is divisible by 4, the integer is even. **True**.

23. If I am neat and well dressed, then I can get a date. **False**.

25. If you pass this course, then you get passing grades on all the tests. **False**.

27. If we cannot find a cure for cancer, then the research is inadequately funded.

29. If a person does not want to improve the world, then the person is not a radical.

31. Let p be: "n is even," and q be: "3n is even." Then, we see that the equivalence given in (**c**) makes the desired transformation.

33. Let p be: "The day is cool," and q be: "I will go fishing." Then the given statement is $p \to q$. Statement (**b**) translates into $q \to p$, which is not logically equivalent to $p \to q$.

35.

1	2	3	4
p	q	(p ∧ q)	→ p
T	T	T	T
T	F	F	T
F	T	F	T
F	F	F	T

Column 3 is the conjunction of Columns 1 and 2, so has T only in the first row, where both p and q are T. Therefore, Column 4 is all T's, which shows that **(p ∧ q) → p is a tautology**.

37.
1	3	2
p	p ↔ ~p	~p
T	F	F
F	F	T

Since Column 3 has all F's, the statement **p ↔ ~p is a contradiction**.

39.
1	2	3	4	6	5
p	q	(~p ∧ q)	→	(p → q)	
T	T	F	F	T	T
T	F	F	F	T	F
F	T	T	T	T	T
F	F	T	F	T	T

Since Column 6 is all T's, the **first** statement, ~p ∧ q, **implies** the **second**, p → q.

41. If p is true, then the first statement is true, and if the first statement is true, then p is true. Hence, the two statements are **equivalent**.

43. If p ∧ ~q is true, then ~q is true, so that ~p ∨ ~q is also true. Thus the second statement implies the first. However, the first statement is true if p is true, in which case the second statement is false. Hence, the **first** statement **does not imply** the **second**.

45. "For q to be true, it is necessary for p to be true." means that q cannot be true if p is not true. It does not mean that q is true if p is true. For example, for an integer n to be divisible by 4, it is necessary for n to be even, but that is not sufficient. (6 is even but not divisible by 4.) "For q to be true, it is sufficient for p to be true." means that q is true if p is true. For an integer to be divisible by 4, it is sufficient for the integer to be the square of an even integer. (Look at Problem 45.) However, this is not a necessary condition. (12 is divisible by 4, but 12 is not the square of any integer.)

47. Answers may vary. Here are two examples.
  i. To freeze water, it is necessary that the temperature be below 0° C.
  ii. To be admitted to a college, it is necessary for you to satisfy the entrance requirements.

49. The truth values for p → q are TFTT. If the F case is missing, so that p → q is a tautology, then p ⇒ q.

51. By definition, the statement p → q is true when p is false. Hence a false statement implies any statement.

SECTION 2.4 Variations of the Conditional

53. (a) The truth set of q is Q and the truth set of ~r is R'. Hence, the truth set of q ∧ ~r is **Q ∩ R'**.
    (b) The truth set of p ∧ q is P ∩ Q and the truth set of ~r is R'. Thus, the truth set of (p ∧ q) ∧ ~r is **(P ∩ Q) ∩ R'**.

55. The contrapositive of ~q → ~p is **p → q**.

57. The inverse of p → q is ~p → ~q, and the contrapositive of ~p → ~q is **q → p**.

59. (~r ∧ ~s) ∨ (p ∨ q) ⇔ (r ∨ s) → (p ∨ q) is true because
    (r ∨ s) → (p ∨ q) ⇔ ~(r ∨ s) ∨ (p ∨ q) ⇔ (~r ∧ ~s) ∨ (p ∨ q).

61. The **direct** statement.   63. The **contrapositive**.

EXERCISE 2.5

STUDY TIPS You can show that an argument is valid (when all premises are true, the conclusion is true) by using Euler diagrams. Stop here! Memorize the four diagrams after Definition 2.8. If you do not know how to diagram. The statements "All P's are Q's," "No P's are Q's," "Some P's are Q's" and "Some P's are not Q's" you will not be able to do the problems. After you learn how to diagram these four statements it is just a matter of diagramming the premises (the given part) and making sure that the conclusion, (the part to be shown) is present in the final diagram.

1. Premises: "No misers are generous," and "Some old persons are not generous."
   Conclusion: "Some old persons are misers."

3. Premises: "All diligent students make A's," and "All lazy students are not successful."
   Conclusion: "All diligent students are lazy."

5. Premises: "No kitten that loves fish is unteachable," and "No kitten without a tail will play with a gorilla."
   Conclusion: "No unteachable kitten will play with a gorilla."

42 CHAPTER 2 LOGIC

7. In the diagram, the circle P (professors) must be inside the circle W (wise), and the point B (Ms. Brown) must be inside the circle P, so the argument is **valid**.

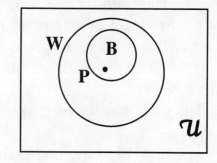

9. In the diagram, the circle D (drinkers) must be outside of both the circle H (healthy) and the circle J (joggers). The premises do not prevent the circles J and H from intersecting, so the argument is **invalid**.

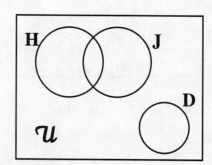

11. In the diagram, the circle M (men) must be inside the circle F (funny), and the dot (Joey) must be inside the circle M, and so inside the circle F. Thus, the argument is **valid**.

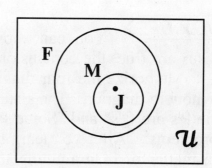

13. In the diagram, the circle F (felines) must be inside the circle M (mammals). The circle D (dogs) must not intersect the circle F, but the premises do not prevent circle D from intersecting circle M, so the argument is **invalid**.

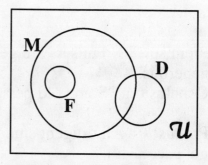

15. In the diagram, the circle M (Math teachers) must not intersect circle W (wealthy). The circle P (panthers) must not intersect the circle M, but may intersect the circle W, so the argument is **invalid**.

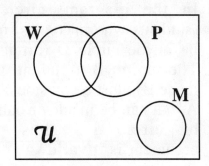

17. In the diagram, the circle M must be inside the circle P (publications). The Ph (Ph. D.) circle must intersect the circle P, but does not have to intersect the circle M, so the argument is **invalid**.

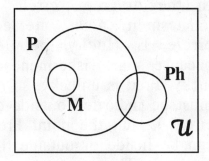

19. In the diagram, the circle H (heavy cars) must be inside the circle C (comfortable to ride in) and the circle C must be outside the circle S (shoddily built). Thus, circle H must lie outside of the circle S, so the argument is **valid**.

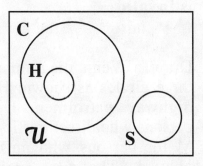

21. In the diagram, circle B (bulldogs) must be inside circle U (ugly). Point D (this dog) must be inside circle U, but not necessarily inside circle B, so the argument is **invalid**.

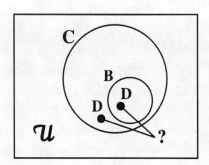

44   CHAPTER 2   LOGIC

23. In the diagram, circle A (students who make A's) must be inside circle D (drudges). The point B (student who made B) must be outside circle A, but can be inside or outside of circle D. Thus, the argument is **invalid**.

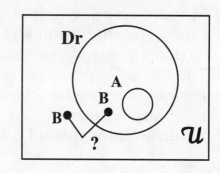

25. In the diagram, circle F (fishermen) must intersect circle L (lucky people) because some fishermen are lucky. Fred is unlucky, so he must be indicated outside of circle L, but the point Fred can be inside or outside the circle F. Thus, the argument is **invalid**.

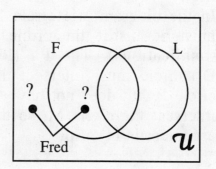

27. Draw a Venn diagram with three circles as shown: H for highway patrolmen, D for persons who direct traffic, and M for persons who must be obeyed. The first premise makes Regions 2 and 5 empty, and the second premise makes Regions 1 and 4 empty. Region 7 is not empty which means that Statement (b) can be logically deduced.

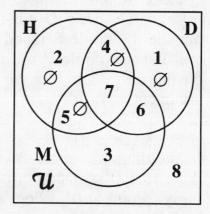

29. (a) An argument is valid if and only if the conclusion is true whenever all the premises are true. Thus, a valid argument reaches a true conclusion from true premises.
    (b) **No.** By the definition stated in (a), if the premises are all true and the conclusion is false, then the argument is invalid.

31. **No**. It may be that the conclusion does not follow from the premises. Example 2 of this section is a good illustration.

33. By the definition of a valid argument, the conclusion is true whenever all the premises are true.

35. Draw a Venn diagram with two circles as shown: M for men and F for funny. The statement. "All men are funny," means that Region 1 is empty. Thus, the point J (for Joey) must be in Region 3. This shows that the argument is **valid**.

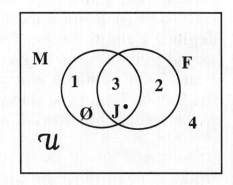

37. Draw a Venn diagram with three circles as shown: F for feline, M for mammals, and D for dogs. Because "All felines are mammals," Regions 1 and 6 are empty. "No dog is a feline," makes Region 7 empty. "This dog" can be in Region 3 or 5, so the argument is **invalid**.

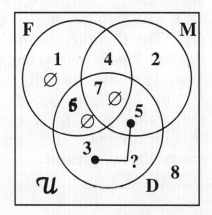

39. Draw a Venn diagram with three circles as shown: M for Math teachers, W for wealthy, and P for panthers. The first premise makes Regions 4 and 7 empty. The second premise makes Region 6 also empty. Since a given panther could be in Region 3 or 5, the argument is **invalid**.

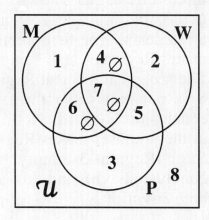

46 CHAPTER 2 LOGIC

41. Draw a Venn diagram with three circles as shown: M for Math teachers, P for publications, and Ph for Ph. D's. The first premise makes Regions 1 and 6 empty. The second premise means Regions 5 and 7 can't both be empty. But Region 7 might be empty, so the argument is **invalid**.

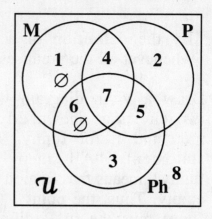

43. Draw a Venn diagram with three circles as shown: H for heavy cars, C for comfortable cars, and S for shoddily built cars. The first premise makes Regions 1 and 6 empty, and the second premise makes Regions 5 and 7 empty. Since the regions common to H and S (Regions 6 and 7) are empty, the argument is **valid**.

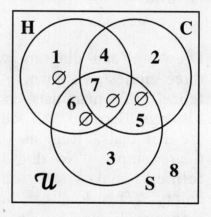

45. Draw a Venn diagram with three circles as shown: W (won't study), A (get A's), and S (students). The first premise makes Regions 4 and 7 empty. The second means that Region 6 is not empty, so the first conclusion is **valid**. Nothing in the premises makes Region 5 or Region 3 empty, so conclusions (b) and (c) are both **invalid**.

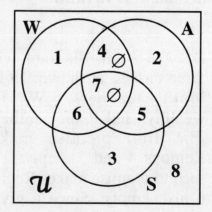

SECTION 2.6 Truth Tables and Validity of Arguments

EXERCISE 2.6

STUDY TIPS   To show that an argument is valid, make a truth table containing all the **premises** (what is assumed) and the **conclusion**. You need to check only the cases in which **all** the premises are true. In these cases, the conclusion must also be **true**. If you find any case in which this is not so (premises are *true* but conclusion is *false*) the argument is **invalid**.

1. $e \rightarrow p$ Using the symbolic form at the left, make a truth table:
   $\sim e$
   $\therefore \sim p$

e	p	e → p	~e	~p	
		Prem.	Prem.	Concl.	
T	T	T	F	F	
T	F	F	F	T	
F	T	T	T	F	← Prem. T, Concl. F
F	F	T	T	T	

In the third row, the premises are true and the conclusion is false. Thus, the argument is **invalid**.

3. $s \rightarrow e$ Use the symbolic form at the left and make a truth
   $\sim e$ table:
   $\therefore \sim s$

s	e	s → e	~e	~s
		Prem.	Prem.	Concl.
T	T	T	F	F
T	F	F	T	F
F	T	T	F	T
F	F	T	T	T

There is no row where the premises are both true and the conclusion is false. Thus, the argument is **valid**.

5. g Use the symbolic form at the left and make a truth
   $\therefore g \wedge r$ table:

g	r	g	g ∧ r	
		Prem.	Concl.	
T	T	T	T	
T	F	T	F	← Prem. T, Concl. F
F	T	F	F	
F	F	F	F	

In the second row, the premise is true and the conclusion is false, so the argument is **invalid**.

48 CHAPTER 2 LOGIC

7. $w \rightarrow m$
    $\sim w \rightarrow g$
    $\therefore m \vee g$

    Use the symbolic form at the left and make a truth table:

w	m	g	Prem. $w \rightarrow m$	Prem. $\sim w \rightarrow g$	Concl. $m \vee g$
T	T	T	T	T	T
T	T	F	T	T	T
T	F	T	F	T	T
T	F	F	F	T	F
F	T	T	T	T	T
F	T	F	T	F	T
F	F	T	T	T	T
F	F	F	T	F	F

    Rows 1, 2, 5, and 7 are the only ones where both premises are true In these rows the conclusion is also true, so the argument is **valid**.

9. $t \rightarrow b$
    t
    $\therefore b$

    Use the symbolic form at the left and make a truth table:

t	b	Prem. $t \rightarrow b$	Prem. t	Concl. b
T	T	T	T	T
T	F	F	T	F
F	T	T	F	T
F	F	T	F	F

    The first row is the only one in which both premises are true. In that row, the conclusion is also true, so the argument is **valid**.

11. $s \rightarrow f$
    s
    $\therefore f$

    Use the symbolic form at the left and make a truth table:

s	f	Prem. $s \rightarrow f$	Prem. s	Concl. f
T	T	T	T	T
T	F	F	T	F
F	T	T	F	T
F	F	T	F	F

    The first row is the only one where both premises are true, In that row, the conclusion is also true, so the argument is **valid**.

13. $m \to e$ Use the symbolic form at the left to make a truth
       $\sim m$ table:
     $\therefore \sim e$

			Prem.	Prem.	Concl.	
m	e	$m \to e$	$\sim m$	$\sim e$		
T	T	T	F	F		
T	F	F	F	T		
F	T	T	T	F	←	Prem.T, Concl.F
F	F	T	T	T		

In the third row, both premises are true and the conclusion is false, so the argument is **invalid**.

15. $f \to s$ Use the symbolic form at the left to make a truth
       $\sim f$ table:
     $\therefore \sim s$

			Prem.	Prem.	Concl.	
f	s	$f \to s$	$\sim f$	$\sim s$		
T	T	T	F	F		
T	F	F	F	T		
F	T	T	T	F	←	Prem.T, Concl.F
F	F	T	T	T		

In the third row, both premises are true and the conclusion is false, so the argument is **invalid.**

17. $p \lor q$ is equivalent to $\sim p \to q$, and if this statement is true and $\sim p$ is also true, then q is true. Thus, the argument is **valid.**

19. $p \to q$ is equivalent to $\sim p \lor q$, a statement that is true if $\sim p$ is true, regardless of whether q is true or false. Hence, the argument is **invalid.**

21. $\sim r \to \sim p$ is equivalent to its contrapositive $p \to r$, so the argument is **valid** since it can be written in exactly the same form as that of Example 6.

23. Use the hint and replace $\sim q \lor r$ by $q \to r$. The premises are then $p \to q$ and $q \to r$, so that a valid conclusion is $\mathbf{p \to r}$.

50 CHAPTER 2 LOGIC

25. Replace the premise $\sim s \to \sim r$ by its contrapositive, $r \to s$. The premises can then be written p, p $\to$ q, q $\to$ r, and r $\to$ s. Thus, a valid conclusion using all the premises is **s**.

27. Replace the disjunction $\sim p \vee r$ by its equivalent, $\sim r \to \sim p$. The premises can then be written $\sim r$, $\sim r \to \sim p$, and $\sim p \to q$. Thus, a valid conclusion using all the premises is **q**.

29. Let n be "I do not recommend them," h be "The books are healthy in tone," b be "The books are bound," w be "The books are well written," and r be "The books are romances." Then the symbolic forms of the given premises are: $n \to \sim h$, $b \to w$, $r \to h$, and $\sim b \to n$. Replace the first and last of these premises by their contrapositives, $h \to \sim n$, and $\sim n \to b$. Then arrange the premises in the order:

$$r \to h,\ h \to \sim n,\ \sim n \to b,\ b \to w.$$

You can now read a conclusion that uses all the premises: **r $\to$ w**, or in words, "**All romances are well written.**"

31. Let p be "You are patriotic," v be "You vote," a be "You are an aardvark," and e be "You have emotions." Then, the symbolic forms of the given premises are: $\sim p \to \sim v$, $a \to \sim e$, and $\sim e \to \sim p$. Now arrange these in the order:
$$a \to \sim e,\ \sim e \to \sim p,\ \sim p \to \sim v.$$

A conclusion that uses all the premises is: **a $\to$ $\sim$v**, or in words, "**Aardvarks do not vote.**"

33. Conclusion (c) makes the argument valid. This can be seen by letting p be "I drive to work," q be "I will not be late," and n be "I do not lose any pay." Then the premises can be written as $p \to q$ and $q \to n$, from which it follows that $p \to n$, which is conclusion **(c)**.

35. Let p be "All persons pay their bills on time," and q be "Some collection agencies are needed." Then the first premise is $p \to \sim q$ and the second premise is q. The contrapositive of the first premise is $q \to \sim p$. Thus, we have q and $q \to \sim p$, so a valid conclusion is $\sim p$. This is statement **(b)**.

37. Let e be "Bill studies economics," m be "Bill makes money," b be "Bill studies business procedures," and g be "Bill makes good money." Then the premises are e → m, b → g, and e ∧ ~b. If the third premise is true, then b is false. If b is false, then the second premise allows g to be either true or false. Thus, neither statement (a) nor statement (b) is a logical conclusion. Statement (c) cannot be deduced from the given premises. So **none** of the proposed statements is a logical conclusion.

39. You read X magazine.

41. "Affirming the consequent" means that the "then" statement is affirmed and "if" statement is taken as a valid conclusion. This is a fallacy because the statement p → q is true if p is false and q is either true or false.

43. Let f be "A kitten loves fish," u be "A kitten is unteachable," t be "A kitten has a tail," p be "A kitten will play with a gorilla," w be "A kitten has whiskers," and g be "A kitten has green eyes." Then, the symbolic forms of the given statements are:

    (1) f → ~u (2) ~t → ~p (3) w → f (4) ~u → ~g (5) ~w → ~t

    Now replace statements (2) and (5) by their respective contrapositives:(2') p → t (5') t → w. Then arrange the premises in the order (2'), (5'), (3), (1), (4):   p → t, t → w, w → f, f → ~u, ~u → ~g

    A conclusion using all these premises is: **p → ~g** or, in words, "**Kittens that will play with a gorilla do not have green eyes**. (An equivalent conclusion is, "**No kitten with green eyes will play with a gorilla.**")

52 CHAPTER 2 LOGIC

EXERCISE 2.7

STUDY TIPS If you have a **series** of switches going from A to B, they will be shown in a straight line. The circuit will be closed when all switches in this line are closed. This is why a series circuit is associated with the statement $p \wedge q$. If you have switches in **parallel** they will be shown by parallel lines. Switches in parallel are associated with the statement $p \vee q$. In this case you can go from A to B by using either path.

1. Corresponding to $(p \wedge q)$ the network must have a branch with switches P and Q in series. Corresponding to p, there must be a branch with switch P alone. These two branches must be in parallel because the symbol joining the two statements is $\vee$ (or).

3. Corresponding to $(\sim p \wedge q)$, the network must have a branch with P' and Q in series. Corresponding to $(p \wedge \sim r)$, it must have a branch with P and R' in series. These two branches must be in parallel because the symbol joining the two statements is $\vee$ (or).

5. Corresponding to $(p \vee \sim q)$, the network must have a branch with P and Q' in parallel.. Corresponding to q, it must have a branch with Q alone. Corresponding to $(\sim p \vee q)$, it must have a branch with P' and Q in parallel. These three branches must be in parallel because the three statements are joined by the the symbol $\vee$ (or). The brackets in the given statement mean that the three branches form a separate circuit
In parallel with this circuit, there must be a branch with Q alone, corresponding to the final q in the given statement.

SECTION 2.7 Switching Networks

7. The branch with P and Q in series corresponds to (p ∧ q), and the branch with P' and R in series corresponds to (~p ∧ r). These two branches are in parallel, so the final statement is **(p ∧ q) ∨ (~p ∧ r)**.

9. Here there are three parallel branches, each with two switches in series: P and R, Q and P, and R and Q'. The three corresponding statements are p ∧ r, q ∧ p, and r ∧ ~q. Since the branches are in parallel, the required statement is **(p ∧ r) ∨ (q ∧ p) ∨ (r ∧ ~q)**.

11. Here there are two circuits in series. The first has two parallel branches, one with P and Q in series and the other with R alone. The statement corresponding to this circuit is (p ∧ q) ∨ r. The second circuit also has two parallel branches, one with P' and R and the other with Q only. The statement corresponding to this circuit is (~p ∧ r) ∨ q. Since the two circuits are in series, the required statement is **[(p ∧ q) ∨ r] ∧ [(~p ∧ r) ∨ q]**.

13. Here are the two circuits corresponding to the given statements:

    The first circuit has two parallel branches, each with P alone. The second has one branch with P alone. If switch P is closed, current will flow in both circuits, and if switch P is open, current will not flow in either circuit. Thus, the two circuits are equivalent, so the statements **p ∨ p** and **p** are also **equivalent**.

15. Here are the two circuits corresponding to the given statements:

    ```
      ┌─P─Q─┐
    A─┤     ├─B         A— P— B
      └──P──┘
    ```

    The first circuit has two parallel branches, one with P and Q in series, and the other with P alone. The second has one branch with P alone. If switch P is closed, current will flow in both circuits, and if P is open, current will not flow in either circuit. Whether Q is open or closed does not matter. Thus, the two circuits are equivalent, so that the statements **(p ∧ q) ∨ p** and **p** are **equivalent**.

54 CHAPTER 2 LOGIC

17. In the given circuit, if switch P is closed (P' is open), then current flows if Q is open (Q' closed), and current does not flow if Q is closed (Q' open). If switch P is open (P' closed), current could flow through the upper branch if Q were closed. But in that case, Q' is open so current cannot flow from A to B. Hence, an equivalent circuit has just one branch with P and Q' in series.

    A———P———Q'———B

    Note: The statement $[(\sim p \wedge q) \vee (p \wedge \sim q)] \wedge q$ corresponds to the given circuit. This statement can be shown to be equivalent to the simpler statement $p \wedge \sim q$, which corresponds to the final answer.

19. In the given circuit, current will not flow if switch Q is closed because Q' is open. If Q' is closed, current will flow if P is closed and not if P is open. Thus, an equivalent circuit has just one branch with P and Q' in series.

    A———P———Q'———B

    Note: The statement $[(p \vee q) \vee (q \vee p)] \wedge \sim q$ corresponds to the given circuit. This statement can be shown to be equivalent to the simpler statement $p \wedge \sim q$, which corresponds to the circuit above.

21. In a circuit with switches P and Q in series, current will flow if and only if both P and Q are closed. This corresponds to the statement $p \wedge q$, which is true if and only if both p and q are true.

23. In the diagram, A is an AND gate and B is a NOT gate. Thus, the inputs to A correspond to p and $\sim q$. The final output is symbolized by $p \wedge \sim q$. The table shows that the final output voltage is low (0) except when the input voltage corresponding to P is high (1) and that corresponding to Q is low (0).

1	2	4	3
p	q	p	$\wedge \sim q$
1	1	0	0
1	0	1	1
0	1	0	0
0	0	0	1

SECTION 2.7 Switching Networks

25. In the diagram, A and D are OR gates, B and C are NOT gates, and E is an AND gate. Thus, the output of A corresponds to $p \vee q$, the output of B corresponds to $\sim p$, the output of C to $\sim q$, the output of D to $\sim p \vee \sim q$. So the output of E must correspond to $(p \vee q) \wedge (\sim p \vee \sim q)$. Column 7 of the table shows that the final output voltage is low (0) when the input voltages corresponding to P and Q are both low (0) or both high (1); the output voltage is high in the other cases.

1	2	3	7	4	6	5
p	q	$(p \vee q)$	$\wedge$	$(\sim p$	$\vee$	$\sim q)$
1	1	1	0	0	0	0
1	0	1	1	0	1	1
0	1	1	1	1	1	0
0	0	0	0	1	1	1

PRACTICE TEST 2

STUDY TIPS Remember, there are two extra Practice Tests with answers at the end of this manual   Here are two test taking tips:
1. Get a good night's sleep before taking your test and
2. Keep good notes with all the information you need for your test.
   (3 by 5 cards are great for this.)

1. **(b), (c), (d)**, and **(e)** are statements. Each is either true or false. (a) is not a statement because people do not agree on what is good. (f) is a question, so is neither true nor false.

2. (a) d: The number of years is divisible by 4. p: The year is a presidential election year. The logical connective is **if ... then ...** $d \rightarrow p$.
   (b) b: I love Bill. $\sim$m: Bill does not love me. The logical connective is **and**. $b \wedge \sim m$.
   (c) e: A candidate is elected president of the United States. m: He receives a majority of the electoral college votes. The logical connective is **if and only if.** $e \leftrightarrow m$.
   (d) s: Janet can make sense out of symbolic logic. f: She fails this course. The logical connective is **or.** $s \vee f$.
   (e) s: Janet can make sense out of symbolic logic. The logical connective is **not**. $\sim s$

3. (a) It is not the case that he is a gentleman and a scholar.
   (b) He is not a gentleman, but he is a scholar.

56 CHAPTER 2 LOGIC

4. (a) I will go neither to the beach nor to the movies.
   (b) I will either not stay in my room or not do my homework.
   (c) Pluto is a planet.

5. (a) Some cats are not felines.
   (b) No dog is well trained.
   (c) Some dogs are afraid of a mouse.

6. (a) Joey does not study, and he will not fail this course.
   (b) Sally studies hard, but she does not get an A in this course.

7. (a) $p \leftrightarrow q$ (b) $p \wedge q$ (c) $\sim p$ (d) $p \rightarrow q$ (e) $p \vee q$

8.
1	2	3	7	4	6	5
p	q	(p ∨ q)	∧	(∼p	∨	∼q)
T	T	T	F	F	F	F
T	F	T	T	F	T	T
F	T	T	T	T	T	F
F	F	F	F	T	T	T

9.
1	2	3	5	4
p	q	(p ∨ q)	→	∼p
T	T	T	F	F
T	F	T	F	F
F	T	T	T	T
F	F	F	T	T

10. Make a truth table for the given statement and statements a and b:

1	2	3	4	6	5	8	7
p	q	∼p	∨ q	∼p	∧ ∼q	∼	(p ∧ ∼q)
T	T	F	T	F	F	T	F
T	F	F	F	F	T	F	T
F	T	T	T	F	F	T	F
F	F	T	T	T	T	T	F

Since Columns 4 and 8 show the same truth values, TFTT, statement (b) is equivalent to the given statement.

11. When at least one of the statements "Sally is naturally beautiful," and "Sally knows how to use makeup," is true.

12. The premise, "2 + 2 = 5," is false, so the statement (a conditional) is true.

13. In making the truth table, keep in mind that a biconditional is true when both components have the same truth value, and is false in the other cases.

1 2	3	6	5 4
p q	(p → q)	↔	(q ∨ ~p)
T T	T	T	T F
T F	F	T	F F
F T	T	T	T T
F F	T	T	T T

14. You should keep in mind that, starting with p → q, the converse is q → p, the inverse is ~p → ~q, and the contrapositive is ~q → ~p.
    (a) If you make a golf score of 62 again, then you made it once.
    (b) If you did not make a golf score of 62 once, then you will not make it again.
    (c) If you do not make a golf score of 62 again, then you did not make it once.

15. (a) m → p   (b) p → m   (c) p ↔ m

16. (a) b → c   (b) c → b   (c) b ↔ c

17. b implies a; b implies c; c implies a.

18. Statement (b), p ∨ ~p, is always true, so it is a tautology.
    Statement (a), p ∧ ~p, is always false, so it is not a tautology.
    Statement (c), (p → q) ↔ (~q ∨ p), is not always true. For example, if p is false and q is true, then (p → q) is true, but (~q ∨ p) is false, so the biconditional is false. Thus, it is **not** a tautology.

19. Nothing in the premises tells whether the J (for John) is inside or outside of the circle H. Thus, the argument is **invalid**.

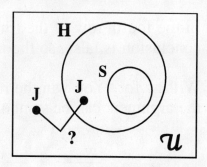

20. Nothing in the premises tells whether S (for Sally) goes inside the circle L or not. Thus, the argument is **invalid**.

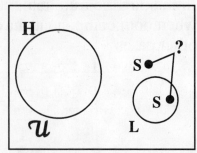

21. With s for "He is a student," and h for "He studies hard," the argument can be symbolized and a truth table constructed.

    s → h
    ~s
    ∴ ~h

    | | | | Prem. | Prem. | Concl. |
s	h	s → h	~s	~h	
T	T	T	F	F	
T	F	F	F	T	
F	T	T	T	F	← Prem.T,Concl.F
F	F	T	T	T	

    In the third row of the table, both premises are true and the conclusion is false, so the argument is **invalid**.

22. With f for "Sally is a loafer," and h for "Sally works hard," the argument can be symbolized and a truth table constructed.

    f → ~h
    ~h
    ∴ f

    | | | Prem. | Prem. | Concl. |
f	h	f → ~h	~h	f
T	T	F	F	T
T	F	T	T	T
F	T	T	F	F
F	F	T	T	F ← Prem.T, Concl.F

    In the fourth row of the truth table, the premises are both true and the conclusion is false, so the argument is **invalid**.

23. With w for "You win the race," and r for "You are a good runner", the argument can be symbolized and a truth table constructed.

    w → r
    w
    ∴ r

    | | | | Prem. | Prem. | Concl. |
r	w	w → r	w	r	
T	T	T	T	T	
T	F	T	F	T	
F	T	F	T	F	
F	F	T	F	F	

    The first row of the table is the only row where the premises are both true. In this row the conclusion is also true, so the argument is **valid**.

24. Suppose that p, q, and r are all true. Then, the first premise, p → q, is true because p and q are both true. The second premise, ~q → ~r is also true, because ~q is false. However, the conclusion, p → ~r is false, because p is true and ~r is false. In this case, the premises are both true and the conclusion is false Thus, the argument is **invalid**. (The first row of the truth table would correspond to the preceding statements.)

25. Since p → q is equivalent to ~p ∨ q, and ~p → -q is equivalent to p ∨ ~q, the given statement is equivalent to (~p ∨ q) ∧ (p ∨ ~q). Thus, the following is the corresponding switching circuit.

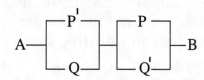

CHAPTER 3
NUMERATION SYSTEMS

EXERCISE 3.1

STUDY TIPS This chapter will give you an insight into the development of our number system by studying the Egyptian, Babylonian, and Roman systems. What are the differences and similarities of these systems? For one thing, the symbols used. You have to study these symbols to be able to write numbers in the corresponding systems. Also, the Egyptian system was an **additive** system but not a **positional** system. Thus, 12 could be written as ∩|| or as ||∩. (In our system 21 is certainly different from 12.) The Babylonians, on the other hand, used a **positional** system. For instance, ▼▼ meant 2 but ▼ ▼ meant 61. The position of the symbol changed its meaning and, to get the final value, the values of the symbols were added. This is why this system is a *positional*, *additive* system. (The base is also different, as the Babylonians used base 60.) The Roman system is also positional and additive, even though the symbols are different from the Babylonian and Egyptian symbols.. Problems 57-60 will acquaint you with numerology (Don't take this too seriously!) as well as give you practice with our number system.

1. ∩ ∩ ||||

3. 9 ∩ ∩ ∩ ||

5. 9 9 9 9 ∩ ∩ ∩ |||
   9 9 9 9 ||

7. 113   9. 322   11. 11,232

13. ∩ ∩ ∩ | | | |
   + ∩ ∩ | | |
   ─────────────────
     ∩ ∩ ∩ ∩ ∩ | | | | | | |

15. Step 1. Subtract one scroll, two heelbones, and two strokes as indicated by the cancellation marks.

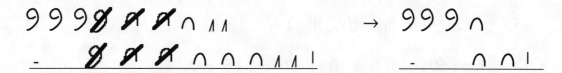

Step 2. In the top line, replace one scroll by ten heelbones and one heelbone by ten strokes. This lets you do the remainder of the subtraction. Subtract two heelbones and one stroke as shown by the cancellation marks. The answer is given as two scrolls, eight heelbones, and nine strokes.

17.	\1	40	19.	1	51	21.	18	32
	\2	80		\2	102		9	64
	\4	160		\4	204		4	128
	\8	320		8	408		2	256
	15	**600**		\16	816		1	512
				22	**1122**			**576**

23.	12	51
	6	102
	3	204
	1	408
		612

25. ▼▼▼▼▼▼

27. $32 = 3 \times 10 + 2 =$ <<< ▼▼

29. $123 = 2 \times 60 + 3 =$ ▼▼   ▼▼▼

31. $258 = 4 \times 60 + 1 \times 10 + 8$

$=$ ▼▼▼▼   <▼▼▼▼▼▼▼▼

62 CHAPTER 3 NUMERATION SYSTEMS

33. $3733 = 1 \times 3600 + 2 \times 60 + 1 \times 10 + 3$
    = ▼ ▼▼ <▼▼▼

35. One 60 plus three 10's plus two 1's = 92.

37. Three 60's plus one 10 plus two 1's = 192.

39. One 3600 plus twelve 60's plus two 1's = 4322.

41.   <<< ▼▼   First add, then replace six 10's
    + <<<< ▼▼▼   by one 60.
    —————————————
    <<<<<<<< ▼▼▼▼▼ = ▼   <▼▼▼▼▼

43.   ▼▼ <▼▼▼   First add, then replace ten
    + ▼ ▼▼▼▼▼▼▼   1's by a ten.
    —————————————
    ▼▼▼   <▼▼▼▼▼▼▼▼▼▼ = ▼▼▼   <<▼

45. CXXVI = $100 + 2 \times 10 + 5 + 1 = 126$

47. $\overline{\text{XLII}} = (\overline{\text{XL}})\overline{\text{II}} = [(50 - 10) + 2] \times 1000 = 42{,}000$

49. $\overline{\text{XCCDV}} = \overline{\text{XC}}(\text{CD})\text{V} = (100 - 10) \times 1000 + (500 - 100) + 5 = 90{,}405$

51. $72 = 50 + 10 + 10 + 2 = $ LXXII

53. $145 = 100 + (50 - 10) + 5 = $ CXLV

55. $32{,}503 = (3 \times 10 + 2) \times 1000 + 500 + 3 = \overline{\text{XXXII}}\,\text{DIII}$

57. J o h n
    $1 + 6 + 8 + 5 = 20 \to 2 + 0 = \mathbf{2}$
    F i t z g e r a l d
    $6 + 9 + 2 + 8 + 7 + 5 + 9 + 1 + 3 + 4 = 54 \to 5 + 4 = 9$
    K e n n e d y
    $2 + 5 + 5 + 5 + 5 + 4 + 7 = 33 \to 3 + 3 = \mathbf{6}$

    $\mathbf{2} + \mathbf{9} + \mathbf{6} = 17 \to 1 + 7 = \mathbf{8} \to \to$ Lonely, misunderstood

SECTION 3.1   Egyptian, Babylonian and Roman Systems   63

59. R   i   n   g   o
$9 + 9 + 5 + 7 + 6 = 36 \rightarrow 3 + 6 = \mathbf{9}$

S   t   a   r   r
$1 + 2 + 1 + 9 + 9 = 22 \rightarrow 2 + 2 = \mathbf{4}$

$\mathbf{4 + 9} = 13 \rightarrow 1 + 3 = \mathbf{4} \rightarrow \rightarrow$ Rebels, unconventional

61. No. C is more than two steps larger than I, so this subtraction is not allowed. I may be subtracted from V or X only.

63. The Babylonian system is a base 60 system and our decimal system is a base 10 system. Another important difference is the lack of a symbol for zero in the Babylonian system. The Babylonian system was not a good place system, it depended on spacing. The symbol for 1 was the same as that for 60, and only the spacing could show which was intended.

65. The Egyptian system was based on 10 and the Babylonian on 60. The Egyptian system was not a positional system; it depended essentially on the addition of the symbol values. The Babylonian system used spacing to change symbol values.

67. Assume the answer is **6**. $6 + (\frac{1}{6})(\mathbf{6}) = 7$, and $21 \div 7 = 3$. Hence, the correct answer is $3 \times \mathbf{6} = 18$.

69. Assume the answer is **3**. $3 + (\frac{2}{3})(\mathbf{3}) = 5$. $5 - (\frac{1}{3})(5) = \frac{10}{3}$ and $10 \div \frac{10}{3} = 3$. Therefore, the correct answer is $3 \times \mathbf{3} = \mathbf{9}$.

71. If you calculated $n^3 + n^2$ for $n = 8$, you found the result to be $512 + 64 = 576$. Thus, the answer is $n = \mathbf{8}$.

64 CHAPTER 3 NUMERATION SYSTEMS

EXERCISE 3.2

STUDY TIPS If you know how much the U.S. owes (the deficit) you know big numbers! To understand how to read, write and operate with numbers you need to know how to write numbers in **expanded form** using **exponents**. What is the meaning of an exponent? It is a number (say **n**) that will tell you how many times another number (say **b**) is to be used as a factor. Thus, b^n means that the number **b** is to be used as a factor **n** times, that is, $b^n = \underbrace{b \times b \times b \times \ldots \times b}_{n \text{ times}}$

Make sure you memorize the laws of exponents: $a^n \times b^m = a^{b+m}$ and $a^n \div b^m = a^{b-m}$. This means that to **multiply** numbers *with the same base*, you **add** the exponents; to **divide** numbers *with the same base*, you **subtract** the exponents. There is one more rule for exponents: raising a power to a power, that is, $(a^m)^n = a^{m \times n}$

1. $(4 \times 10^2) + (3 \times 10) + (2 \times 10^0)$

3. $(2 \times 10^3) + (3 \times 10^2) + (7 \times 10^0)$

5. $(1 \times 10^4) + (2 \times 10^3) + (3 \times 10^2) + (4 \times 10) + (9 \times 10^0)$

7. 1   9. 45   11. 9071   13. 748,308   15. 4,000,031

17. $\begin{array}{r} 23 \\ +13 \\ \hline 36 \end{array}$ $\begin{array}{l} (2 \times 10) + (3 \times 10^0) \\ (1 \times 10) + (3 \times 10^0) \\ \hline (3 \times 10) + (6 \times 10^0) \end{array}$ 19. $\begin{array}{r} 71 \\ +23 \\ \hline 94 \end{array}$ $\begin{array}{l} (7 \times 10) + (1 \times 10^0) \\ (2 \times 10) + (3 \times 10^0) \\ \hline (9 \times 10) + (4 \times 10^0) \end{array}$

21. $\begin{array}{r} 76 \\ -54 \\ \hline 22 \end{array}$ $\begin{array}{l} (7 \times 10) + (6 \times 10^0) \\ (-) (5 \times 10) + (4 \times 10^0) \\ \hline (2 \times 10) + (2 \times 10^0) \end{array}$ 23. $\begin{array}{r} 84 \\ -31 \\ \hline 53 \end{array}$ $\begin{array}{l} (8 \times 10) + (4 \times 10^0) \\ (-) (3 \times 10) + (1 \times 10^0) \\ \hline (5 \times 10) + (3 \times 10^0) \end{array}$

25. $7^{8+3} = 7^{11}$ 27. $6^{19+21} = 6^{40}$ 29. $6^{10-3} = 6^7$

31. $6^{12-0} = 6^{12}$ 33. $(5^4)^3 = 5^{4 \times 3} = 5^{12}$

35. $(10^3)^{10} = 10^{3 \times 10} = 10^{30}$

SECTION 3.2 The Hindu-Arabic (Decimal) System

37. 25 $(2\times 10) + (5\times 10^0)$
     $\underline{\times\; 51}$ $\underline{\times\; (5\times 10) + (1\times 10^0)}$
     25 $(2\times 10) + (5\times 10^0)$
     $\underline{125}$ $\underline{(10\times 10^2) + (25\times 10)}$
     1275 $10^3 + (27\times 10) + (5\times 10^0)$

   $= (1\times 10^3) + (2\times 10^2) + (7\times 10) + (5\times 10^0) = 1275$

39. 62 $(6\times 10) + (2\times 10^0)$
     $\underline{\times\; 25}$ $\underline{\times\; (2\times 10) + (5\times 10^0)}$
     310 $(30\times 10) + (10\times 10^0)$
     $\underline{124}$ $\underline{(12\times 10^2) + (4\times 10)}$
     1550 $(12\times 10^2) + (34\times 10) + (1\times 10)$

   $= (1\times 10^3) + (5\times 10^2) + (5\times 10) = 1550$

41. 8 (8×10^0)
     $8\overline{)64}$ $8\times 10^0\; \overline{)\; (6\times 10) + (4\times 10^0)}$
     $\underline{64}$ $\underline{(6\times 10) + (4\times 10^0)}$
     0 0

43. 12 $(1\times 10) + (2\times 10^0)$
     $6\overline{)72}$ $(6\times 10^0)\overline{)\;(7\times 10) + (2\times 10^0)}$
     $\underline{6}$ $\underline{(6\times 10)}$
     12 $(1\times 10) + (2\times 10^0)$
     $\underline{12}$ $\underline{(1\times 10) + (2\times 10^0)}$
     0 0

45. $4\times 75{,}000 = 300{,}000 = 3\times 10^5$ trees would be saved.

47. $300\times (3\times 10^7) = (3\times 10^2)\times (3\times 10^7)$
     $ = 3\times 3\times 10^2\times 10^7 = 9\times 10^9$

49. $(240\times 10^7) \div 12 = 20\times 10^7 = 2\times 10^8$

51. You must add the exponents to obtain a^{m+n}.

53. You must multiply the exponent m by n to obtain a^{mn}.

55. 7 women, $7^2 = 49$ mules, $7^3 = 343$ sacks, $7^4 = 2401$ loaves, $7^5 = 16{,}807$ knives, $7^6 = 117{,}649$ sheats. Thus, there were $7 + 49 + 343 + 2401 + 16{,}807 + 117{,}649 = 137{,}256$ in all on the road to Rome.

EXERCISE 3.3

STUDY TIPS Numbers in bases other than 10 are used in science and computers. The **base** of a system tells you how the numbers are grouped in that system, so in binary base we group by 2's but in base 5 we group by 5's. Thus, to change a number from base 10 to base 8 you simply find how many groups of 8 there are in the number. The most efficient way to do this is by dividing by 8. (See Example 6) Just remember that to change a number from base 10 to any base b, you must divide the base 10 number by b. These ideas are used in the **bar codes** used to price and identify merchandise in stores and the **postnet** code used by the U.S. Postal Service in delivering your mail.

1. 22_{three} 3. 31_{four} 5. $15_{ten} = 17_{eight}$

7. $15_{ten} = 21_{seven}$ 9. $42_{five} = 4 \times 5 + 2 = 22$

11. $213_{eight} = 2 \times 8^2 + 1 \times 8 + 3 = 128 + 8 + 3 = 139$

13. $11011_{two} = 2^4 + 2^3 + 2 + 1 = 16 + 8 + 2 + 1 = 27$

15. $123_{sixteen} = 1 \times 16^2 + 2 \times 16 + 3 = 256 + 32 + 3 = 291$

17. 5)15
       3 | 0 $15 = 30_{five}$

19. 2)28
    2)14 | 0 $28 = 11100_{two}$
    2) 7 | 0
    2) 3 | 1
       1 | 1

21. 16)25
        1 | 9 $25 = 19_{sixteen}$

23. 6)25
       4 | 1 $25 = 41_{six}$

25. 7)64
    7) 9 | 1
       1 | 2 $64 = 121_{seven}$

27. 8)38
       4 | 6 $38 = 46_{eight}$

29. 16) 1467
        16) 91 | 11
               5 | 11   1467 = $5BB_{sixteen}$

31. 2) 73   8) 73
        2) 36 | 1   8) 9 | 1
        2) 18 | 0   1 | 1
        2) 9 | 0
        2) 4 | 1
        2) 2 | 0   73 = 1001001_{two} = 111_{eight}
        2) 1 | 0

In working problems 33-39, make sure you refer to Table 3.5

33. 00110   01010   00101   00011   01100
          3   5   2   1   6

35. 01001   00101   00011   00110   01010
          4   2   1   3   5

37. |||||||||||||||||||||||||

39. |||||||||||||||||||||||||

41. No. If the Zip Code + 4 consists of 9 numbers and each number is represented by 5 bars, there should only be 45 bars and the two long bars at the beginning and the end of the Zip Code + 4 numbers for a total of 47 bars. See Problem 42 for the solution of this mystery!

43. One of the meanings of binary is "based on two". The prefix bi means "two".

45. Hexadecimal means "based on 16". The prefix hexa means "six".

47 The trick works because the columns correspond to the binary digits in the number. For instance, 6 = 110_{two} and this corresponds to the number 6 = 2 + 4, the numbers that head columns B and C. Note that 6 occurs in columns B and C, but not in A.

68 CHAPTER 3 NUMERATION SYSTEMS

49 Use the same procedure as for the numbers from 1 to 7, but with five columns instead of three. If you label the columns A, B, C, D, E, then the numbers 1, 2, 4, 8, and 16 would head these columns. The number 13, for example, would be put into columns A, C, and D because 13 = 01101_{two}, which means that 13 = 1 + 4 + 8, the numbers that head these three columns.

CALCULATOR CORNER 3.3

1. Key in the following and the calculator will show the answer 13:

   $\boxed{1}\ \boxed{\times}\ \boxed{2}\ \boxed{+}\ \boxed{1}\ \boxed{=}\ \boxed{\times}\ \boxed{2}\ \boxed{+}\ \boxed{0}\ \boxed{=}\ \boxed{\times}\ \boxed{2}\ \boxed{+}\ \boxed{1}\ \boxed{=}$

3. Key in the following and the calculator will show the answer 113:

   $\boxed{4}\ \boxed{\times}\ \boxed{5}\ \boxed{+}\ \boxed{2}\ \boxed{=}\ \boxed{\times}\ \boxed{5}\ \boxed{+}\ \boxed{3}\ \boxed{=}$

5. Key in the following and the calculator will show the answer 1914:

   $\boxed{3}\ \boxed{\times}\ \boxed{8}\ \boxed{+}\ \boxed{5}\ \boxed{=}\ \boxed{\times}\ \boxed{8}\ \boxed{+}\ \boxed{7}\ \boxed{=}\ \boxed{\times}\ \boxed{8}\ \boxed{+}\ \boxed{2}\ \boxed{=}$

EXERCISE 3.4

STUDY TIPS If you have a computer you may know that computer characters (like the letter A) are written in ASCII code using binary values. For example, the letter A has the value $0100\ 0001_2$. In this section we study **binary** arithmetic. To do so, you have to go back in time and recall how you learned the addition tables for base 10 so you could do arithmetic. It's the same idea here. You must know the addition and multiplication facts (Tables 3.6 and 3.7) but this time, instead of doing base 10, we will be doing binary arithmetic (base 2). Start this section by studying and even memorizing these two tables.

1.
$$\begin{array}{r} 1 \\ 111_2 \\ +\ 10_2 \\ \hline 1001_2 \end{array}$$

3.
$$\begin{array}{r} 1 \\ 1101_2 \\ +\ 110_2 \\ \hline 10011_2 \end{array}$$

5.
$$\begin{array}{r} 11 \\ 110_2 \\ +\ 101_2 \\ +\ 111_2 \\ \hline 10010_2 \end{array}$$

7.
$$\begin{array}{r} 111_2 \\ -\ 10_2 \\ \hline 101_2 \end{array}$$

9.
$$\begin{array}{r} 111 \\ 1000_2 \\ -\ 111_2 \\ \hline 1_2 \end{array}$$

11.
$$\begin{array}{r} 1111_2 \\ -\ 101_2 \\ \hline 1010_2 \end{array}$$

13.
$$\begin{array}{r} 110_2 \\ \times\ 11_2 \\ \hline 110 \\ 110 \\ \hline 10010_2 \end{array}$$

15.
$$\begin{array}{r} 1111_2 \\ \times\ 11_2 \\ \hline 1111 \\ 1111 \\ \hline 101101_2 \end{array}$$

17.
$$\begin{array}{r} 1011_2 \\ \times\ 101_2 \\ \hline 1011 \\ 10110 \\ \hline 110111_2 \end{array}$$

19.
$$\begin{array}{r} 110_2\ R\ 1_2 \\ 10_2 \overline{)1101_2} \\ \underline{10} \\ 10 \\ \underline{10} \\ 01 \end{array}$$

21.
$$\begin{array}{r} 100_2\ R\ 10_2 \\ 11_2 \overline{)1110_2} \\ \underline{11} \\ 010 \end{array}$$

23.
$$\begin{array}{r} 1011_2\ R\ 100_2 \\ 101_2 \overline{)1110111_2} \\ \underline{101} \\ 1001 \\ \underline{101} \\ 1001 \\ \underline{101} \\ 100 \end{array}$$

70 CHAPTER 3 NUMERATION SYSTEMS

25. The message is HELLO. The decimal numerals corresponding to the letters are:

$H \leftrightarrow 72, E \leftrightarrow 69, L \leftrightarrow 76, O \leftrightarrow 79$

		Binary	Decimal	Hexadecimal
27.	A	01000001	65	41
29.	Q	01010001	81	51
31.	X	01011000	88	58

33.

Hexadecimal	Binary	Letter
48	01001000	H
45	01000101	E
4C	01001100	L
50	01010000	P

35. $110111_2 = 2^5 + 2^4 + 2^2 + 2 + 1 = 32 + 16 + 4 + 2 + 1 = 55$

37. Since white is represented by 000000_2, the next smallest number, 000001_2 would represent the lightest shade of gray that is not white.

39. $31 = 16 + 8 + 4 + 2 + 1 = 2^4 + 2^3 + 2^2 + 2 + 1 = 011111_2$ (Note that this problem can also be done by successive divisions as shown in Section 3.3.)

PRACTICE TEST 3

STUDY TIPS For your convenience, a list of most of the symbols used in the Babylonian, Egyptian and Roman numeration systems appears in the Summary. Consult your instructor to see if you have to memorize all these symbols or if they will be available to you when you are taking the test. Make sure you are familiar with exponents and their uses, since they are needed in many of the problems in the Practice Test.

1. (a) ∩ ∩ ∩ ||| (b) 9999 ∩ ∩ |||
         ∩ ∩ ∩   999 ∩ ||

2. (a) $10 + 10 + 1 + 1 + 1 = 23$ (b) $100 + 10 + 10 + 1 = 121$

3. (a) ▼ ▼▼▼   (b) ◁ ▼▼   ◁ ▼▼▼▼

4. (a) $60 + 10 + 10 + 1 + 1 = 82$ (b) $60 + 60 + 10 + 1 = 131$

5. (a) \1   21   (b)   23   ☐21☐
   \2   42
   \4   84   11   ☐42☐
    8   168
   \16   336   5   ☐84☐
         483
                                    2   168

                                    1   ☐336☐
                                              483

6. (a) LIII   (b) XLII   (c) $\overline{\text{XXII}}$

7. (a) 67   (b) 48,000

8. (a) $(2 \times 10^3) + (5 \times 10^2) + (0 \times 10) + (7 \times 10^0)$

   (b) $(1 \times 10^2) + (8 \times 10) + (9 \times 10^0)$

72 CHAPTER 3 NUMERATION SYSTEMS

9. (a) 3702 (b) 59,040

10. (a) $\;\;\;75\;\;\;\;\;(7\times10)+5$
    $\;\;\;\;\;\underline{+32}\;\;\;\underline{+(3\times10)+2}$
    $\;\;\;\;107\;\;\;\;(10\times10)+7$
    $\;\;\;\;\;\;\;\;\;\;\;\;=(1\times10^2)+7$
    $\;\;\;\;\;\;\;\;\;\;\;\;=107$

    (b) $\;\;\;56\;\;\;\;\;(5\times10)+6$
    $\;\;\;\;\underline{-24}\;\;\;\underline{(-)(2\times10)+4}$
    $\;\;\;\;32\;\;\;\;\;(3\times10)+2$
    $\;\;\;\;\;\;\;\;\;\;\;=32$

11. (a) $3^4 \times 3^8 = 3^{4+8} = 3^{12}$ (b) $2^9 \div 2^3 = 2^{9-3} = 2^6$

12. (a) $\;\;\;\;83\;\;\;\;\;\;\;\;\;\;\;\;\;\;\;\;(8\times10)+3$
    $\;\;\underline{\times\;21}\;\;\;\;\;\;\;\;\;\;\;\;\underline{\times\;\;\;(2\times10)+1}$
    $\;\;\;\;\;83\;\;\;\;\;\;\;\;\;\;\;\;\;\;\;\;(8\times10)+3$
    $\;\;\underline{166}\;\;\;\;\;\;\;\;\;\;\underline{16\times10^2+(6\times10)}$
    $\;1743\;\;\;\;\;\;\;\;(16\times10^2)+(14\times10)+3$
    $\;\;\;\;\;\;\;\;\;\;\;\;\;\;=(1\times10^3)+(6\times10^2)+(1\times10^2)+(4\times10)+3$
    $\;\;\;\;\;\;\;\;\;\;\;\;\;\;=(1\times10^3)+(7\times10^2)+(4\times10)+3=1743$

    (b) $\;\;\;\;\;\;\;\;\;\;7\;R\;5\;(7\times10^0)\;R\;(5\times10^0)$
    $\;\;\;\;\;\;7)\overline{54}\;\;\;\;\;\;\;\;\;7\times10^0)\overline{(5\times10)+(4\times10^0)}$
    $\;\;\;\;\;\;\;\;\underline{49}\;\;\;\;\;\;\;\;\;\;\;\;\;\;\;\;\;\underline{(4\times10)+(9\times10^0)}$
    $\;\;\;\;\;\;\;\;\;\;5\;(5\times10^0)$

13. (a) $203_4 = (2\times4^2)+(0\times4)+3 = 32+3 = 35$
    (b) $143_5 = (1\times5^2)+(4\times5)+3 = 25+20+3 = 48$
    (c) $11101_2 = (1\times2^3)+(1\times2^2)+(0\times2)+1 = 8+4+1 = 13$

14. (a) $152_8 = (1\times8^2)+(5\times8)+2 = 64+40+2 = 106$
    (b) $A2C_{16} = (10\times16^2)+(2\times16)+12 = 2560+32+12 = 2604$

15. (a) $5)\underline{33\;\;\;|\;\;\;}$
    $\;\;\;\;5)\;\underline{\;\;6\;}|\;3$
    $\;\;\;\;\;\;\;\;\;\;1\;|\;1$
    Thus, $33 = 113_5$.

    (b) $6)\underline{33\;\;\;|\;\;\;}$
    $\;\;\;\;\;\;\;\;\;\;5\;|\;3$
    Thus, $33 = 53_6$.

16. (a)
```
2) 39
2) 19  1
2)  9  1
2)  4  1
2)  2  0
2)  1  0
```
Thus, $39 = 100111_2$.

(b)
```
2) 527
2) 263  1
2) 131  1
2)  65  1
2)  32  1
2)  16  0
2)   8  0
2)   4  0
2)   2  0
     1  0
```
Thus, $527 = 1000001111_2$.

17. (a)
```
8) 47
   5 | 7
```
Thus, $47 = 57_8$.

(b)
```
16) 47
    2 | 15
```
Thus, $47 = 2F_{16}$.

18. (a)
$$\begin{array}{r} 1\,1 \\ 1101_2 \\ +\ 101_2 \\ \hline 10010_2 \end{array}$$

(b)
$$\begin{array}{r} 0\,1 \\ 1101_2 \\ -\ 111_2 \\ \hline 110_2 \end{array}$$

19. (a)
$$\begin{array}{r} 1101_2 \\ \times\ 11_2 \\ \hline 1101 \\ 1101 \\ \hline 100111_2 \end{array}$$

(b)
$$\begin{array}{r} 10110_2 \\ \times\ 101_2 \\ \hline 10110 \\ 101100 \\ \hline 1101110_2 \end{array}$$

20. (a)
```
            111₂ R 1₂
   11₂) 10110₂
         11
         101
          11
         100
          11
           1
```

(b)
```
            1001₂ R 1₂
  110₂) 110111₂
        110
        0111
         110
           1
```

CHAPTER 4
NUMBER THEORY AND THE REAL NUMBERS

EXERCISE 4.1

STUDY TIPS How do we use the **natural** numbers? Let us count the ways! At least three. Since the natural numbers are also called the counting numbers, we use them for counting, that is, as **cardinal** numbers. For example, how many students are in your class? 20, 35, 50? All these numbers are cardinal numbers. We also use natural numbers to assign **order**. Which section of Chapter 4 is this? The **first** section. And finally, we use natural numbers for **identification**. The course you are taking is Math ____. Whatever number goes in that blank is for identification. We also study a special type of natural number called **prime** numbers, that is, numbers that are divisible by exactly two numbers, themselves and 1. If numbers are not prime, they are **composite**, except for the number 1 which is neither prime nor composite (because it has only one divisor, itself). To find the prime numbers, you can use the Sieve of Eratosthenes given in the text. Next, the **Fundamental Theorem of Arithmetic** tells us that any composite number can be written (factored) as a product of primes. You try dividing the given number by successive primes (2, 3, 5, 7, and so on). Caution: To know if a given number is divisible by 2, 3, 5 and so on you need the divisibility rules in Table 4.1. There are other divisibility rules on page 162. As far as prime numbers are concerned, we study two important applications:

(1) **reducing** fractions (you have to learn how to get the GCF, page 156).

(2) **adding** fractions (here you need to find the LCD, page 157).

Good news: there is a simple way of finding both the GCF and LCD (See page 159).

1. For **identification** only.

3. A **cardinal** number (used for counting).

5. The "First" is for **identification**; the "one" is an **ordinal** number.

7. The sieve is constructed in the same way as it was for the numbers from 1 to 50. We cross out all the multiples of 2, of 3, of 5, and of 7. The remaining numbers are all primes. We don't have to go beyond 7, because the next prime is 11 and its square is greater than 100. The primes in the table are circled.

$$51\ 52\ \widehat{53}\ 54\ 55\ 56\ 57\ 58\ \widehat{59}\ 60$$
$$\widehat{61}\ 62\ 63\ 64\ 65\ 66\ \widehat{67}\ 68\ 69\ 70$$
$$\widehat{71}\ 72\ \widehat{73}\ 74\ 75\ 76\ 77\ 78\ \widehat{79}\ 80$$
$$81\ 82\ \widehat{83}\ 84\ 85\ 86\ 87\ 88\ \widehat{89}\ 90$$
$$91\ 92\ 93\ 94\ 95\ 96\ \widehat{97}\ 98\ 99\ 100$$

9. There are **6** primes between 25 and 50: 29, 31, 37, 41, 43, 47.

11. There are **4** primes between 75 and 100: 79, 83, 89, 97.

13. (a) 2 and 3 are both primes and are consecutive counting numbers.
    (b) If any pair of consecutive counting numbers greater than 2 is selected, one of the pair must be an even number (divisible by 2) and hence, not a prime. Thus, there cannot be a second pair of primes that are consecutive counting numbers.

15. (a) The product part of m is exactly divisible by 2, so that m divided by 2 would have a remainder of 1.
    (b) The product part of m is exactly divisible by 3, so that m divided by 3 would have a remainder of 1.
    (c) and (d) Exactly the same reasoning as in Parts (a) and (b) applies here. If m is divided by any prime from 2 to P, there is remainder of 1.
    (e) Because P was assumed to be the largest prime.
    (f) Because m is not divisible by any of the primes from 2 to P.

17. $50 = 2 \times 5 \times 5$. Thus, the divisors of 50 are 1, 2, 5, 2×5 or 10, 5×5 or 25, and $2 \times 5 \times 5$ or 50.

19. $128 = 2 \times 2 \times 2 \times 2 \times 2 \times 2 \times 2 = 2^7$ Thus, the divisors of 128 are 1, 2, 2×2 or 4, 2×4 or 8, 2×8 or 16, 2×16 or 32, 2×32 or 64, and 2×64 or 128.

21. $1001 = 7 \times 11 \times 13$. Thus, the divisors of 1001 are 1, 7, 11, 13, 7×11 or 77, 7×13 or 91, 11×13 or 143, and $7 \times 11 \times 13$ or 1001.

23. 41 is a prime.   25. $91 = 7 \times 13$ 27. $148 = 2^2 \times 37$

29. $2 \times 5 \times 7^2 = 490$ 31. $2^4 \times 3 \times 5^2 = 1200$

76 CHAPTER 4 NUMBER THEORY AND THE REAL NUMBERS

33. (a) 6345 is not an even number, so is not divisible by 2. The sum of the digits is 18, so the number is divisible by 3. The number ends in 5, so is divisible by 5.
   (b) 8280 ends in 0, so is divisible by 2 and by 5. The sum of the digits is 18, so the number is divisible by 3.
   (c) 11,469,390 ends in 0, so is divisible by 2 and by 5. The sum of the first three digits is 6, which is divisible by 3, and the other digits are all multiples of 3; thus, the number is divisible by 3.

35. In order to leave a remainder of 1 when divided into 23, the divisor must be an exact divisor of 22: 2, 11 or 22. All three of these leave a remainder of 1 when divided into 45. Thus, there are three whole numbers, 2, 11 and 22, that satisfy the given condition.

37. $135 = 3^3 \times 5$, $351 = 3^3 \times 13$. GCF $= 3^3 = 27$.

39. $147 = 3 \times 7^2$, $260 = 2^2 \times 5 \times 13$. GCF $= 1$, so 147 and 260 are relatively prime.

41. $282 = 2 \times 3 \times 47$, $329 = 7 \times 47$. GCF $= 47$.

43. $12 = 2^2 \times 3$, $15 = 3 \times 5$, $20 = 2^2 \times 5$. GCF $= 1$, so 12, 15 and 20 are relatively prime.

45. $100 = 2^2 \times 5^2$, $200 = 2^3 \times 5^2$, $320 = 2^6 \times 5$. GCF $= 2^2 \times 5 = 20$.

47. $\dfrac{62}{88} = \dfrac{2 \times 31}{2 \times 44} = \dfrac{31}{44}$

49. $156 = 2^2 \times 3 \times 13$ and $728 = 2^3 \times 7 \times 13$. GCF is $2^2 \times 13$ or 52. So
$$\dfrac{156}{728} = \dfrac{52 \times 3}{52 \times 14} = \dfrac{3}{14}$$

51. $96 = 2^5 \times 3$ and $384 = 2^7 \times 3$, so the GCF is $2^5 \times 3$ or 96. Thus,
$$\dfrac{96}{384} = \dfrac{96 \times 1}{96 \times 4} = \dfrac{1}{4}$$

53. $15 = 3 \times 5$ and $55 = 5 \times 11$, so the LCM is $3 \times 5 \times 11 = 165$.
$$\dfrac{1}{15} + \dfrac{1}{55} = \dfrac{11}{165} + \dfrac{3}{165} = \dfrac{14}{165}$$

SECTION 4.1 Number Theory 77

55. $32 = 2^5$ and $124 = 2^2 \times 31$, so the LCM is $2^5 \times 31 = 992$.
$$\frac{3}{32} + \frac{1}{124} = \frac{3 \times 31}{992} + \frac{8 \times 1}{992}$$
$$= \frac{93}{992} + \frac{8}{992} = \frac{101}{992}$$

57. $180 = 2^2 \times 3^2 \times 5$ and $240 = 2^4 \times 3 \times 5$, so the LCM is $2^4 \times 3^2 \times 5 = 720$.
$$\frac{1}{180} + \frac{1}{240} = \frac{4}{720} + \frac{3}{720} = \frac{7}{720}$$

59. $12 = 2^2 \times 3$, $18 = 2 \times 3^2$, $30 = 2 \times 3 \times 5$, so the LCM is $2^2 \times 3^2 \times 5$ or 180.
$$\frac{1}{12} + \frac{1}{18} + \frac{1}{30} = \frac{15}{180} + \frac{10}{180} + \frac{6}{180} = \frac{31}{180}$$

61. $285 = 3 \times 5 \times 19$, $315 = 3^2 \times 5 \times 7$, and $588 = 2^2 \times 3 \times 7^2$, so the LCM is $2^2 \times 3^2 \times 5 \times 7^2 \times 19 = 167{,}580$.
$$\frac{1}{285} + \frac{1}{315} + \frac{1}{588} = \frac{2^2 \times 3 \times 7^2}{167580} + \frac{2^2 \times 7 \times 19}{167580} + \frac{3 \times 5 \times 19}{167580}$$
$$= \frac{588 + 532 + 285}{167580} = \frac{1405}{167580} = \frac{281}{33516}$$

63. $200 = 2^2 \times 5 \times 10$
$300 = 2 \times 3 \times 5 \times 10$
$420 = 2 \times 3 \times 7 \times 10$
LCM $= 2^2 \times 3 \times 5 \times 7 \times 10 = 4200$

$$\frac{1}{200} + \frac{1}{300} + \frac{1}{420} = \frac{21 + 14 + 10}{4200} = \frac{45}{4200} = \frac{3}{280}$$

65. $1 - \frac{1}{4} - \frac{3}{20} - \frac{1}{10} = 1 - \frac{5}{20} - \frac{3}{20} - \frac{2}{20} = \frac{10}{20} = \frac{1}{2}$

67. $1 - \frac{3}{5} - \frac{1}{4} = 1 - \frac{12}{20} - \frac{5}{20} = \frac{3}{20}$

69. $1 - \frac{3}{10} - \frac{1}{5} - \frac{1}{20} - \frac{3}{20} = 1 - \frac{6}{20} - \frac{4}{20} - \frac{1}{20} - \frac{3}{20} = 1 - \frac{14}{20} = 1 - \frac{7}{10} = \frac{3}{10}$

78 CHAPTER 4 NUMBER THEORY AND THE REAL NUMBERS

71. (a) This can be done in several ways:
    $100 = 3 + 97 = 11 + 89 = 17 + 83 = 29 + 71 = 41 + 59$.
    (b) This can also be done in several ways:
    $200 = 3 + 197 = 7 + 193 = 19 + 181 = 37 + 163 = 43 + 157$
    $= 61 + 139 = 73 + 127 = 97 + 103$

73. The number 1 has only one divisor, itself. It is not a prime because a prime must have exactly **two distinct** divisors, 1 and itself. It is not a composite number because it has only one divisor.

75. The largest prime that you need to try is **13**, because the next prime is 17 and $17^2 = 289$ which is greater than 211.

77. All the other digits are multiples of 3, so their sum is divisible by 3. Hence, only the sum of **2** and **7** needs to be checked.

79. Since 999 and 99 and 9 are all divisible by 9, only the sum
    $$2 \times 1 + 8 \times 1 + 5 \times 1 + 3,$$
    which is exactly the sum of the digits, needs to be checked. If this sum is divisible by 9, the original number is divisible by 9 and not otherwise.

81. (a) 1436 is divisible by 4 because 36 is divisible by 4. 1436 is not divisible by 8 because 436 is not divisible by 8.
    (b) 21,408 is divisible by 8 because 408 is divisible by 8. Being divisible by 8, the number is also divisible by 4.
    (c) 347,712 is divisible by 8 because 712 is divisible by 8. Since the number is divisible by 8, it is also divisible by 4.
    (d) 40,924 is divisible by 4 because 24 is divisible by 4. The number is not divisible by 8 because 924 is not divisible by 8.

83. None of the numbers 1, 2, 3, 4, 5 is the sum of its proper divisors. Therefore, **6** is the smallest perfect number.

85.
    ```
    2 )496
    2 )248
    2 )124
    2 ) 62
        31
    ```
    The division shows that $496 = 2^4 \times 31$. Thus the proper divisors of 496 are 1, 2, 4, 8, 16, 31, 62, 124, and 248. The sum of these divisors is 496, so 496 is a perfect number.

87. The only proper divisor of a prime is the number 1. Thus, a prime is always greater than the sum of its proper divisors. Consequently, all primes are deficient.

SECTION 4.2 The Whole Numbers and the Integers

EXERCISE 4.2

STUDY TIPS Two new sets of numbers are introduced: the set of whole numbers {0, 1, 2, 3...} (almost the same as the natural numbers, only **0** has been added), and the set of integers {...-2, -1, 0, 1, 2... }. The set of integers consists of the whole numbers and their **additive inverses.** If you mark a number and its additive inverse on the number line, they are always on opposite sides of the 0 point. Two consequences of this fact: (1) - (-a) = a and (2) additive inverses are sometimes called **opposites.** Note that integers can be **positive** (1, 2, 3 and so on), **negative** (-1, -2, -3 and so on) or 0, which is neither positive, nor negative. Caution: When we say **non-negative** it means 0, 1, 2, 3 and so on, since these number are **not** negative. Three more items: learn the rule of signs (Table 4.2), the properties on pages 169-170 and the order in which operations must be performed. (**PEMDAS** is a good way to remember that you work from left to right and remove **p**arentheses, do **e**xponents, do **m**ultiplications and **d**ivisions as they occur from left to right and do **a**ddition and **s**ubtraction also as they occur from left to right.)

1.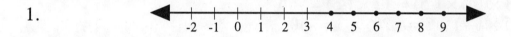

3.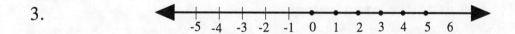

5. Because 3 + (-3) = 0, the additive inverse of 3 is **-3**.

7. Because (-8) + 8 = 0, the additive inverse of -8 is **8**.

9.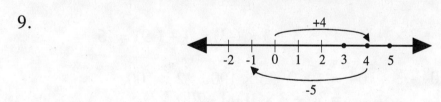

11.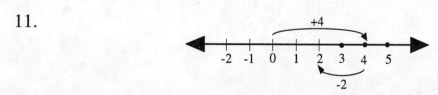

80 CHAPTER 4 NUMBER THEORY AND THE REAL NUMBERS

13.
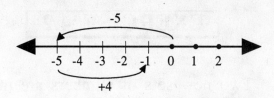

15.
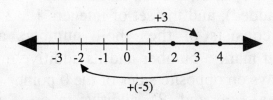

17. $3 - 8 = 3 + (-8) = $ **-5**   19. $3 - 4 = 3 + (-4) = $ **-1**

21. $-5 - 2 = -5 + (-2) = $ **-7**   23. $5 - (-6) = 5 + (+6) = $ **11**

25. $-3 - (-4) = -3 + (+4) = $ **1**   27. $-5 - (-3) = -5 + (+3) = $ **-2**

29. (a) $(-5) \times 3 = $ **-15**   (b) $(-8) \times 9 = $ **-72**

31. (a) $4 \times (-5) = $ **-20**   (b) $3 \times (-13) = $ **-39**

33. (a) $3 \times (4) \times (-5) = 12 \times (-5) = $ **-60**
    (b) $5 \times (-4) \times 3 = (-20) \times 3 = $ **-60**

35. If n is a negative odd integer, then n - 1 is a negative even integer and $(n - 1)^3$ is a **negative even integer**.

37. (a) $-3(4+5) = -3 \times 9 = $ **-27**   (b) $-4(4 - 5) = -4 \times -1 = $ **4**

39. (a) $-5 + (-5 + 1) = -5 + (-4) = $ **-9**   (b) $-8 + (-2 + 5) = -8 + (+3) = $ **-5**

41. $(-2 - 4)(-3) - 8(5 - 4) = (-6)(-3) - 8(1) = 18 - 8 = $ **10**

43. $6 \times 2 \div 3 + 6 \div 2 \times (-3) = 12 \div 3 + 3 \times (-3) = 4 + (-9) = $ **-5**

45. $4 \times 9 \div 3 \times 10^3 - 2 \times 10^2 = 4 \times 9 \div 3 \times 1000 - 2 \times 100$
    $= 36 \div 3 \times 1000 - 200$
    $= 12 \times 1000 - 200$
    $= 12{,}000 - 200$
    $= $ **11,800**

47. $9 - (-11) = 9 + 11 = $ **20 km**

SECTION 4.2   The Whole Numbers and the Integers   81

49. Step 1. m is assumed to be a multiplicative identity, so m•a = a.
Step 3. 1 is a multiplicative identity, so m•1 = m.
Step 4. Both m and 1 equal m•1, so m = 1.

51. Step 1. By the definition of subtraction.
Step 3. By the associative property of addition,
$$[a + (-0)] + 0 = a + [(-0) + 0]$$
Step 5. a + 0 = a because 0 is the additive identity.

53. (b) adding a   (c) identity   (e) q   (f) identity ... unique

55. Step 1. 0 is the additive identity, so 0 + 0 = 0.
Step 3. By the distributive property, a•(0 + 0) = a•0 + a•0
Step 5. Since the additive identity is unique, if a•0 + x = a•0, then x must be 0. Here, x = a•0.

57. The product of two positive numbers is a **positive** number.

59. The product of two negative numbers is a **positive** number.

61. 5 + 4 × (-2) = 5 - 8 = **-3**   63. 1 + 1 + 4 + 3 × (-2) = 6 + (-6) = **0**

EXERCISE 4.3

STUDY TIPS   List all the kinds of rational numbers that you know. The list should include integers, fractions, decimals and percents. The key here is that a rational number **can be** written in the form $\frac{a}{b}$, where a and b are integers. We do addition, subtraction and multiplication of rationals. (Remember that we told you earlier that you would need LCD's, LCM's and the laws of signs?) Be consistent with answers. If the problem has proper fractions, give the answer as a proper fraction when possible. If it has mixed numbers, use mixed numbers for the answer.

1. The numerator is **3**; the denominator is **4**.

3. The numerator is **3**; the denominator is **-5**.

5. 17 × 697 = 11,849 = 41 × 289. Thus, $\frac{17}{41} = \mathbf{\frac{289}{697}}$.

82 CHAPTER 4 NUMBER THEORY AND THE REAL NUMBERS

7. $11 \times 2093 = 23{,}023 = 91 \times 253$. Thus, $\dfrac{11}{91} = \dfrac{\mathbf{253}}{\mathbf{2093}}$.

9. $\dfrac{95}{38} = \dfrac{5 \times 19}{2 \times 19} = \dfrac{\mathbf{5}}{\mathbf{2}}$ 　　　11. $\dfrac{21}{48} = \dfrac{3 \times 7}{3 \times 16} = \dfrac{\mathbf{7}}{\mathbf{16}}$

13. $\dfrac{30}{28} = \dfrac{2 \times 15}{2 \times 14} = \dfrac{\mathbf{15}}{\mathbf{14}}$ 　　　15. $\dfrac{52}{78} = \dfrac{2 \times 26}{3 \times 26} = \dfrac{\mathbf{2}}{\mathbf{3}}$

17. $\dfrac{2}{9} + \dfrac{1}{6} + \dfrac{7}{18} = \dfrac{4}{18} + \dfrac{3}{18} + \dfrac{7}{18} = \dfrac{4+3+7}{18} = \dfrac{\mathbf{14}}{\mathbf{18}}$

19. $\dfrac{1}{3} + \dfrac{1}{6} + \dfrac{1}{9} = \dfrac{6}{18} + \dfrac{3}{18} + \dfrac{2}{18} = \dfrac{6+3+2}{18} = \dfrac{\mathbf{11}}{\mathbf{18}}$

21. $\dfrac{1}{7} + \dfrac{1}{9} = \dfrac{9}{63} + \dfrac{7}{63} = \dfrac{\mathbf{16}}{\mathbf{63}}$ 　　　23. $\dfrac{3}{4} + \dfrac{5}{6} = \dfrac{9}{12} + \dfrac{10}{12} = \dfrac{\mathbf{19}}{\mathbf{12}}$

25. $\dfrac{3}{17} + \dfrac{7}{19} = \dfrac{57}{323} + \dfrac{119}{323} = \dfrac{\mathbf{176}}{\mathbf{323}}$ 　　　27. $\dfrac{1}{7} - \dfrac{1}{9} = \dfrac{9}{63} - \dfrac{7}{63} = \dfrac{\mathbf{2}}{\mathbf{63}}$

29. $\dfrac{3}{4} - \dfrac{5}{6} = \dfrac{9}{12} - \dfrac{10}{12} = -\dfrac{\mathbf{1}}{\mathbf{12}}$ 　　　31. $\dfrac{7}{19} - \dfrac{3}{17} = \dfrac{119}{323} - \dfrac{57}{323} = \dfrac{\mathbf{62}}{\mathbf{323}}$

33. $\dfrac{2}{5} \times \dfrac{5}{3} = \dfrac{\mathbf{2}}{\mathbf{3}}$ 　　　35. $\dfrac{3}{4} \div \dfrac{2}{7} = \dfrac{3}{4} \times \dfrac{7}{2} = \dfrac{\mathbf{21}}{\mathbf{8}}$

37. $\dfrac{7}{9} \div \dfrac{3}{8} = \dfrac{7}{9} \times \dfrac{8}{3} = \dfrac{\mathbf{56}}{\mathbf{27}}$ 　　　39. $\left(-\dfrac{6}{7}\right) \times \left(-\dfrac{3}{11}\right) = \dfrac{\mathbf{18}}{\mathbf{77}}$

41. $\left(-\dfrac{3}{4}\right) \div \left(-\dfrac{7}{6}\right) = \left(-\dfrac{3}{4}\right) \times \left(-\dfrac{6}{7}\right) = \dfrac{\mathbf{9}}{\mathbf{14}}$

43. $\dfrac{1}{8} \div \left(-\dfrac{3}{4}\right) = \dfrac{1}{8} \times \left(-\dfrac{4}{3}\right) = -\dfrac{\mathbf{1}}{\mathbf{6}}$ 　　　45. $\left(-\dfrac{1}{8}\right) + \dfrac{1}{4} = -\dfrac{1}{8} + \dfrac{2}{8} = \dfrac{\mathbf{1}}{\mathbf{8}}$

47. $\dfrac{3}{8} - \left(\dfrac{1}{4} - \dfrac{1}{8}\right) = \dfrac{3}{8} - \left(\dfrac{2}{8} - \dfrac{1}{8}\right) = \dfrac{3}{8} - \dfrac{1}{8} = \dfrac{2}{8} = \dfrac{\mathbf{1}}{\mathbf{4}}$

49. $\dfrac{1}{2} \times \left(\dfrac{7}{8} \times \dfrac{7}{5}\right) = \dfrac{1 \times 7 \times 7}{2 \times 8 \times 5} = \dfrac{\mathbf{49}}{\mathbf{80}}$

51. $(\frac{1}{2} \div \frac{1}{8}) \div \frac{1}{4} = (\frac{1}{2} \times \frac{8}{1}) \times \frac{4}{1} = 4 \times 4 =$ **16**

53. $\frac{2}{3}(\frac{1}{2} + \frac{3}{4}) + \frac{2}{3} = \frac{2}{3}(\frac{2}{4} + \frac{3}{4}) + \frac{2}{3} = \frac{2}{3} \times \frac{5}{4} + \frac{2}{3} = \frac{5}{6} + \frac{4}{6} = \frac{9}{6} = \mathbf{\frac{3}{2}}$

55. $\frac{1}{3}(\frac{3}{2} - \frac{1}{5}) - \frac{1}{30} = \frac{1}{3}(\frac{15}{10} - \frac{2}{10}) - \frac{1}{30} = \frac{1}{3} \times \frac{13}{10} - \frac{1}{30} = \frac{13}{30} - \frac{1}{30} = \frac{12}{30} = \mathbf{\frac{2}{5}}$

57. $1\frac{1}{2} + \frac{1}{7} = \frac{3}{2} + \frac{1}{7} = \frac{21}{14} + \frac{2}{14} = \frac{23}{14} = \mathbf{1\frac{9}{14}}$

59. $\frac{1}{4} \times 1\frac{1}{7} = \frac{1}{4} \times \frac{8}{7} = \mathbf{\frac{2}{7}}$

61. $3\frac{1}{4} + \frac{1}{6} = 3 + \frac{1}{4} + \frac{1}{6} = 3 + \frac{3}{12} + \frac{2}{12} = \mathbf{3\frac{5}{12}}$

63. $\frac{1}{5} \times 2\frac{1}{7} = \frac{1}{5} \times \frac{15}{7} = \mathbf{\frac{3}{7}}$

65. $-3 + 2\frac{1}{4} = -3 + 2 + \frac{1}{4} = -1 + \frac{1}{4} = \mathbf{-\frac{3}{4}}$

67. $(-8) \times 2\frac{1}{4} = (-8) \times \frac{9}{4} = (-2) \times 9 =$ **-18**

69. $-2 + 1\frac{1}{5} = -2 + 1 + \frac{1}{5} = -1 + \frac{1}{5} = \mathbf{-\frac{4}{5}}$

71. $(-9) \times 3\frac{1}{3} = (-9) \times \frac{10}{3} = (-3) \times 10 =$ **-30**

73. $7\frac{1}{4} + (-\frac{1}{8}) = 7 + \frac{1}{4} - \frac{1}{8} = 7 + \frac{2}{8} - \frac{1}{8} = 7 + \frac{1}{8} = \mathbf{7\frac{1}{8}}$

75. $(-1\frac{1}{4}) \times (-2\frac{1}{10}) = (-\frac{5}{4}) \times (-\frac{21}{10}) = (-\frac{1}{4}) \times (-\frac{21}{2}) = \frac{21}{8} = \mathbf{2\frac{5}{8}}$

77. $\frac{1}{2} \times \frac{1}{6} - \frac{1}{3} + \frac{1}{4} = \frac{1}{12} - \frac{4}{12} + \frac{3}{12} = \mathbf{0}$

79. $\dfrac{1}{3} - \dfrac{1}{3} \times \dfrac{2}{3} \div \dfrac{2}{5} = \dfrac{1}{3} - \dfrac{2}{9} \div \dfrac{2}{5} = \dfrac{1}{3} - \dfrac{2}{9} \times \dfrac{5}{2} = \dfrac{3}{9} - \dfrac{5}{9} = -\dfrac{2}{9}$

81. $(2\dfrac{1}{2}) \times (-3\dfrac{1}{4}) - (-7\dfrac{1}{8}) \div 3 = (\dfrac{5}{2}) \times (-\dfrac{13}{4}) - (-\dfrac{57}{8}) \div 3$

$\qquad\qquad\qquad\qquad\qquad = -\dfrac{65}{8} - (-\dfrac{19}{8}) = -\dfrac{65}{8} + \dfrac{19}{8}$

$\qquad\qquad\qquad\qquad\qquad = -\dfrac{46}{8} = -\dfrac{23}{4} = -5\dfrac{3}{4}$

83. $101\dfrac{6}{10} - 98\dfrac{6}{10} = $ **3** degrees **above** normal

85. $4\dfrac{7}{16} - 3\dfrac{1}{8} = 4 - 3 + \dfrac{7}{16} - \dfrac{2}{16} = 1 + \dfrac{5}{16} = 1\dfrac{5}{16}$ **lb above average**.

87. $1 - \dfrac{3}{10} - \dfrac{1}{5} = 1 - \dfrac{3}{10} - \dfrac{2}{10} = 1 - \dfrac{5}{10} = \dfrac{5}{10} = \dfrac{1}{2}$

89. $3\dfrac{1}{2} + 1\dfrac{2}{5} = 3 + 1 + \dfrac{1}{2} + \dfrac{2}{5} = 4 + \dfrac{5}{10} + \dfrac{4}{10} = 4\dfrac{9}{10}$ billion dollars per week.

91. $7\dfrac{1}{2} + 2\dfrac{3}{5} + 2 = 7 + 2 + 2 + \dfrac{1}{2} + \dfrac{3}{5} = 11 + \dfrac{5}{10} + \dfrac{6}{10} = 11 + \dfrac{11}{10} = 12\dfrac{1}{10}$ hr

93. The symbol $\dfrac{0}{0}$ cannot be uniquely defined. You can write any number you wish for the n in the formula $\dfrac{0}{0} = n$. This gives $0 = 0 \times n = 0$ no matter what value you assign to n.

95. $36 \times 3\dfrac{1}{2} = 36 \times \dfrac{7}{2} = 18 \times 7 = $ **126 mi**   97. $108 \div 36 = $ **3 in.**

99. $\dfrac{5}{1} \; \dfrac{5}{2} \; \dfrac{5}{3} \; \dfrac{5}{4} \; \dfrac{5}{5} \; \dfrac{5}{6} \; \dfrac{5}{7} \; \dfrac{5}{8} \; \dfrac{5}{9} \; \ldots$

$ \dfrac{6}{1} \; \dfrac{6}{2} \; \dfrac{6}{3} \; \dfrac{6}{4} \; \dfrac{6}{5} \; \dfrac{6}{6} \; \dfrac{6}{7} \; \dfrac{6}{8} \; \dfrac{6}{9} \; \ldots$

101. $\dfrac{2}{2} = 1, \dfrac{4}{2} = 2, \dfrac{3}{3} = 1, \dfrac{2}{4} = \dfrac{1}{2}$ and these have already been caught in the one-to-one correspondence.

103.   **Neither**. The two sets have the same cardinal number.

SECTION 4.4 Decimals

EXERCISE 4.4

STUDY TIPS All terminating decimals are rational numbers. (If you read 0.3 correctly as "three tenths", you can immediately write $0.3 = \frac{3}{10}$.) Writing decimals in **expanded form** requires negative exponents. Just remember that $a^{-n} = \frac{1}{a^n}$, the exponent n remains the same! Negative exponents are also used with **scientific notation.** Note: to write 345.85 in scientific notation, you need to write the given number as a number between 1 and 10, so write 3.4585 first. But now, you have changed the problem! 3.4585 must be made equivalent to the original number, 345.85. Do that by multiplying by 10^2, which moves the decimal point two places right. Thus, $345.85 = 3.485 \times 10^2$. What about 0.00345? Make it 3.45 and remember to move the decimal point three places left by multiplying by 10^{-3}. (Multiplying by 10 with a negative exponent moves the decimal point **left**.) The rest of the chapter may be a review for you. Two tips: when adding or subtracting decimals make sure you align the decimal point and **round** the answer to the correct number of places (page 192). When multiplying decimals, do the multiplication first as if they were whole numbers. Count how many decimal places are in the multiplicand, add the number of decimal places in the multiplier, go to the answer and count that many places, starting at the right. This gets you to where the decimal point goes! Then, round to the correct number of decimal places. (See the rule on page 193.)

1. $692.087 = 6 \times 10^2 + 9 \times 10 + 2 + 8 \times 10^{-2} + 7 \times 10^{-3}$

3. $0.00107 = 1 \times 10^{-3} + 7 \times 10^{-5}$

5. $5 \times 10^3 + 2 \times 10 + 3 \times 10^{-1} + 9 \times 10^{-2} = 5020.39$

7. $4 \times 10^{-3} + 7 \times 10^{-4} + 2 \times 10^{-6} = 0.004702$

9. $935 = 9.35 \times 10^2$ 11. $0.0012 = 1.2 \times 10^{-3}$

13. $8.64 \times 10^4 = 86{,}400$ 15. $6.71 \times 10^{-3} = 0.00671$

17. $0.0346 \div 1{,}730{,}000 = (3.46 \times 10^{-2}) \div (1.73 \times 10^6)$
     $= (3.46 \div 1.73) \times (10^{-2} \div 10^6) = 2 \times 10^{-8}$

19. $(3.1 \times 10^5) \times (2.2 \times 10^{-6}) = (3.1 \times 2.2) \times (10^5 \times 10^{-6}) = 6.82 \times 10^{-1}$

21. $\dfrac{(2 \times 10^6)(6 \times 10^{-5})}{4 \times 10^3} = \dfrac{(2 \times 6) \times (10^6 \times 10^{-5})}{4 \times 10^3}$
$= \dfrac{12 \times 10}{4 \times 10^3} = 3 \times 10^{-2}$

23. 4×10^{25} (There are 25 zeros after the 4.)

25. $\dfrac{2.8 \times 10^8}{1.4 \times 10^5} = 2 \times 10^3$ (Note that $2.8 = 2 \times 1.4$, and $10^8 = 10^3 \times 10^5$.)

27. $\dfrac{1.47 \times 10^{11}}{490} = \dfrac{1470 \times 10^8}{490} = 3 \times 10^8$ (Note that $1470 = 3 \times 490$.)

29. $\dfrac{6.28 \times 10^{11}}{2.0 \times 10^{10}} = 3.14 \times 10 = 31.4$, so the answer to the nearest year is 31 yr.

31. (a) 3.81   (b)   -3.81
        + 0.93   + (-0.93)
          4.74   - 4.74

33. (a) 2.08   (b)   3.07
        - 6.238   - 8.934
        - 4.158   - 5.864

35. (a) $(-0.03) \times (-1.5) = 0.045$ (b) $(-3.2) \times (-0.04) = 0.128$

37. (a) Write in fractional form: $\dfrac{-0.07}{1.4} = -\dfrac{0.7}{14}$. Then divide 0.7 by 14.

```
       0.05
14 )0.70      Thus, the answer is - 0.05.
    0.70
       0
```

(b) Write in fractional form: $\dfrac{-0.09}{-4.5} = \dfrac{0.9}{45}$. Then divide 0.9 by 45.

```
       0.02
45 )0.90      Thus, the answer is 0.02.
    0.90
       0
```

39. In $ millions:
    926.3
    721.21
    488.34
    674
    792.45
    229.58
    $3831.88 (millions)

41. 36 oz of the 6 oz cans would cost
    $6 \times 0.48 = \$2.88$
    36 oz of the 12 oz cans would cost
    $3 \times 0.89 = \$2.67$
    Buying the 12 oz cans would save
    $\$2.88 - \$2.67 = \$0.21$

43. To find how much Harry lost, subtract $7.25 from the total $25.75 and divide the result by 2. Thus:
    $\$25.75 - \$7.25 = \$18.50$ and $\$18.50 \div 2 = \$9.25$
    Harry lost $9.25 and George lost $9.25 + $7.25 = $16.50.

45. 28 books were sold for $31.50 each. The gross sales amounted to $28 \times \$31.50 = \$882.00$. The cost of the books sold was $28 \times \$25.00$ or $700 and the service charge on the two books returned was $6.00. Thus, the total cost was $706.00 and the gross profit was $882.00 - $706.00 = $176

47. 3265 oz rounded to the nearest 100 oz gives 3300 oz.

49. 18,719.7
    - 18,327.2
    392.5 mi

51. $19,230,000 \times 0.75 = \$14,422,500$

53. $(2.15 \times 10^6) \times (5.878 \times 10^{12}) = 2.15 \times 5.878 \times 10^{6+12}$
    $= 1.26 \times 10 \times 10^{18}$
    $= 1.26 \times 10^{19}$

55. $1150.3 \div 9.41 = 59.26$

57. $3600 \div 46 = 78$ (to the nearest whole number)

59. Answers will vary. Some of the advantages are that operations can be done just as with the whole numbers and following the rules for the placement of the decimal point.

61. Round each cost to the nearest dollar. This gives the best estimate as
    $\$2 + \$2 + \$3 + \$4 = \$11$.

EXERCISE 4.5

STUDY TIPS Percents are a special type of rational number. Remember that rational numbers are written as $\frac{a}{b}$ where a and b are integers, but you can change them to decimals or percents by dividing a by b. The result may be a **terminating** decimal (If the number is $\frac{3}{4}$ divide 3 by 4 and get 0.75.) or **nonterminating** (Divide 1 by 3 and use a bar to show that the 3 repeats, like this $\frac{1}{3} = 0.\overline{3}$.) How do you work it the other way, that is, how do you write $0.\overline{37}$ as a fraction? Here is a quick way: $0.\overline{37}$ has two digits 3 and 7 under the bar, so write $0.\overline{37} = \frac{37}{99}$. (The digits under the bar go in the numerators and the denominator is two nines.) If the number were $0.\overline{371}$, which has three digits under the bar, write $0.\overline{371} = \frac{371}{999}$, three nines in the denominator this time.

Back to percents. To change a percent to a decimal move the decimal point in the number two places left and omit the percent symbol. To change a decimal to a percent reverse the procedure. Finally, to write the percent increase or decrease use the memory device PC's are a C/ON, that is, **Percent Change** = $\frac{\text{Change}}{\text{Original Number}}$.

1. 0.9
3. 1.1
5. 0.17
7. 1.21
9. 0.003
11. 1.243
13. 0.6
15. 0.5625
17. 0.625
19. 0.714285...
21. 0.2666...
23. 7.142857...
25. 0.1875
27. 0.015625
29. 0.00992
31. $0.\overline{5}$
33. $0.\overline{64}$
35. $0.\overline{235}$
37. $0.21\overline{5}$
39. $0.07\overline{935}$
41. $5.\overline{07}$
43. $\frac{8}{9}$
45. $\frac{31}{99}$
47. $\frac{114}{999} = \frac{38}{333}$
49. $2 + \frac{31}{99} = \frac{229}{99}$

SECTION 4.5 Rational Numbers as Decimals: Percents

51. $1 + \frac{234}{999} = 1 + \frac{26}{111} = \frac{137}{111}$

53. $1 + \frac{27}{99} = 1 + \frac{3}{11} = \frac{14}{11}$

55. $\frac{45}{100} + \frac{75}{9900} = \frac{45}{100} + \frac{25}{3300} = \frac{1510}{3300} = \frac{151}{330}$

57. $\frac{2016}{9999} = \frac{224}{1111}$

59. 0.29   61. 0.009   63. 0.4569   65. 0.3415

67. 0.000234   69. 345%   71. 56.7%   73. 900.3%

75. 0.45%   77. 60%   79. 83.3%

81. 13% of 70 = 0.13 × 70 = 9.1   83. $\frac{24}{72} = \frac{1}{3} = 33\frac{1}{3}\%$

85. $\frac{100}{80} = 1\frac{1}{4} = 125\%$

87. 8,214,671 ÷ 16,300,000 = 0.50397 = 50.4% (to one decimal place)

89. (a) 0.43 × 150 = 64.5   (b) 0.26 × 150 = 39

91. 0.055 × 196.50 = $10.81

93. 168,607,000 ÷ 0.379 = 445,000,000 (to the nearest million)

95. 5 ÷ 25 = 0.20 = 20%

97. 25% of 28 = 0.25 × 28 = 7. 28 + 7 = 35.

99. 8000 ÷ 28,000 = 29% (to the nearest percent)

101. 2000 ÷ 20,000 = 10%

103. 115% of 22 = 1.15 x 22 = 25.3 mi/gal

105. For the first set, let c_1 be the cost. Then $391 = 1.15c_1$ so that c_1 = 391 ÷ 1.15 = 340. Thus, the first set cost the dealer $340 and he made a profit of $51. For the second set, let c_2 be the cost. Then $391 = 0.85c_2$ so that c_2 = 340 ÷ 0.85 = 460.   Thus, the second set cost the dealer $460 and he lost $69. His net loss on the two sets was $69 - $51 or $18.

107. $0.\overline{4} = 0.444\ldots = \frac{4}{9}$

CHAPTER 4 NUMBER THEORY AND THE REAL NUMBERS

EXERCISE 4.6

STUDY TIPS Some numbers, like $\sqrt{2}$ and π are not rational. If you write them as decimals, they are nonterminating and nonrepeating. Such numbers are called **irrational numbers.** In particular, the irrational number π is used in the formula for the circumference of a circle, which we study here. We shall study operations with radicals in the next section.

1. $\sqrt{120}$ is an **irrational** number. (120 is not a perfect square.)

3. $\sqrt{125}$ is an **irrational** number. (125 is not a perfect square.)

5. $\sqrt{\frac{9}{16}} = \frac{3}{4}$, so $\sqrt{\frac{9}{16}}$ is a **rational** number.

7. $\frac{3}{5}$ is a **rational** number.   9. $-\frac{5}{3}$ is a **rational** number

11. $0.232323\ldots = \frac{23}{99}$ is a **rational** number.

13. $0.121231234\ldots$ is a nonterminating, nonrepeating decimal, so is an **irrational** number.

15. $6\frac{1}{4} = \frac{25}{4}$ is a **rational** number.

17. $0.24681012\ldots$ is a nonterminating, nonrepeating decimal, so is an **irrational** number.

19. 3.1415 is a terminating decimal, so is a **rational** number.

21. $\sqrt{16} = 4$ 23. $\sqrt{64} = 8$ 25. $\sqrt{81} = 9$

27. $-\sqrt{169} = -13$ 29. $\sqrt{196} = 14$ 31. $-\sqrt{81} = -9$

33. $3 < 4$ 35. $\frac{1}{5} < \frac{1}{4}$ 37. $\frac{5}{7} = \frac{10}{14}$

39. $(4.5)^2 = 20.25$, so $\sqrt{20} < 4.5$ 41. $0.333\ldots < 0.333444\ldots$

43. $0.999\ldots = 1$ 45. $3(0.333\ldots) = 0.999\ldots = 1$

SECTION 4.6 The Real Numbers

NOTE: IN PROBLEMS 47-55, OTHER ANSWERS ARE POSSIBLE

47. 0.315

49. 0.311212345 . . .

51. 0.1011

53. 0.101101001000 . . .

55. $\frac{7}{22}$ This is the average (one-half the sum) of the given numbers.

57. $\frac{4}{9} = 0.444\ldots$ and $\frac{5}{9} = 0.555\ldots$, so $0.5101001000\ldots$ is a possible answer.

59. $0.\overline{5} = 0.555\ldots = \frac{5}{9} = \frac{10}{18}$ and $\frac{2}{3} = \frac{12}{18}$, so $\frac{11}{18}$ is a possible answer.

61. $0.21 < 0.2121 < 0.21211 < 0.212112111\ldots < 0.21212$

63. $C = \pi d = 3.14 \times \frac{5200}{5280} = 3.09$ mi

65. $C = \pi d = 3.14 \times 52 = 163$ mi

67. $\pi d_2 - \pi d_1 = \pi(d_2 - d_1) = \pi(100 - 90) = 3.14 \times 10 = 31.4 = 31$ yd (to the nearest yard).

69. Recall that a rational number can be expressed as a terminating decimal or as a nonterminating, repeating decimal. An irrational number cannot be expressed this way.

71. $(OB)^2 = 1^2 + (\sqrt{5})^2 = 1 + 5 = 6$, so that $OB = \sqrt{6}$.

73. $h^2 = 1^2 + (\sqrt{8})^2 = 1 + 8 = 9$, so that $h = \sqrt{9} = 3$.

CHAPTER 4 NUMBER THEORY AND THE REAL NUMBERS

EXERCISE 4.7

STUDY TIPS To be able to do operations with radicals you need to know that $\sqrt{a \cdot b} = \sqrt{a} \cdot \sqrt{b}$ and $\sqrt{\dfrac{a}{b}} = \dfrac{\sqrt{a}}{\sqrt{b}}$ Memorize these facts!

Then, use them for simplifying radical expressions. After simplification, it is as easy as adding apples. For example, 3 apples + 4 apples = 7 apples. Well $3\sqrt{10} + 4\sqrt{10} = 7\sqrt{10}$. Make sure that, by now, you can classify any number using the categories given in the text.

1. $\sqrt{90} = \sqrt{9 \times 10} = \mathbf{3\sqrt{10}}$

3. $\sqrt{122}$ is in **simplest** form.

5. $\sqrt{180} = \sqrt{36 \times 5} = \mathbf{6\sqrt{5}}$

7. $\sqrt{200} = \sqrt{100 \times 2} = \mathbf{10\sqrt{2}}$

9. $\sqrt{384} = \sqrt{64 \times 6} = \mathbf{8\sqrt{6}}$

11. $\sqrt{588} = \sqrt{196 \times 3} = \mathbf{14\sqrt{3}}$

13. $\dfrac{3}{\sqrt{7}} = \dfrac{3\sqrt{7}}{\sqrt{7} \times \sqrt{7}} = \mathbf{\dfrac{3\sqrt{7}}{7}}$

15. $\dfrac{-\sqrt{2}}{\sqrt{5}} = -\dfrac{\sqrt{2} \times \sqrt{5}}{\sqrt{5} \times \sqrt{5}} = \mathbf{-\dfrac{\sqrt{10}}{5}}$

17. $\dfrac{4}{\sqrt{8}} = \dfrac{4\sqrt{2}}{\sqrt{8} \times \sqrt{2}} = \dfrac{4\sqrt{2}}{\sqrt{16}} = \dfrac{4\sqrt{2}}{4} = \mathbf{\sqrt{2}}$

19. $\sqrt{\dfrac{3}{49}} = \mathbf{\dfrac{\sqrt{3}}{7}}$

21. $\sqrt{\dfrac{4}{3}} = \dfrac{2}{\sqrt{3}} = \dfrac{2\sqrt{3}}{\sqrt{3} \times \sqrt{3}} = \mathbf{\dfrac{2\sqrt{3}}{3}}$

23. $\sqrt{\dfrac{8}{49}} = \mathbf{\dfrac{2\sqrt{2}}{7}}$

25. $\sqrt{\dfrac{18}{50}} = \sqrt{\dfrac{9}{25}} = \mathbf{\dfrac{3}{5}}$

27. $\sqrt{\dfrac{32}{125}} = \dfrac{4\sqrt{2}}{5\sqrt{5}} = \dfrac{4\sqrt{2} \times \sqrt{5}}{5\sqrt{5} \times \sqrt{5}} = \mathbf{\dfrac{4\sqrt{10}}{25}}$

29. $\sqrt{5} \cdot \sqrt{50} = \sqrt{250} = \sqrt{25 \cdot 10} = \mathbf{5\sqrt{10}}$

31. $\dfrac{\sqrt{28}}{\sqrt{2}} = \sqrt{\dfrac{28}{2}} = \mathbf{\sqrt{14}}$

33. $\dfrac{\sqrt{10}}{\sqrt{250}} = \sqrt{\dfrac{10}{250}} = \sqrt{\dfrac{1}{25}} = \mathbf{\dfrac{1}{5}}$

35. $\dfrac{\sqrt{33}}{\sqrt{22}} = \sqrt{\dfrac{33}{22}} = \sqrt{\dfrac{3}{2}} = \sqrt{\dfrac{3 \times 2}{2 \times 2}} = \dfrac{\sqrt{6}}{2}$

37. $\sqrt{3} + \sqrt{12} = \sqrt{3} + \sqrt{4 \times 3} = \sqrt{3} + 2\sqrt{3} = \mathbf{3\sqrt{3}}$

39. $\sqrt{125} + \sqrt{80} = \sqrt{25 \times 5} + \sqrt{16 \times 5} = 5\sqrt{5} + 4\sqrt{5} = \mathbf{9\sqrt{5}}$

41. $\sqrt{3^2 + 4^2} = \sqrt{9 + 16} = \sqrt{25} = \mathbf{5}$

43. $\sqrt{(13)^2 - (12)^2} = \sqrt{169 - 144} = \sqrt{25} = \mathbf{5}$

45. $6\sqrt{7} + \sqrt{7} - 2\sqrt{7} = (6 + 1 - 2)\sqrt{7} = \mathbf{5\sqrt{7}}$

47. $5\sqrt{7} - 3\sqrt{28} - 2\sqrt{63} = 5\sqrt{7} - 3\sqrt{4 \times 7} - 2\sqrt{9 \times 7}$
$= 5\sqrt{7} - 3 \times 2\sqrt{7} - 2 \times 3\sqrt{7}$
$= 5\sqrt{7} - 6\sqrt{7} - 6\sqrt{7} = \mathbf{-7\sqrt{7}}$

49. $-3\sqrt{45} + \sqrt{20} - \sqrt{5} = -3\sqrt{9 \times 5} + \sqrt{4 \times 5} - \sqrt{5}$
$= -3 \times 3\sqrt{5} + 2\sqrt{5} - \sqrt{5}$
$= -9\sqrt{5} + 2\sqrt{5} - \sqrt{5} = \mathbf{-8\sqrt{5}}$

51. $\sqrt{(80)^2 + (100)^2} = 10\sqrt{64 + 100} = 10\sqrt{164} = 10\sqrt{4 \times 41} = \mathbf{20\sqrt{41}}$ m

53. $t = \sqrt{\dfrac{50}{16}} = \sqrt{\dfrac{25 \times 2}{16}} = \dfrac{\mathbf{5\sqrt{2}}}{\mathbf{4}}$ sec

55. $r = \sqrt{\dfrac{144}{100}} - 1 = \dfrac{12}{10} - 1 = 1.2 - 1 = 0.\mathbf{20} = \mathbf{20\%}$

57. Check Rational Numbers and Real Numbers

59. Check Whole Numbers, Integers, Rational Numbers and Real Numbers

61. Check Natural Numbers, Whole Numbers, Integers, Rational Numbers and Real Numbers

63. Check Rational Numbers and Real Numbers

65. Check Irrational Numbers and Real Numbers

67. Try an example. Suppose a = -2 and b = -3. Then, by definition, $\sqrt{-2}$ and $\sqrt{-3}$ are not real numbers but $\sqrt{-2} \cdot \sqrt{-3} = \sqrt{(-2)(-3)} = \sqrt{6}$, which is a real number. How can the product of two undefined numbers yield a real number?

69. 40 - 36 = 4
$\sqrt{36} = 6$
$\sqrt{40} = 6\frac{4}{13} = \mathbf{6.31}$
$\sqrt{49} = 7$
49 - 36 = 13
(Calculator gives 6.32.)

71. 85 - 81 = 4
$\sqrt{81} = 9$
$\sqrt{85} = 9\frac{4}{19} = \mathbf{9.21}$
$\sqrt{100} = 10$
100 - 81 = 19
(Calculator gives 9.22.)

EXERCISE 4.8

STUDY TIPS We study two types of sequences: **arithmetic** and **geometric**. How do you tell them apart? Look at successive terms. Take the *difference*. If the answer is always the same, the sequence is an *arithmetic* sequence. If not, try *dividing* each term by the preceding one. If the answer is always the same, the sequence is *geometric*. To find the sum of a sequence you must learn the formulas given on page 225 for an arithmetic sequence and on page 226 for a geometric sequence. An important application of the sum of an infinite geometric sequence is to write a given repeating decimal as a fraction.

1. (a) $a_1 = \mathbf{7}$ (b) $d = \mathbf{6}$ (c) $a_{10} = a_1 + 9d = 7 + 54 = \mathbf{61}$
   (d) $a_n = a_1 + (n-1)d = 7 + (n-1)(6) = 7 + 6n - 6 = \mathbf{6n + 1}$

3. (a) $a_1 = \mathbf{43}$ (b) $d = \mathbf{-9}$ (c) $a_{10} = a_1 + 9d = 43 + 9(-9) = 43 - 81 = \mathbf{-38}$
   (d) $a_n = a_1 + (n-1)d = 43 + (n-1)(-9) = 43 - 9n + 9 = \mathbf{52 - 9n}$

5. (a) $a_1 = \mathbf{2}$ (b) $d = \mathbf{-5}$ (c) $a_{10} = a_1 + 9d = 2 + 9(-5) = 2 - 45 = \mathbf{-43}$
   (d) $a_n = a_1 + (n-1)d = 2 + (n-1)(-5) = 2 - 5n + 5 = \mathbf{7 - 5n}$

7. (a) $a_1 = -\dfrac{5}{6}$ (b) $d = \dfrac{1}{2}$

   (c) $a_{10} = a_1 + 9d = -\dfrac{5}{6} + 9(\dfrac{1}{2}) = -\dfrac{5}{6} + \dfrac{9}{2} = \dfrac{22}{6} = \dfrac{11}{3}$

   (d) $a_n = a_1 + (n-1)d = -\dfrac{5}{6} + (n-1)(\dfrac{1}{2}) = -\dfrac{5}{6} + \dfrac{n}{2} - \dfrac{1}{2} = \dfrac{n}{2} - \dfrac{4}{3}$, or $\dfrac{\mathbf{3n-8}}{\mathbf{6}}$

9. (a) $a_1 = \mathbf{0.6}$ (b) $d = \mathbf{-0.4}$ (c) $a_{10} = a_1 + 9d = 0.6 + 9(-0.4) = \mathbf{-3}$

   (d) $a_n = a_1 + (n-1)d = 0.6 + (n-1)(-0.4) = 0.6 - 0.4n + 0.4 = \mathbf{1 - 0.4n}$

11. $S_{10} = \dfrac{n(a_1 + a_n)}{2} = \dfrac{10(7+61)}{2} = \dfrac{680}{2} = \mathbf{340}$

    $S_n = \dfrac{n(7 + 6n + 1)}{2} = \mathbf{n(3n + 4)}$

13. $S_{10} = \dfrac{n(a_1 + a_n)}{2} = \dfrac{10(43-38)}{2} = \dfrac{50}{2} = \mathbf{25}$

    $S_n = \dfrac{n(43 + 52 - 9n)}{2} = \dfrac{\mathbf{n}}{\mathbf{2}}\mathbf{(95 - 9n)}$

15. $S_{10} = \dfrac{n(a_1 + a_n)}{2} = \dfrac{10(2-43)}{2} = -\dfrac{410}{2} = \mathbf{-205}$

    $S_n = \dfrac{n(2 + 7 - 5n)}{2} = \dfrac{\mathbf{n}}{\mathbf{2}}\mathbf{(9 - 5n)}$

17. $S_{10} = \dfrac{n(a_1 + a_n)}{2} = \dfrac{10}{2}(-\dfrac{5}{6} + \dfrac{11}{3}) = 5(\dfrac{17}{6}) = \dfrac{85}{6} = \mathbf{14\dfrac{1}{6}}$

    $S_n = \dfrac{n}{2}(-\dfrac{5}{6} + \dfrac{3n-8}{6}) = \dfrac{\mathbf{n}}{\mathbf{12}}\mathbf{(3n - 13)}$

19. $S_{10} = \dfrac{n(a_1 + a_n)}{2} = \dfrac{10}{2}(0.6 - 3) = (5)(-2.4) = \mathbf{-12}$

    $S_n = \dfrac{n}{2}(0.6 + 1 - 0.4n) = \dfrac{n}{2}(1.6 - 0.4n) = \dfrac{n}{2}(0.4)(4-n) = \dfrac{\mathbf{n}}{\mathbf{5}}\mathbf{(4 - n)}$

21. (a) $a_1 = \mathbf{3}$ (b) $r = \mathbf{2}$ (c) $a_{10} = a_1 r^{n-1} = 3 \cdot 2^9 = 3 \cdot 512 = \mathbf{1536}$

    (d) $a_n = \mathbf{3 \cdot 2^{n-1}}$

23. (a) $a_1 = \dfrac{1}{3}$ (b) $r = \mathbf{3}$ (c) $a_{10} = a_1 r^{n-1} = \dfrac{1}{3}(3^9) = 3^8 = \mathbf{6561}$

    (d) $a_n = \dfrac{1}{3}(3^{n-1}) = \mathbf{3^{n-2}}$

96 CHAPTER 4 NUMBER THEORY AND THE REAL NUMBERS

25. (a) $a_1 = 16$ (b) $r = -\frac{1}{4}$ (c) $a_{10} = a_1 r^{n-1} = 16(-\frac{1}{4})^9 = -\frac{1}{4^7} = -\frac{1}{16384}$

   (d) $a_n = a_1 r^{n-1} = 16(-\frac{1}{4})^{n-1} = \frac{(-1)^{n-1}}{4^{n-3}}$

27. $S_{10} = \frac{a_1(1-r^n)}{1-r} = \frac{3(1-2^{10})}{1-2} = 3(2^{10} - 1)$

   $S_n = \frac{3(1-2^n)}{1-2} = 3(2^n - 1)$

29. $S_{10} = \frac{a_1(1-r^n)}{1-r} = \frac{\frac{1}{3}(1-3^{10})}{1-3} = \frac{1}{6}(3^{10} - 1)$

   $S_n = \frac{\frac{1}{3}(1-3^n)}{1-3} = \frac{1}{6}(3^n - 1)$

31. $S_{10} = \frac{a_1(1-r^n)}{1-r} = \frac{16(1-(-\frac{1}{4})^{10})}{1+\frac{1}{4}} = \frac{64}{5}(1 - \frac{1}{4^{10}}) = \frac{64(4^{10}-1)}{5 \cdot 4^{10}} = \frac{4^{10}-1}{5 \cdot 4^7}$

   $S_n = \frac{16(1-(-\frac{1}{4})^n)}{1+\frac{1}{4}} = \frac{64}{5}[1-(-\frac{1}{4})^n] = \frac{64[4^n - (-1)^n]}{5 \cdot 4^n} = \frac{4^n - (-1)^n}{5 \cdot 4^{n-3}}$

33. $a_1 = 6$, $r = \frac{1}{2}$, so $S = \frac{a_1}{1-r} = \frac{6}{1-\frac{1}{2}} = 12$

35. $a_1 = -8$, $r = \frac{1}{2}$, so $S = \frac{a_1}{1-r} = \frac{-8}{1-\frac{1}{2}} = -16$

37. $a_1 = 0.7$, $r = 0.1$, so $S = \frac{a_1}{1-r} = \frac{0.7}{1-0.1} = \frac{0.7}{0.9} = \frac{7}{9}$

39. First write this as $2 + 0.101010...$. Then for the decimal part, $a_1 = 0.10$, $r = 0.01$, so $S = \frac{a_1}{1-r} = \frac{0.10}{1-0.01} = \frac{0.10}{0.99} = \frac{10}{99}$. Thus, the number is $2 + \frac{10}{99}$ or $\frac{208}{99}$.

SECTION 4.8 Number Sequences

41. This results in an arithmetic sequence with $a_1 = 1380$ and $d = -40$.
    (a) $a_{10} = a_1 + (n-1)d = 1380 + 9(-40) = 1380 - 360 = 1020$. Thus, the depreciation the tenth year will be **$1020**.
    (b) The total depreciation for the ten years is obtained as the sum of the first ten terms of the sequence:
    $$S_{10} = \frac{n}{2}(a_1 + a_{10}) = \frac{10}{2}(1380 + 1020) = 5(2400) = 12{,}000.$$
    Thus, the value of the property at the end of ten years will be 30,000 - $12,000 = **$18,000**.

43. The costs per foot form an arithmetic progression with $a_1 = 50$ and $d = 5$.
    (a) The cost of the tenth foot is $a_{10} = a_1 + (n-1)d = 50 + 9 \times 5 = 50 + 45 = 95$. So the cost is **$95**.
    (b) $a_{50} = 50 + 49 \cdot 5 = 50 + 245 = 295$. The total cost for the 50 ft is obtained from the sum
    $$S_{50} = \frac{n}{2}(a_1 + a_{50}) = \frac{50}{2}(50 + 295) = 25 \times 345 = 8625.$$
    Thus, the total cost for a 50 foot well is **$8625**.

45. Here, the sequence is geometric with $a_1 = 100$ and $r = 1.10$. The sum of the first five terms is
    $$S_5 = a_1 \frac{1 - r^n}{1 - r} = 100 \frac{1 - (1.10)^5}{1 - 1.10} = 100 \frac{(1.10)^5 - 1}{0.10} = 1000[(1.10)^5 - 1]$$
    $$= 1000(1.61051 - 1) = 610.51$$
    Thus, the final amount is **$610.51**.

47. Assume that n is an even number. Then there are $\frac{n}{2}$ pairs and the sum of each pair is $n + 1$. Thus, the sum of the sequence is $\frac{n}{2}(n+1)$. If n is an odd number, first consider the sequence
    $$1 + 2 + 3 + \cdots + (n-3) + (n-2) + (n-1)$$
    Here there are $\frac{n-1}{2}$ pairs with n as the sum of each pair. So the sum is $\frac{n-1}{2}(n) = \frac{n(n-1)}{2}.$ Adding the term n that was omitted, the final sum is $\frac{n(n-1)}{2} + n = \frac{n(n-1) + 2n}{2} = \frac{n(n-1+2)}{2} = \frac{n}{2}(n+1)$ as before.

98 CHAPTER 4 NUMBER THEORY AND THE REAL NUMBERS

49. In an arithmetic sequence, each term after the first is obtained by adding the constant difference d to the preceding term. In a geometric sequence, each term after the first is obtained by multiplying the preceding term by the constant ratio r.

51. There is a common difference of 85, so it **is** an arithmetic sequence.

53. $127 + (n - 1)85$ or **$42 + 85n$**

PRACTICE TEST 4

STUDY TIPS If you have memorized the formulas and tips we have given you, write them on your test paper as soon as you receive it and before you forget them. Remember, two practice tests are given at the end of this manual.

1. (a) The <u>third</u> is an ordinal number.
   (b) The <u>two</u> is a cardinal number.
   (c) The <u>270-891</u> is used for identification only.

2. (a) The **commutative** property of addition. $(9 + 2 = 2 + 9)$
   (b) The **distributive** property of multiplication over addition.
   $$8(10 + 2) = 8 \times 10 + 8 \times 2$$
   (c) The **commutative** property of multiplication. $(27 \times 4 = 4 \times 27)$

3. $2 \overline{)1220}$
   $2 \overline{)610}$
   $5 \overline{)305}$ Thus, $1220 = 2^2 \times 5 \times 61$.
   61

4. **53, 59, 61** and **67**

5. $143 = 11 \times 13$, so 143 is **composite**.

6. (a) 436 and 1530 are divisible by **two**.
   (b) $3 + 8 + 7 = 18$, which is divisible by 3, so 387 is divisible by **3**.
       $1 + 5 + 3 = 18$, which is divisible by 3, so 1530 is divisible by **3**.
   (c) 2345 ends in a 5 and 1530 ends in a 0, so these are both divisible by **5**.

7. $2 \overline{)216 \quad 254}$
    $\quad\;\; 108 \quad 127$

    108 and 127 are relatively prime, so that GCF(216, 254) = 2. Consequently,
    $$\frac{216}{254} = \frac{2 \times 108}{2 \times 127} = \frac{\mathbf{108}}{\mathbf{127}}$$

8. $2 \overline{)18 \quad 54 \quad 60}$
    $3 \overline{)\;\,9 \quad 27 \quad 30}$
    $\quad\;\; 3 \quad\;\, 9 \quad 10$

    This shows that LCM(18, 54, 60) = $2 \times 3 \times 9 \times 10$ = 540

    Thus, $\dfrac{1}{18} + \dfrac{1}{54} - \dfrac{1}{60} = \dfrac{30}{540} + \dfrac{10}{540} - \dfrac{9}{540} = \dfrac{\mathbf{31}}{\mathbf{540}}$

9. $1 - \dfrac{1}{4} - \dfrac{1}{2} - \dfrac{1}{8} = \dfrac{8 - 2 - 4 - 1}{8} = \dfrac{\mathbf{1}}{\mathbf{8}}$

10. (a) $8 - 19 = 8 + (-19) = \mathbf{-11}$ (b) $8 - (-19) = 8 + (+19) = \mathbf{27}$
    (c) $-8 - 19 = -8 + (-19) = \mathbf{-27}$ (d) $-8 - (-19) = -8 + (+19) = \mathbf{11}$

11. $4 \times 12 \div 3 \times 10^3 - 2(-6 + 4) \times 10^4 = 48 \div 3 \times 10^3 - 2(-2) \times 10^4$
    $\qquad\qquad\qquad\qquad\qquad\qquad\qquad = 16 \times 10^3 + 4 \times 10^4$
    $\qquad\qquad\qquad\qquad\qquad\qquad\qquad = 16{,}000 + 40{,}000$
    $\qquad\qquad\qquad\qquad\qquad\qquad\qquad = \mathbf{56{,}000}$

12. Since $16 = 4 \times 4$, $\dfrac{3}{4} = \dfrac{4 \times 3}{16} = \dfrac{\mathbf{12}}{\mathbf{16}}$

13. (a) $\dfrac{\mathbf{3}}{\mathbf{2}}$ (b) $-\dfrac{\mathbf{7}}{\mathbf{4}}$ (c) $2\dfrac{5}{8} = \dfrac{21}{8}$, so the reciprocal is $\dfrac{\mathbf{8}}{\mathbf{21}}$ (d) $-\dfrac{\mathbf{1}}{\mathbf{8}}$

14. (a) $\dfrac{7}{8} \times \left(-\dfrac{5}{16}\right) = -\dfrac{7 \times 5}{8 \times 16} = -\dfrac{\mathbf{35}}{\mathbf{128}}$

    (b) $-\dfrac{7}{8} \div \left(-\dfrac{5}{16}\right) = -\dfrac{7}{8} \times \left(-\dfrac{16}{5}\right) = \dfrac{7 \times 2}{1 \times 5} = \dfrac{\mathbf{14}}{\mathbf{5}}$

15. (a) $23.508 = (2 \times 10) + 3 + (5 \times 10^{-1}) + (8 \times 10^{-3})$
    (b) This is $800 + 3 + 4$ one-hundredths, that is, **803.04**.

16. $(6 \times 10^4) \times (4 \times 10^{-6}) = 6 \times 8 \times 10^{4-6} = 48 \times 10^{-2}$
    $\qquad\qquad\qquad\qquad\qquad\;\; = 4.8 \times 10 \times 10^{-2} = \mathbf{4.8 \times 10^{-1}}$

17.
(a)
```
  6.73
 +2.8
  9.53
```
(b)
```
  9.34
 -4.71
  4.63
```
(c)
```
   6.7
 ×0.29
   603
   134
 1.943
```
(d)
```
      5.6
 31)173.6
    155
     186
     186
       0
```

18. $18.7 + 6.25 + 19.63 = 44.58$. Thus, the perimeter is **44.6 cm**.

19. $A = 93.5 \times 50.6 = $ **4731.1 ft²**

20. (a)
```
    0.75
 4)3.00
   2 8
     20
     20
      0
```
Thus, the answer is **0.75**.

(b)
```
    0.066..
 15)1.000
     90
     100
      90
      10
```
Thus, the answer is the repeating decimal **0.0666...**

21. (a) $0.\overline{12} = \frac{12}{99} = \frac{4}{33}$ (b) $2.6555\ldots = \frac{26}{10} + \frac{5}{90} = \frac{234 + 5}{90} = \frac{239}{90}$

22. (a) $21\% = $ **0.21**   (b) $9.35\% = $ **0.0935**   (c) $0.26\% = $ **0.0026**

23. (a) $0.52 = $ **52%**   (b) $2.765 = $ **276.5%**   (c) $\frac{3}{5} = 0.6 = $ **60%**

(d) $\frac{2}{11} = 0.182 = $ **18.2%** (to one decimal place)

24. $86 - 48 = 38$ and $\frac{38}{48} = 0.791666\ldots$ Thus, the percent of profit is **79.17%**.

25. $4 \div 0.15 = 26.666\ldots$ Thus, the answer is about **26.7** million.

26. (a) $\sqrt{49} = 7$, a **rational** number.
    (b) $\sqrt{45}$ is an **irrational** number because 45 is not a perfect square.
    (c) $\sqrt{121} = 11$, a **rational** number.
    (d) 0.41252525... is a nonterminating, repeating decimal, so is a **rational** number.
    (e) 0.212112111... is a nonterminating, nonrepeating decimal, so is an **irrational** number.
    (f) 0.246810... is a nonterminating, nonrepeating decimal, so is an **irrational** number.

27. (a) $0.\overline{2} = 0.222...$, so **0.24** is a rational number between $0.\overline{2}$ and 0.25. (Other answers are possible.)
    (b) **0.23456...** is a nonterminating, nonrepeating decimal, so is an irrational number. Also, it is between 0.222... and 0.25. (Other answers are possible.)

28. $C = \pi d = 3.14 \times 4 = $ **12.6 in**.

29. $d = \dfrac{C}{\pi} = \dfrac{29.5}{3.14} = 9.3949$. Thus, the diameter is **9.39 cm**.

30. (a) $\sqrt{96} = \sqrt{16 \times 6} = \mathbf{4\sqrt{6}}$ (b) $\sqrt{58}$ is in **simplest form**.

31. (a) $\dfrac{4}{\sqrt{20}} = \dfrac{4}{2\sqrt{5}} = \dfrac{2}{\sqrt{5}} = \mathbf{\dfrac{2\sqrt{5}}{5}}$ (b) $\sqrt{\dfrac{48}{49}} = \sqrt{\dfrac{16 \times 3}{49}} = \mathbf{\dfrac{4\sqrt{3}}{7}}$

32. (a) $\sqrt{8} \times \sqrt{6} = \sqrt{48} = \sqrt{16 \times 3} = \mathbf{4\sqrt{3}}$ (b) $\dfrac{\sqrt{56}}{\sqrt{7}} = \sqrt{\dfrac{56}{7}} = \sqrt{8} = \mathbf{2\sqrt{2}}$

33. (a) $\sqrt{90} - \sqrt{40} = \sqrt{9 \times 10} - \sqrt{4 \times 10} = 3\sqrt{10} - 2\sqrt{10} = \mathbf{\sqrt{10}}$
    (b) $\sqrt{32} + \sqrt{18} - \sqrt{50} = \sqrt{16 \times 2} + \sqrt{9 \times 2} - \sqrt{25 \times 2}$
       $= 4\sqrt{2} + 3\sqrt{2} - 5\sqrt{2} = \mathbf{2\sqrt{2}}$

34. $c = \sqrt{4^2 + 6^2} = \sqrt{16 + 36} = \sqrt{52} = \sqrt{4 \times 13} = \mathbf{2\sqrt{13}}$

35. (a) Check **Natural numbers, Integers, Rational numbers** and **Real Numbers**.
    (b) Check **Rational numbers** and **Real numbers**.
    (c) Check **Irrational numbers** and **Real numbers**.
    (d) Check **Rational numbers** and **Real numbers**.
    (e) Check **Rational numbers** and **Real numbers**.

36. (a) Each term is 2 **times** the preceding term, so this is a geometric sequence.
    (b) Each term is 3 **plus** the preceding term, so this is an arithmetic sequence.

37. This is an arithmetic sequence with $a_1 = 9$ and $d = 4$. So
$$a_{10} = a_1 + 9d = 9 + 9 \times 4 = \mathbf{45}$$

$$S_{10} = \frac{10}{2}(a_1 + a_{10}) = 5(9 + 45) = 5 \times 54 = \mathbf{270}$$

38. This is a geometric sequence with $a_1 = 1$ and $r = \frac{1}{2}$. Thus,

$$S_5 = \frac{a_1(1 - r^5)}{1 - r} = \frac{1 - (\frac{1}{2})^5}{1 - \frac{1}{2}} = \frac{1 - \frac{1}{32}}{\frac{1}{2}} = 2(1 - \frac{1}{32}) = 2 - \frac{1}{16} = \mathbf{\frac{31}{16}}$$

39. (a) $a_5 = (\frac{1}{3})(\frac{1}{2})^4 = \mathbf{\frac{1}{48}}$

    (b) $S_5 = \frac{a_1(1 - r^5)}{1 - r} = \frac{\frac{1}{3}[1 - (\frac{1}{2})^5]}{1 - \frac{1}{2}} = \frac{2}{3}(1 - \frac{1}{32}) = \mathbf{\frac{31}{48}}$

40. (a) $S = \frac{0.4}{1 - 0.1} = \frac{0.4}{0.9} = \mathbf{\frac{4}{9}}$ (b) $S = \frac{0.21}{1 - 0.01} = \frac{0.21}{0.99} = \mathbf{\frac{7}{33}}$

    (c) $S = 2 + \frac{0.5}{1 - 0.1} = 2 + \frac{0.5}{0.9} = 2 + \frac{5}{9} = \mathbf{\frac{23}{9}}$

CHAPTER 5
EQUATIONS, INEQUALITIES AND PROBLEM SOLVING

EXERCISE 5.1

STUDY TIPS You must learn how to distinguish between the *replacement set* and the *solution set* for an equation. The **replacement set** is given and you replace the variable in the equation by the elements of this set until a true statement results. (The equation is **satisfied.**) When this happens, the element becomes part of the **solution set.** Note that linear equations can have **one** solution (the solution set has one element), **infinitely many** solutions (the solution set may be the set of real numbers or the set of integers), or **no** solution (the solution set is the empty set.) But how do you actually **solve** equations? The operations you can perform with equations are given on page 241, while the procedure to do so is on page 243. *Study and use this procedure in every problem.* Fortunately, the procedure for solving **inequalities** is the same as for equations with *one* exception: **If you multiply or divide an inequality by a negative number, you must reverse the sense of the inequality.**

For example, to solve -2x < 6, divide both sides by -2 (which is negative) and reverse the sign of the inequality to obtain x > - 3. Similarly, if $-\frac{1}{3}$ x > 4, multiply both sides by -3 (which is negative) and reverse the sign of the inequality to obtain x < - 12.

1. 2 + x ≤ 2 - x
   (a) Substitute **2** for x: 2 + **2** = 4, 2 - **2** = 0 and 4 > 0. So 2 is not a solution.
   (b.) Substitute **- 2** for x: 2 + (**-2**) = 0, 2 - (**-2**) = 4 and 0 < 4. So **- 2** is a solution.
   (c) Substitute **0** for x: 2 + **0** = 2, 2 - **0** = 2 and 2 = 2. So **0** is a solution.
   (d) Substitute **5** for x: 2 + **5** = 7, 2 - **5** = -3 and 7 > - 3. So 5 is not a solution.

3. $3x - 2 \geq 2x - 1$
   (a) Substitute **0** for x: $3 \cdot 0 - 2 = -2$, $2 \cdot 0 - 1 = -1$ and $-2 < -1$. So **0** is not a solution.
   (b) Substitute **3** for x: $3 \cdot 3 - 2 = 7$, $2 \cdot 3 - 1 = 5$ and $7 > 5$. So **3** is a solution.
   (c) Substitute **-2** for x: $3 \cdot (-2) - 2 = -8$, $2 \cdot (-2) - 1 = -5$ and $-8 < -5$. So **-2** is not a solution.
   (d) Substitute **1** for x: $3 \cdot 1 - 2 = 1$, $2 \cdot 1 - 1 = 1$ and $1 = 1$. So **1** is a solution.

5. **2** is the only solution of $2x - 1 = 3$.

7. The given equation is: $\qquad x + 10 = 15$
   Subtract **10** from both sides: $\qquad x + 10 - \mathbf{10} = 15 - \mathbf{10}$
   Simplify: $\qquad x = 5$

9. The given equation is: $\qquad 2x - 1 = 5$
   Add **1** to both sides $\qquad 2x - 1 + \mathbf{1} = 5 + \mathbf{1}$
   Simplify: $\qquad 2x = 6$
   Divide both sides by 2: $\qquad x = 3$

11. The given equation is: $\qquad 2x + 2 = x + 4$
    Subtract **x** from both sides: $\qquad 2x + 2 - \mathbf{x} = x + 4 - \mathbf{x}$
    Simplify: $\qquad x + 2 = 4$
    Subtract 2 from both sides to get $\qquad x = 2$

13. The given equation is: $\qquad 3x + 1 = 4x - 8$
    Subtract **3x** from both sides: $\qquad 3x + 1 - \mathbf{3x} = 4x - 8 - \mathbf{3x}$
    Simplify: $\qquad 1 = x - 8$
    Add **8** to both sides: $\qquad 1 + \mathbf{8} = x - 8 + \mathbf{8}$
    Simplify: $\qquad 9 = x$ (or $x = 9$)

15. The given equation is: $\qquad 7 = 3x + 4$
    Subtract **4** from both sides: $\qquad 7 - \mathbf{4} = 3x + 4 - \mathbf{4}$
    Simplify: $\qquad 3 = 3x$
    Divide both sides by **3**: $\qquad \dfrac{3}{3} = \dfrac{3x}{3}$
    Simplify: $\qquad 1 = x$ (or $x = 1$)

17. The given equation is: $\qquad 4 = 3x - 2$
    Add 2 to both sides: $\qquad 6 = 3x$
    Divide both sides by 3: $\qquad 2 = x$ (or $x = 2$)

SECTION 5.1 Solution of First-Degree Sentences

19. The given equation is: $7n + 10 - 2n = 4n - 2 + 3n$
    Simplify: $5n + 10 = 7n - 2$
    Subtract 5n from both sides: $10 = 2n - 2$
    Add 2 to both sides: $12 = 2n$
    Divide both sides by 2: $6 = n$ (or $n = 6$)

21. The given equation is: $2(x + 5) = 13$
    Simplify: $2x + 10 = 13$
    Subtract 10 from both sides: $2x = 3$
    Divide both sides by 2: $x = \frac{3}{2}$

23. The given equation is: $\frac{1}{2}(x - 2) = 5$
    Multiply both sides by 2: $x - 2 = 10$
    Add 2 to both sides: $x = 12$

25. The given equation is: $3(x + 1) - x = 2(9 - x)$
    Do the multiplications: $3x + 3 - x = 18 - 2x$
    Simplify: $2x + 3 = 18 - 2x$
    Add 2x to both sides: $4x + 3 = 18$
    Subtract 3 from both sides: $4x = 15$
    Divide both sides by 4: $x = \frac{15}{4} = 3\frac{3}{4}$

27. The given equation is: $8(x - 1) = x + 2$
    Do the multiplication: $8x - 8 = x + 2$
    Subtract x from both sides: $7x - 8 = 2$
    Add 8 to both sides: $7x = 10$
    Divide both sides by 7: $x = \frac{10}{7} = 1\frac{3}{7}$

29. The given equation is: $\frac{1}{4}(x - 2) = \frac{1}{3}(x - 4)$
    Multiply both sides by 12: $3x - 6 = 4x - 16$
    Subtract 3x from both sides: $-6 = x - 16$
    Add 16 to both sides: $10 = x$ (or $x = 10$)

106 CHAPTER 5 EQUATIONS, INEQUALITIES, PROBLEM SOLVING

31. The given equation is: $\quad\quad\quad\quad\quad 3 \times 2p + 5 = 37 - \dfrac{4p^2}{2p}$

   Do the multiplication and the division: $\quad 6p + 5 = 37 - 2p$
   Add 2p to both sides: $\quad\quad\quad\quad\quad\quad\quad 8p + 5 = 37$
   Subtract 5 from both sides: $\quad\quad\quad\quad\quad\quad 8p = 32$
   Divide both sides by 8: $\quad\quad\quad\quad\quad\quad\quad\;\; p = 4$

33. The given inequality is: $\quad\quad\quad\quad\quad\; x - 3 < 1$
   Add 3 to both sides: $\quad\quad\quad\quad\quad\quad\quad\;\; x < 4$
   The solution set is $\{x \mid x < 4\}$.

35. The given inequality is: $\quad\quad\quad\quad\quad\; x - 4 > -1$
   Add 4 to both sides: $\quad\quad\quad\quad\quad\quad\quad\;\; x > 3$
   The solution set is $\{x \mid x > 3\}$.

37. The given inequality is: $\quad\quad\quad\quad\quad\; 2x - 1 > x + 2$
   Subtract x from both sides: $\quad\quad\quad\quad\quad x - 1 > 2$
   Add 1 to both sides: $\quad\quad\quad\quad\quad\quad\quad\;\; x > 3$
   The solution set is $\{x \mid x > 3\}$.

39. The given inequality is: $\quad\quad\quad\quad\quad\; 2x + 3 \leq 9 + 5x$
   Subtract 2x from both sides: $\quad\quad\quad\quad\; 3 \leq 9 + 3x$
   Subtract 9 from both sides: $\quad\quad\quad\quad\quad -6 \leq 3x$
   Divide both sides by 3 $\quad\quad\quad\quad\quad\quad\quad -2 \leq x$
   Since the last result is equivalent to $x \geq -2$, the solution set is $\{x \mid x \geq -2\}$.

41. The given inequality is: $\quad\quad\quad\quad\quad\; x + 1 > \dfrac{1}{2}x - 1$

   Multiply both sides by 2: $\quad\quad\quad\quad\quad\quad 2x + 2 > x - 2$
   Subtract x from both sides: $\quad\quad\quad\quad\quad x + 2 > -2$
   Subtract 2 on both sides: $\quad\quad\quad\quad\quad\quad x > -4$
   The solution set is $\{x \mid x > -4\}$.

43. The given inequality is: $\quad\quad\quad\quad\quad\; x \geq 4 + 3x$
   Subtract x from both sides: $\quad\quad\quad\quad\quad 0 \geq 4 + 2x$
   Subtract 4 from both sides: $\quad\quad\quad\quad\quad -4 \geq 2x$
   Divide both sides by 2: $\quad\quad\quad\quad\quad\quad\;\; -2 \geq x$
   Since the last result is equivalent to $x \leq -2$, the solution set is $\{x \mid x \leq -2\}$.

SECTION 5.1   Solution of First-Degree Sentences

45. The given inequality is: $\quad\quad\quad\quad\quad\quad\quad\quad \frac{1}{3}x - 2 \geq \frac{2}{3}x + 1$

   Multiply both sides by 3: $\quad\quad\quad\quad\quad\quad\quad x - 6 \geq 2x + 3$
   Subtract x from both sides: $\quad\quad\quad\quad\quad\quad -6 \geq x + 3$
   Subtract 3 from both sides: $\quad\quad\quad\quad\quad\quad -9 \geq x$ or $x \leq -9$
   Thus, the solution set is $\{x \mid x \leq -9\}$.

47. The given inequality is: $\quad\quad\quad\quad\quad\quad\quad\quad 2x - 2 > x + 1$
    Subtract x from both sides: $\quad\quad\quad\quad\quad\quad x - 2 > 1$
    Add 2 to both sides: $\quad\quad\quad\quad\quad\quad\quad\quad x > 3$
    Thus, the solution set is $\{x \mid x > 3\}$.

49. The given inequality is: $\quad\quad\quad\quad\quad\quad\quad\quad x + 3 > \frac{1}{2}x + 1$

    Multiply both sides by 2: $\quad\quad\quad\quad\quad\quad\quad 2x + 6 > x + 2$
    Subtract x from both sides: $\quad\quad\quad\quad\quad\quad x + 6 > 2$
    Subtract 6 from both sides: $\quad\quad\quad\quad\quad\quad x > -4$
    The solution set is $\{x \mid x > -4\}$.

51. The given inequality is: $\quad\quad\quad\quad\quad\quad\quad\quad x \geq 2 + 4x$
    Subtract x on both sides: $\quad\quad\quad\quad\quad\quad\quad 0 \geq 2 + 3x$
    Subtract 2 on both sides: $\quad\quad\quad\quad\quad\quad\quad -2 \geq 3x$

    Divide both sides by 3: $\quad\quad\quad\quad\quad\quad\quad\quad -\frac{2}{3} \geq 1$

    Since the final result is equivalent to $x \leq -\frac{2}{3}$ the solution set is

    $\{x \mid x \leq -\frac{2}{3}\}$.

53. The given inequality is: $\quad\quad\quad\quad\quad\quad\quad\quad 2x + 1 < 2x$
    Subtracting 2x on both sides gives: $\quad\quad\quad\quad 1 < 0$
    which is false. There is no value of
    x for which the given inequality is true, so the solution set is $\emptyset$.

55. The given inequality is: $\quad\quad\quad\quad\quad\quad\quad\quad 8x + 2 \leq 3(x + 4)$
    Simplify: $\quad\quad\quad\quad\quad\quad\quad\quad\quad\quad\quad\quad 8x + 2 \leq 3x + 12$
    Subtract 3x from both sides: $\quad\quad\quad\quad\quad\quad 5x + 2 \leq 12$
    Subtract 2 from both sides: $\quad\quad\quad\quad\quad\quad\quad 5x \leq 10$
    Divide both sides by 5: $\quad\quad\quad\quad\quad\quad\quad\quad x \leq 2$
    Thus, the solution set is $\{x \mid x \leq 2\}$.

108 CHAPTER 5 EQUATIONS, INEQUALITIES, PROBLEM SOLVING

57. The given inequality is: $3(x + 4) > -5x - 4$
 Simplify: $3x + 12 > -5x - 4$
 Add 5x to both sides: $8x + 12 > -4$
 Subtract 12 from both sides: $8x > -16$
 Divide both sides by 8: $x > -2$
 The solution set is $\{x \mid x > -2\}$.

59. The given inequality is: $-2(x + 1) \geq 3x - 4$
 Simplify: $-2x - 2 \geq 3x - 4$
 Add 2x on both sides: $-2 \geq 5x - 4$
 Add 4 on both sides $2 \geq 5x$
 Divide both sides by 5: $\frac{2}{5} \geq x$ or $x \leq \frac{2}{5}$

 Thus, the solution set is $\{x \mid x \leq \frac{2}{5}\}$.

61. The given inequality is: $a(x - 1) \leq a(2x + 3)$ with $a < 0$
 Divide by a, and since a is negative,
 reverse the inequality sign: $x - 1 \geq 2x + 3$
 Subtract x from both sides: $-1 \geq x + 3$
 Subtract 3 from both sides: $-4 \geq x$ or $x \leq -4$
 The solution set is $\{x \mid x \leq -4\}$.

63. Use the equation **$0.01rn = p$**. In this problem, $r = 40$, $n = 80$, and p is unknown. Thus, the equation becomes
 $$0.01 \times 40 \times 80 = p \text{ or } 32 = p$$
 Hence, $p = $ **32**.

65. Use the same equation as in Problem 63 but with $n = 3150$, $p = 315$, and r as the unknown. The equation becomes
 $$31.50r = 315$$
 To find r, divide both sides by 31.50. This gives $r = 10$, so the answer is **10%**.

67. Use the same equation as in Problem 63, but with $n = 40$, $p = 5$, and r as the unknown. The equation becomes
 $$0.01 \times r \times 40 = 5 \text{ or } 0.4 r = 5$$
 To find r, divide both sides by 0.4. This gives $r = 12.5$, so the answer is **12.5%**.

69. Use the same equation as in Problem 63, but with $r = 30$, $p = 60$, and n as the unknown. The equation becomes:
 $$0.3n = 60$$
 To find n, divide both sides by 0.3. This gives n = **200**.

SECTION 5.1 Solution of First-Degree Sentences

71. Use the same equation as in Problem 63, but with r = 7, p = 47, and n as the unknown. The equation becomes
$$0.7n = 47$$
To find n, divide both sides by 0.7. This gives 671 to the nearest unit. Thus, the answer is **671** billion barrels.

73. Use the same equation as in Problem 63, but with n = 60, p = 40, and r as the unknown. The equation becomes
$$0.6r = 40$$
To find r, divide both sides by 0.6. This gives r = 66.666 . . .; so to the nearest percent, the answer is **67%**.

75. Use the same equation as in Problem 63. For store A, n = 140, r = 25, and p is the unknown. The equation becomes
$$0.25 \times 140 = p \text{ or } 35 = p$$
Thus, $35 is the discount and store A's sale price is $105. Store B's price of **$100** is the lower.

77. 20 < t < 40   79. s > 23,000   81. 2 ≤ e ≤ 7

83. The solution set for an equation is the set of numbers taken from the replacement set that will make an open sentence a true statement when the number is substituted for the variable in the open sentence.

85. Yes. In order to be in the solution set a number must be part of the replacement set.

87. Since 50% of the calories is 125 of the total calories c, we have
$$0.50c = 125$$
Divide by 0.50:   c = 250
Thus, there are **250** calories in the apple pie.

89. Since 55% of the calories is 343.75 of the total calories c, we have
$$0.55c = 343.75$$
Divide by 0.55:   c = 625
Thus, there are **625** calories in a Whopper.

91. C = 4(77.5 - 40) = 4 × 37.5 = **150**.

93. $S = \frac{1}{6}(10 - 4) = $ **1** cm per sec

110 CHAPTER 5 EQUATIONS, INEQUALITIES, PROBLEM SOLVING

EXERCISE 5.2

STUDY TIPS Sometimes the elements of the solution set of an equation are not easy to write but a picture of the solution set, its **graph**, is easily obtainable. (See the Getting Started.) Note that the **graph** of the solution set is similar to the solution set itself, that is, if the solution set is a point, the graph is a point: if the solution set consists of all real numbers such that $x > 1$, the graph is -1 0 1 2 with the arrow pointing right just like in the > sign. The 1 is by marked by an open circle and is not included. Similarly, for $x \leq 2$, the graph is -1 0 1 2 pointing left, like the < and with 2 included. When the replacement set is the set of real numbers and the compound sentence has an **and**, the answer is either **the empty set** or **a single line segment**. If the compound sentence contains an **or** the answer consists of **two pieces** called **rays**.

If the replacement set is *not* the set of real numbers, the answer is **the intersection** of the two sets when **and** is used or **the union** of the two sets when **or** is used.

1. Subtract 2 from both sides to find the solution $x = 2$. The solution set is thus $\{2\}$, which is shown on the graph.

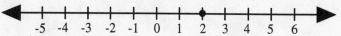

3. Subtract 1 from both sides to get $x \geq 1$. Thus, the solution set consists of all the positive integers as indicated on the graph.

5. Add 3 to both sides to get $x \neq 4$. This means that the solution consists of all the integers except 4, as indicated on the graph.

7. Since x is to be an integer, the inequality $-2 \leq x \leq 4$ has the solution set $\{-2, -1, 0, 1, 2, 3, 4\}$, as shown on the graph.

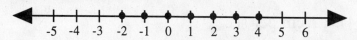

SECTION 5.2 Graphs of Algebraic Sentences   111

9. The solution set of x < 4 is the set of all real numbers less than 4, as shown on the graph.

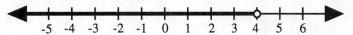

11. Adding 2 to both sides gives the inequality x ≤ 2 so the solution set consists of all the real numbers less than or equal to 2.

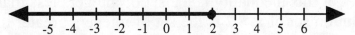

13. The solution set of -2 ≤ x ≤ 4 consists of all the real numbers between -2 and +4, inclusive.

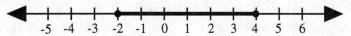

15. Subtract 2 from both sides to get x > 3. The solution set is the set of all real numbers greater than 3.

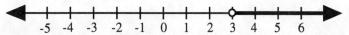

17. The solution set of -1 < x < 2 consists of all the real numbers between -1 and +2 with neither endpoint included.

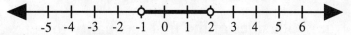

19. Subtracting 4 from both sides gives x < 1, so that the solution set consists of all the real numbers less than 1.

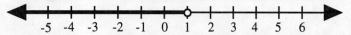

21. x + 1 < x is not true for any real number. Thus, the solution set is ∅.

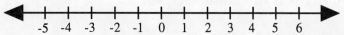

23. Subtract x from both sides to get x + 3 < 1. Then subtract 3 from both sides to get x < -2. Thus, the solution set consists of all the real numbers less than -2.

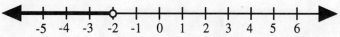

25. Add 7 to both sides to get 3x ≥ 0. Then divide both sides by 3 to get x ≥ 0. The solution set consists of all the real numbers that are greater than or equal to 0.

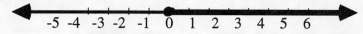

27. Add 1 to both sides of the second inequality to get x ≥ -1. The solution set of this inequality is the set of integers greater than or equal to -1. The solution set of x ≤ 4 is the set of integers less than or equal to 4. So the solution set of x ≤ 4 **and** x ≥ -1 is the set of integers greater than or equal to -1 and less than or equal to 4, that is, {-1, 0, 1, 2, 3, 4}.

29. Subtract 1 from both sides of the first inequality to get x ≤ 6. The solution set of this inequality is the set of integers less than or equal to 6. The solution set of x > 2 is the set of integers greater than 2. Thus, the solution set of x > 2 **and** x ≤ 6 is the set of integers greater than 2 but less than or equal to 6, that is, {3, 4, 5, 6}.

31. Add 1 to both sides of the first inequality to get 2x > 2. Then divide both sides by 2 to obtain x > 1. The solution set of this inequality is the set of integers greater than 1. Now, subtract 1 from both sides of the second inequality to get x < 3. The solution set of this inequality is the set of integers less than 3. Thus, the solution set of x > 1 **and** x < 3 consists of the integer 2, that is, {2}.

33. The solution set of x < -5 or x > 5 consists of all the integers except those between -5 and +5, inclusive, that is the set
{. . ., -8, -7, -6, 6, 7, 8, . . .}.

35. Subtract 1 from both sides of the first inequality to get x ≥ 1. The solution set of x ≥ 1 and x ≤ 4 is the set of all real numbers between 1 and 4, inclusive. This is shown on the graph.

37. There are no numbers that are both greater than 2 and less than -2, so the solution set is the empty set, ∅.

39. Add 1 to both sides of the first inequality to get x > 1. Then, subtract 1 from both sides of the second inequality to get x < 4. The solution set of x > 1 **and** x < 4 is the set of all real numbers between 1 and 4, not including the 1 and the 4. This is shown on the graph.

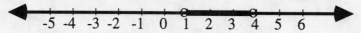

SECTION 5.2   Graphs of Algebraic Sentences   113

41. The first inequality, x < x + 1, is satisfied by all real numbers. Therefore, only the inequality x ≥ 2 must be satisfied, and the solution set of this inequality is the set of all real numbers greater than or equal to 2. The graph shows this set.

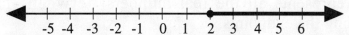

43. Add 2 to both sides of the first inequality to get x ≥ 4. Since there is no real number that is both greater than or equal to 4 and less than 0, the solution set is the empty set, ∅.

45. Add 1 to both sides of the second inequality to get x ≥ 3. Then, the solution set of x ≥ 0 **and** x ≥ 3 is the set of all real numbers that are greater than or equal to 3. This set is shown on the graph.

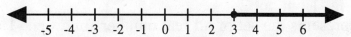

47. Add 1 to both sides of the second inequality to get x < 3. Then, the solution set of x < 0 **or** x < 3 is the set of all real numbers that are less than 3. (Note that each number of this set satisfies at least one of the two inequalities.) The graph is shown below.

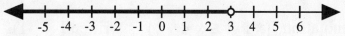

49. Subtract 1 from both sides of the first inequality to get x > 1. Then, add 2 to both sides of the second inequality to get x < 5. The solution set of x > 1 **and** x < 5 is the set of all real numbers between 1 and 5, not including the 1 or the 5. The graph shows this set.

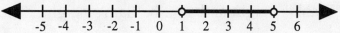

51. Add 1 to both sides of the first inequality and subtract 2 from both sides of the second inequality to get the system x > 1 **or** x < 2. Every real number is greater than 1 or less than 2, so the solution set consists of all the real numbers, as indicated by the graph.

53. (a) This represents all the real numbers **between** -1 and +2, including the 2 but not the -1.
    (b) This represents all the real numbers **between** -1 and +2 including the -1 but not the 2.
    (c) This represents all the real numbers **between** -1 and +2 not including the endpoints.
    (d) This represents all the real numbers **between** -1 and +2 including the endpoints.

55. (a) This represents the set of all real numbers between, but not including -1 and 2.
    (b) This represents the set of all real numbers between -1 and 2, including 2 but not -1.
    (c) This represents the set of all real numbers between -1 and 2, including -1 but not 2.
    (d) This represents the set of all real numbers between -1 and 2, including -1 and 2.

57. In interval notation, $\{x \mid x < 5\}$ is written $(-\infty, 5)$.

59. In interval notation, $\{x \mid x > 9\}$ is written $(9, +\infty)$.

61. In interval notation, $\{x \mid -4 \leq x < -1\}$ is written $[-4, -1)$.

63. In interval notation, $\{x \mid -1 \leq x \leq 10\}$ is written $[-1, 10]$.

65. Let j in. be Joe's height. Then $j = 60$.

67. Let f in. be Frank's height and s in. be Sam's height. Then, $f = s - 3$

69. Let s in. be Sam's height. Then $s = 77$.

71. Bill is taller than 74 in. (6 ft 2 in.)

SECTION 5.3 Sentences Involving Absolute Values

EXERCISE 5.3

STUDY TIPS The best way to understand this section is with examples you can construct yourself. First, graph $|x| = 1$. The graph consists of two points one unit away from 0. Then graph $|x| < 1$. The graph consists of all the points less than one unit away from 0, that is, all real numbers x such that $-1 < x < 1$. Finally, $|x| > 1$ has to consist of all elements more than one unit from 0, that is, $|x| > 1$ consists of all real numbers x such that $x > 1$ or $x < -1$. Note that the graph of $|x| = a$ (a positive) consists of **two** points, $|x| < a$ is a **line segment** and $|x| > a$ consists of **two** pieces called **rays**.

1. **10**

3. $\dfrac{1}{8}$

5. $|5 - 8| = |-3| = 3$

7. $|0| + |-2| = 0 + 2 = \mathbf{2}$

9. **-8**

11. (a) For $x = 2$, $|1 - 3x| = |1 - 6| = |-5| = 5 > 3$, a solution.

    (b) For $x = -\dfrac{1}{2}$, $|1 - 3x| = |1 + \dfrac{3}{2}| = |\dfrac{5}{2}| = \dfrac{5}{2} < 3$, not a solution

    (c) For $x = \dfrac{5}{3}$, $|1 - 3x| = |1 - 5| = |-4| = 4 > 3$, a solution.

    (d) For $x = 0$, $|1 - 3x| = |1| = 1 < 3$, not a solution.

13. **0** is the only integer for which $|x| < 1$ is true.

15. $\{-5, 5\}$

17. $\{\ldots, -3, -2, -1, 1, 2, 3, \ldots\}$

19. $|x| = 1$ is true for $x = -1$ and for $x = 1$. These points are shown on the graph. There is no interval that satisfies this equation.

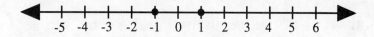

21. $|x| \le 4$ is true for all the real numbers between -4 and 4, inclusive. These numbers are shown on the graph. In interval notation, the solution is $[-4, 4]$.

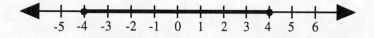

23. |x + 1| < 3 is equivalent to -3 < x + 1 < 3. Thus, by subtracting 1 from each side of this inequality, you get -4 < x < 2. The last inequality is true for all real numbers between -4 and 2, not including the -4 or the 2. This gives the required solution set, which is shown on the graph. In interval notation, the solution is (-4, 2).

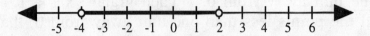

25. |x| ≥ 1 is true if x ≥ 1 or if x ≤ -1. Thus, the solution set consists of all real numbers that are less than or equal to -1 or that are greater than or equal to 1. This set is shown on the graph. In interval notation, the solution is (-∞, -1] and [1, +∞).

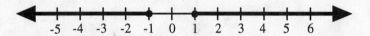

27. |x - 1| > 2 is true if x - 1 > 2 or if x - 1 < -2. By adding 1 to both sides of the last two inequalities, you get x > 3 or x < -1. Thus, the solution set of the given inequality consists of all the real numbers that are less than -1 or are greater than 3. This set is shown on the graph. In interval notation, the solution is (-∞, -1) and (3, +∞).

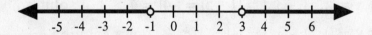

29. Divide both sides of the inequality |2x| < 4 by 2 to get |x| < 2. This inequality is satisfied by all the real numbers between -2 and 2, not including the -2 or the 2. This set is shown on the graph. In interval notation, the solution is (-2, 2).

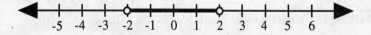

31. Divide both sides of the inequality |3x| ≥ 6 by 3 to get |x| ≥ 2. This inequality is satisfied by all the real numbers that are less than or equal to -2 or that are greater than or equal to 2. The graph shows this set. In interval notation, the solution is (-∞, -2] and [2, +∞).

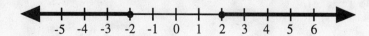

SECTION 5.3 Sentences Involving Absolute Values

33. $|2x - 3| \leq 3$ is equivalent to $-3 \leq 2x - 3 \leq 3$. Add 3 to each member of this inequality to get
$$-3 + 3 \leq 2x - 3 + 3 \leq 3 + 3$$
or $\qquad 0 \leq 2x \leq 6$

Now divide by 2 to get the inequality $0 \leq x \leq 3$. This inequality is satisfied by all the real numbers between 0 and 3, inclusive. The solution set is shown on the graph. In interval notation, the solution is [0, 3].

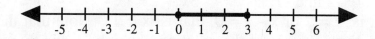

35. $|2x - 3| > 3$ is equivalent to the system $2x - 3 < -3$ or $2x - 3 > 3$. Add 3 to both sides of these inequalities to get the system $2x < 0$ **or** $2x > 6$. Now divide by 2 to get $x < 0$ **or** $x > 3$. The solution set of this system of inequalities is the set of all real numbers that are less than 0 or that are greater than 3. The graph shows this set. In interval notation, the solution is $(-\infty, 0)$ and $(3, +\infty)$.

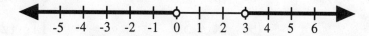

37. Substitute $b = 500$ and $c = 50$ to get $|500 - a| \leq 50$. Since the absolute value is being used, we may write the equivalent as $|a - 500| \leq 50$, and this is equivalent to $-50 \leq a - 500 \leq 50$. Add 500 to each member to get
$$450 \leq a \leq 550$$
Thus, the company can spend any amount between $450 and $500, inclusive.

39. Substitute $b = 300$ and $a = 290$ to get $|300 - 290| = 10 \leq c$. Since 5% of the budgeted $300 = 0.05 \times 300 = 15$, we take $c = 15$ and see that George is within the 5% variance.

41. "The absolute value of x is less than a." is equivalent to "x is between -a and a."

118 CHAPTER 5 EQUATIONS, INEQUALITIES, PROBLEM SOLVING

EXERCISE 5.4

STUDY TIPS Here we learn two ways to solve quadratic equations. Before you attempt to solve a quadratic, however, it must be in the standard form $ax^2 + bx + c = 0$. If a is 1, then you need two numbers whose product is c and whose sum is b to **factor** the equation. To factor $x^2 - 3x + 2 = 0$ we need two numbers that add to -3 and whose product is 2. The numbers are -2 and -1, so $x^2 - 3x + 2 = (x - 2)(x - 1) = 0$. Thus, x - 2 = 0, that is, x = 2 or x - 1 = 0, that is, x = 1. Note that the -2 and the -1 are the negative inverses of the solutions, which means that if you know how to factor the equation, you know the solutions automatically!

The other method used to solve quadratics is by using the **quadratic formula.** After the equation is in the standard form $ax^2 + bx + c = 0$, select a, b and c and use the formula on page 264.

1. $(x + 2)(x + 4)$ 3. $(x - 4)(x + 3)$ 5. $(x + 9)(x - 2)$

7. $(x - 5)^2$ 9. $(x + 5)^2$ 11. $(2x + 3)(x - 1)$

13. $6x^2 - 5x + 1$ is not factorable.

15. This equation is satisfied if and only if x - 2 = 0 or x - 4 = 0, that is, if and only if x = 2 or x = 4. Thus, the solution set is {2, 4}.

17. This equation is satisfied if and only if x + 2 = 0 or x - 3 = 0, that is, if and only if x = -2 or x = 3. Thus, the solution set is {-2, 3}.

19. This equation is satisfied if and only if x = 0, or x - 1 = 0, or x + 1 = 0, that is, if and only if x = 0, or x = 1, or x = -1. Thus, the solution set is {-1, 0, 1}.

21. This equation is satisfied if and only if 2x - 1 = 0, or x + 2 = 0. Add 1 to both sides of the first equation to get 2x = 1. Then, divide both sides by 2 to get x = 1/2. The second equation is satisfied by x = -2. Thus, the solution set of the given equation is $\{-2, \frac{1}{2}\}$.

23. $x^2 - 16$ factors into $(x - 4)(x + 4)$, so the equation $x^2 - 16 = 0$ is satisfied by x = -4 and by x = 4. Thus, the solution set is {-4, 4}.

SECTION 5.4   Quadratic Equations   119

25. Divide both sides of the given equation by 5 to get $x^2 = 25$. The solution set of this equation is $\{-5, 5\}$, because $(-5)^2 = 25$ and $5^2 = 25$.

27. The equation $(3x - 6)(2x + 3)(5x - 8) = 0$ is satisfied if and only if $3x - 6 = 0$, or $2x + 3 = 0$, or $5x - 8 = 0$. The equation $3x - 6 = 0$ can be solved by adding 6 to both sides and then dividing both sides by 3 to get $x = 2$. Similarly, the second equation can be solved by subtracting 3 from both sides and then dividing by 2 to get $x = -3/2$. The third equation is solved by adding 8 to both sides and then dividing by 5 to get $x = 8/5$. Thus, the solution set of the given equation is $\{-\frac{3}{2}, \frac{8}{5}, 2\}$.

29. To solve the equation $6x^2 - 1 = 215$, add 1 to both sides to get $6x^2 = 216$, and then divide both sides by 6 to get $x^2 = 36$. This equation is satisfied by -6 and by 6, because $(-6)^2 = 36$ and $6^2 = 36$. Thus, the solution set is $\{-6, 6\}$.

31. Since $x^2 - 12x + 27 = (x - 3)(x - 9)$, the given equation can be rewritten as $(x - 3)(x - 9) = 0$, which is satisfied if and only if $x = 3$ or $x = 9$. Thus, the solution set of the given equation is $\{3, 9\}$.

33. Since $x^2 - 8x - 20 = (x + 2)(x - 10)$, the given equation can be rewritten as $(x + 2)(x - 10) = 0$, which is satisfied if and only if $x = -2$ or $x = 10$. Thus, the solution set is $\{-2, 10\}$.

35. Since $10x^2 + 7x + 1 = (2x + 1)(5x + 1)$, the given equation can be rewritten as $(2x + 1)(5x + 1) = 0$, which is satisfied if and only if $x = -\frac{1}{2}$ or $x = -\frac{1}{5}$. Thus, the solution set is $\{-\frac{1}{2}, -\frac{1}{5}\}$.

37. First subtract 5 from both sides to get Since $3x^2 + 2x - 5 = 0$. Since $3x^2 + 2x - 5 = (3x + 5)(x - 1)$, the given equation can be rewritten as $(3x + 5)(x - 1) = 0$, which is satisfied if and only if $x = -\frac{5}{3}$ or $x = 1$. Thus, the solution set is $\{-\frac{5}{3}, 1\}$.

39. For the equation $2x^2 + 3x - 5 = 0$, $a = 2$, $b = 3$, and $c = -5$. By substituting these numbers in the quadratic formula

$$x = \frac{-b \pm \sqrt{b^2 - 4ac}}{2a}$$

we get $x = \dfrac{-3 \pm \sqrt{9 - (-40)}}{4} = \dfrac{-3 \pm \sqrt{49}}{4} = \dfrac{-3 \pm 7}{4} = -\dfrac{10}{4}$ or $\dfrac{4}{4}$, that is $-\dfrac{5}{2}$ or 1. Thus, the solution set is $\{-\dfrac{5}{2}, 1\}$.

41. For the equation $2x^2 + 5x - 7 = 0$, $a = 2$, $b = 5$, and $c = -7$. By substituting these numbers in the quadratic formula (see Problem 39), we get

$$x = \frac{-5 \pm \sqrt{25 - (-56)}}{4} = \frac{-5 \pm \sqrt{81}}{4} = \frac{-5 \pm 9}{4} = \frac{-14}{4} \text{ or } \frac{4}{4} = -\frac{7}{2} \text{ or } 1.$$

Thus, the solution set is $\{-\dfrac{7}{2}, 1\}$.

43. For the equation $x^2 + 5x + 3 = 0$, $a = 1$, $b = 5$, and $c = 3$. By substituting these numbers in the quadratic formula (see Problem 39), we get

$$x = \frac{-5 \pm \sqrt{25 - 12}}{2} = \frac{-5 \pm \sqrt{13}}{2}$$

so that the solution set is $\{\dfrac{-5 - \sqrt{13}}{2}, \dfrac{-5 + \sqrt{13}}{2}\}$.

45. For the equation $5x^2 - 8x + 2 = 0$, $a = 5$, $b = -8$, and $c = 2$. By substituting these numbers in the quadratic formula (see Problem 39), we get

$$x = \frac{8 \pm \sqrt{64 - 40}}{10} = \frac{8 \pm \sqrt{24}}{10} = \frac{8 \pm 2\sqrt{6}}{10} = \frac{4 \pm \sqrt{6}}{5}$$

Thus, the solution set is $\{\dfrac{4 - \sqrt{6}}{5}, \dfrac{4 + \sqrt{6}}{5}\}$.

47. By adding 1 to both sides, we can write the given equation in the standard form: $7x^2 - 6x + 1 = 0$. Then, we have $a = 7$, $b = -6$, and $c = 1$. Substituting these values in the quadratic formula (see Problem 39), we get

$$x = \frac{6 \pm \sqrt{36 - 28}}{14} = \frac{6 \pm \sqrt{8}}{14} = \frac{6 \pm 2\sqrt{2}}{14} = \frac{3 \pm \sqrt{2}}{7}$$

Thus, the solution set is $\{\dfrac{3 - \sqrt{2}}{7}, \dfrac{3 + \sqrt{2}}{7}\}$.

SECTION 5.4 Quadratic Equations

49. For the equation $9x^2 - 6x - 2 = 0$, $a = 9$, $b = -6$, and $c = -2$. By substituting these numbers in the quadratic formula (see Problem 39), we get

$$x = \frac{6 \pm \sqrt{36 + 72}}{18} = \frac{6 \pm \sqrt{108}}{18} = \frac{6 \pm 6\sqrt{3}}{18} = \frac{1 \pm \sqrt{3}}{3}$$

Thus, the solution set is $\{\frac{1 - \sqrt{3}}{3}, \frac{1 + \sqrt{3}}{3}\}$.

51. First subtract 1 from both sides to get $2x^2 + 2x - 1 = 0$, for which $a = 2$, $b = 2$ and $c = -1$. Substituting these numbers in the quadratic formula (see Problem 39 above), we get

$$x = \frac{-2 \pm \sqrt{4 + 8}}{4} = \frac{-2 \pm \sqrt{12}}{4} = \frac{-2 \pm 2\sqrt{3}}{4} = \frac{-1 \pm \sqrt{3}}{2}$$

Thus, the solution set is $\{\frac{-1 - \sqrt{3}}{2}, \frac{-1 + \sqrt{3}}{2}\}$.

53. First, add $8x - 5$ to both sides to get $4x^2 + 8x - 5 = 0$, for which $a = 4$, $b = 8$, and $c = -5$. Substitute these numbers in the quadratic formula (see Problem 39 above) to get

$$x = \frac{-8 \pm \sqrt{64 + 80}}{8} = \frac{-8 \pm \sqrt{144}}{8} = \frac{-8 \pm 12}{8} = -\frac{5}{2} \text{ or } \frac{1}{2}.$$

Thus, the solution set is $\{-\frac{5}{2}, \frac{1}{2}\}$.

55. $(2ax + b)^2 = (2ax + b)(2ax + b) = 2ax(2ax + b) + b(2ax + b)$
    $= (4a^2x^2 + 2abx) + (2abx + b^2)$
    $= 4a^2x^2 + 4abx + b^2$

which is the left side of the equation, as stated. If we subtract b from both sides of the equation,

$$2ax + b = \pm\sqrt{b^2 - 4ac}$$

we get $\quad 2ax = -b \pm \sqrt{b^2 - 4ac}$

If we divide both sides of this last equation by 2a, we get the quadratic formula as given.

57. Let the shortest side be x cm long. Then the hypotenuse is (x + 4) cm long and the other leg is (x + 2) cm long. By the Pythagorean theorem,
$$(x + 4)^2 = x^2 + (x + 2)^2$$
Multiply: $x^2 + 8x + 16 = x^2 + x^2 + 4x + 4$
Subtract $x^2 + 8x + 16$: $0 = x^2 - 4x - 12$
Factor: $0 = (x - 6)(x + 2)$
Solve: $x = 6$ or -2.
Since this is to be the side of a triangle, we discard the negative solution. The sides of the triangle are, respectively, 6 cm, 8 cm and 10 cm long.

59. Let the shortest side be x in. long. Then the hypotenuse is (x + 8) in. long and the other leg is (x + 7) in. long. By the Pythagorean theorem
$$(x + 8)^2 = x^2 + (x + 7)^2$$
Multiply: $x^2 + 16x + 64 = x^2 + x^2 + 14x + 49$
Subtract $x^2 + 16x + 64$: $0 = x^2 - 2x - 15$
Factor: $0 = (x - 5)(x + 3)$
Solve: $x = 5$ or -3
Since this is to be a side of a triangle, we discard the negative solution. The sides of the triangle are, respectively, 5 in., 12 in. and 13 in. long.

61. If $b^2 - 4ac = 0$, the two answers coincide. The only solution is $\frac{-b}{2a}$.

63. If $b^2 - 4ac < 0$, there are no real number solutions.

65. Substitute into the formula $h = 16t^2$ to get
$$28 = 16t^2$$
$$t^2 = \frac{28}{16} = \frac{7}{4}$$
$$t = \frac{\sqrt{7}}{2} \approx 1.3 \text{ sec}$$

67. In the equation $x^2 = y^2 + 13^2$, we want y as large as possible relative to x, so we try putting $y = x - 1$. The equation becomes
$$x^2 = (x - 1)^2 + 13^2$$
or $x^2 = x^2 - 2x + 1 + 169$
Subtract x^2 $0 = -2x + 170$
Add 2x $2x = 170$
Divide by 2: $x = 85$.
We can check this by noting that $85^2 = 7225$ and $84^2 = 7056$. Thus, $85^2 - 84^2 = 7225 - 7056 = 169 = 13^2$. So the maximum number of letters that Charlie could have received is $85^2 = 7225$.

SECTION 5.5 Modeling and Problem Solving 123

EXERCISE 5.5

STUDY TIPS Before you start this section it is a good idea to review the **RSTUV** procedure covered in Chapter 1. Try covering the solution in the right hand column of the **Problem Solving** section with a 3 by 5 card and see if you can figure out how to proceed. Use the same idea in the rest of the section paying particular attention to the key words used in mathematics. Learn these words and their translation by studying Table 5.1.

1. $4m = m + 18$

3. If x is the tens digit, then the units digit is x - 3. The number is $10x + (x - 3)$. The required equation is
$$10x + (x - 3) = 26(x - 3)$$
We can solve this equation as follows
$$11x - 3 = 26x - 78$$
$$75 = 15x$$
$$5 = x$$
So the tens digit is **5**. You can check that the number is $52 = 26 \times 2$.

5. If x is the number, then the required equation is $4x + 5 = 29$.
Subtract 5 from both sides: $4x = 24$
Divide both sides by 4: $x = 6$
Thus, the number is **6**.

7. If x is the number, then the required equation is $3x + 8 = 29$.
Subtract 8 from both sides: $3x = 21$
Divide both sides by 3: $x = 7$
Thus, the number is **7**.

9. If x is the number, then the required equation is $3x - 2 = 16$.
Add 2 to both sides: $3x = 18$
Divide both sides by 3: $x = 6$
Thus, the number is **6**.

11. If x is the number, then the required equation is $2x^2 = 2x + 12$.
    Subtract $2x + 12$ from both sides: $\qquad 2x^2 - 2x - 12 = 0$
    Divide both sides by 2: $\qquad\qquad\qquad x^2 - x - 6 = 0$
    Factor the left side: $\qquad\qquad\qquad (x - 3)(x + 2) = 0$
    The last equation is satisfied by $x = -2$ and by $x = 3$. Thus, the number can be **-2**, or it can be **3**.

13. If the number is x, then the required equation is $\frac{1}{3}x^2 - 2 = 10$.

    Add 2 to both sides: $\qquad\qquad \frac{1}{3}x^2 = 12$
    Multiply both sides by 3: $\qquad x^2 = 36$
    Take square roots: $\qquad\qquad x = \pm 6$
    Thus, the number can be **-6** or **6**.

15. Let x million lb be the weight of the orbiter and fuel. Then the weight of the tank and the boosters is $(x - 1.26)$ million lb. The total weight of these components is 4.16 million lb, so the required equation is
    $$x + (x - 1.26) = 4.16$$
    Simplify: $\qquad\qquad\qquad 2x - 1.26 = 4.16$
    Add 1.26 to both sides: $\qquad 2x = 5.42$
    Divide both sides by 2: $\qquad x = 2.71$
    Check: $2.71 - 1.26 = 1.45$, and $2.71 + 1.45 = 4.16$.
    Thus, the weight of the orbiter and fuel is **2.71** million lb.

17. Let x be the number of ships that Russia has. Then $x + 2276$ is the number of ships that Japan has. The total of these two numbers is 15,426, so the required equation is
    $$x + (x + 2276) = 15{,}426$$
    Simplify: $\qquad\qquad\qquad\qquad 2x + 2276 = 15{,}426$
    Subtract 2276 from both sides: $\quad 2x = 13{,}150$
    Divide both sides by 2: $\qquad\qquad x = 6575$
    Check: $6575 + 2276 = 8851$, and $6575 + 8851 = 15{,}426$.
    Thus, Russia has **6575** ships and Japan has **8851** ships.

19. Let m be the number of miles that Margie traveled. Then, $0.20m$ is the number of dollars that her travel cost. Since the total cost was $44, the required equation is
    $$0.20m + 18 = 44$$
    Subtract 18 from both sides: $\qquad 0.20m = 26$
    Multiply both sides by 10: $\qquad 2m = 260$
    Divide both sides by 2: $\qquad\qquad m = 130$
    Check: $(0.20)(130) = 26$ and $26 + 18 = 44$.
    Thus, Margie traveled **130** miles.

SECTION 5.5   Modeling and Problem Solving   125

21. Let r be the annual interest rate. Then, the interest on $10,000 for 3 months (one-fourth of a year) is $\frac{10000r}{4}$ or 2500r. Since the interest was $350, the required equation is
$$2500r = 350$$
Divide both sides by 2500: $r = 0.14$
Check: $(10,000)(0.14) = 1400$, and $1400 \div 4 = 350$.
Thus, the interest rate was **14%** per year.

23. (a) If the total cost is $71, then the required equation is
$$0.25m + 20 = 71$$
Subtract 20 from both sides: $0.25m = 51$
Multiply both sides by 100: $25m = 5100$
Divide both sides by 25: $m = 204$
Check: $(0.25)(204) = 51$, and $51 + 20 = 71$.
Thus, the answer is **204** miles.
(b) Using the mileage rate, the cost would be $(0.25)(60) + 20$, which comes to 35 dollars. The flat rate would be $40, so use the **mileage** rate.

25. Substitute $P = 1000$ and $A = 1210$ into the equation $A = P(1 + r)^2$ to get the required equation
$$1210 = 1000(1 + r)^2$$
Divide both sides by 1000: $\frac{121}{100} = (1 + r)^2$

Take the positive square root: $\frac{11}{10} = 1 + r$

Subtract 1 from both sides: $\frac{1}{10} = r$

Thus, the rate is 0.10 or **10%**.
You can check this answer as follows: The interest for 1 yr would be 10% of $1000, which is $100. So the total amount at the end of the first year would be $1000 + $100 = 1100. The interest for the second year would be 10% of $1100, that is, $110. Adding this interest to the $1100 gives $1210, which shows that the answer is correct.

27. Substitute $b = 54$ into the given formula to get the required equation: $54 = 0.06v^2$
Multiply both sides by 100 $5400 = 6v^2$
Divide both sides by 6: $900 = v^2$
Take the positive square root: $30 = v$
Thus, the answer is **30** mph

126 CHAPTER 5 EQUATIONS, INEQUALITIES, PROBLEM SOLVING

29. Substitute v = 30 and t = 0.5 into the given formula to get
$$d = (1.5)(0.5)(30) + (0.06)(30)^2$$
$$= 22.5 + 54 = 76.5$$
Thus, the stopping distance is **76.5** ft.

31. Substitute v = 20 and d = 42 into the given formula to get
$$42 = 30t + (0.06)(20^2)$$
or
$$42 = 30t + 24$$
Subtract 24 from both sides: 18 = 30t
Divide both sides by 30:   0.6 = t
Thus, the driver's reaction time is **0.6** sec.

33. Substitute $t = \frac{2}{3}$ and d = 44 into the given formula to get the equation
$44 = v + 0.06v^2$.
Rewrite in standard quadratic form: $0.06v^2 + v - 44 = 0$
Multiply both sides by 100: $6v^2 + 100v - 4400 = 0$
Divide both sides by 2: $3v^2 + 50v - 2200 = 0$
In the last equation, you can read off: a = 3, b = 50, and c = -2200.
Substitute these numbers into the quadratic formula to get
$$v = \frac{-50 \pm \sqrt{2500 + 26400}}{6} = \frac{-50 \pm \sqrt{28900}}{6} = \frac{-50 \pm 170}{6}$$
Since a positive number is needed for the answer, use the plus sign to find $v = \frac{-50 + 170}{5} = \frac{120}{6} = 20$. Thus, Loren was going **20** mph at the instant of the stop signal.

35. Let the two consecutive integers be x and x + 1. Then the given condition says that 6x < 5(x + 1). Simplify this inequality to get 6x < 5x + 5. Then subtract 5x from both sides to get x < 5. Since 4 is the largest integer less than 5, take x = 4, so that x + 1 = 5. Thus, **4** and **5** are the largest integers for which 6 times the smaller is less than 5 times the larger.

37. Let x billion cans be the number that are recycled. Then $\frac{2}{3}x$ billion cans are thrown away. Since the total number of billion cans is 60, we have the equation
$$x + \frac{2}{3}x = 60 \quad \text{or} \quad \frac{5}{3}x = 60$$
Thus, $x = \frac{3}{5}(60) = 36$. So **36** billion cans are recycled and 24 billion cans are not recycled.

SECTION 5.5 Modeling and Problem Solving

39. (a) Tuition = T

   State Contribution = $\frac{3N}{40}(2800) = 210N$

   Revenue to Institution = **T + 210N**

   (b) Costs to Institution:
   - (s) (40)($100) = $4,000
   - (U) (40)($150) = $6,000
   - (G) (40)($1000) = $40,000
   - (S) (3)($1000) = $3,000
   - Total Cost = **$53,000**

   (c) T + 210N = 53,000

   (d) $N = \dfrac{53000 - T}{210}$

   (e) If N is the number of students, then the total tuition revenue is
   T = 40 × 3 × N = 120 N.
   To break even, T + 210 N = 53000, that is
   $$120N + 210N = 53000$$
   $$330N = 53000$$
   $$N = \frac{53000}{330} = 160.606060\ldots$$

   Thus, to insure no loss, the least number of students required is **161**.

41. (a)

   Identical Row Alignment

Unit	Cost per month	Number per row	Rows	Layers	Total	Units Needed	Cost for 2 months
A	$25	5	5	3	75	5	$250
B	$90	10	5	5	250	2	$360
C	$128	10	10	5	500	1	$256

   Staggered Row Alignment

Unit	Cost per month	Number per row	Rows	Layers	Total	Units Needed	Cost for 2 months
A	$25	2-4's & 3-5's	5	3	69	6	$300
B	$90	3-10s & 2-9's	5	5	240	2	$360
C	$128	6-10s & 5-9-s	11	5	525	1	$256

41. (b) Use **5** units A for a cost of $250.
   (c) Use **one** unit C for a cost of $256.
   (d) Use **one** unit A and **one** unit B for a cost of $230.

128 CHAPTER 5 EQUATIONS, INEQUALITIES, PROBLEM SOLVING

43.

Hours Spent and Grade Earned

Math	Science	English	GPA
9(A)	2(C)	1(C)	$\frac{16 + 8 + 6}{11} = \frac{30}{11}$
6(B)	4(B)	2(B)	$\frac{12 + 12 + 9}{11} = \frac{33}{11}$
3(C)	6(A)	3(A)	$\frac{8 + 16 + 12}{11} = \frac{36}{11}$

The table shows that you should spend 3 hr on Math, 6 hr on Science and 3 hours on English to get the best GPA.

45. (a)

Late	2 days	3 days	4 days
Blockbusters:			
New	$3 + $4 = $7	$3 + $6 = $9	$3 + $8 = $11
Old	$3 + $2 = $5	$3 + $4 = $7	$3 + $6 = $9
Red Rabbit:			
New	$3 + $6 = $9	$3 + $9 = $12	$3 + $12 = $15
Old	$1.60 + $3 = $4.60	$1.60 + $6 = $7.60	$1.60 + $9 = $10.60

(b) Let C_B = cost at Blockbusters, C_R = cost at Red Rabbit.
(In dollars) $C_B = 6 + 4n$, $C_R = 4.60 + 4.60n$

(c) $C_B = \$26$ $C_R = \$27.60$

47. (a) Average stopping distance in feet
$d = (1.5)(0.6)v + 0.06v^2$; $d = 0.9v + 0.06v^2$

(b) $d = (0.9)(45) + (0.06)(45^2) = 40.5 + 121.5 = 161.7$ ft

(c) 161.7 ft + 80 ft = 241.7 ft

(d) $D = RT$: $241.7 = \frac{(45)(5280)T}{3600}$

$T = \frac{(241.7)(3600)}{(45)(5280)} = 3.66$ sec.

49. You should try to determine what is the unknown, that is, what is wanted.

SECTION 5.5 Modeling and Problem Solving 129

51. $y = \dfrac{(32)(10)(7000)}{(22)(15000)(1.3)} = \dfrac{(32)(7)}{(11)(3)(1.3)} = \dfrac{224}{42.9} = \mathbf{5.22}$ yr

53. If M is greater than G, then y would be negative, which is unrealistic. If M = G, then there is a zero denominator, which could be interpreted to mean that you should keep the old car and not buy the new one.

EXERCISE 5.6

STUDY TIPS This section covers ratio, proportion and variation. A ratio is an indicated quotient and can be written in 3 ways: a to b, a:b or the actual quotient $\dfrac{a}{b}$. A proportion is an equality between ratios. Proportions are usually given in words, so be prepared to use the **RSTUV** procedure when solving proportions. Finally, we mention two types of variation: direct and inverse. If y varies directly as x, y = kx, but if y varies inversely as x, $y = \dfrac{k}{x}$. If you think of y as $\dfrac{y}{1}$, and y varies directly as x, (y = kx) they are both in the numerator. If y varies inversely as x, $(y = \dfrac{k}{x})$, y is in the numerator and x is in the denominator.

1. 7000 to 2000;   7000 : 2000; $\dfrac{7000}{2000}$

3. 70 to 4260;   70 : 4260; $\dfrac{70}{4260}$

5. $\dfrac{2000}{600} = \dfrac{10}{3}$ 7. $\dfrac{12000}{700} = \dfrac{120}{7} = 17$ (to the nearest whole number)

9. (a) $\dfrac{1.31}{22} = 0.0595 = 6$ cents (to the nearest cent)

    (b) $\dfrac{1.75}{32} = 0.0547 = 5$ cents (to the nearest cent)

    (c) Based on price per oz alone, White Magic is the better buy.

11. From the proportion $\dfrac{x}{9} = \dfrac{4}{3}$, you get 3x = 36, by using "cross-products". Divide both sides by 3 to find x = 12.

13. Use "cross-products" for the proportion $\frac{8}{x} = \frac{4}{3}$ to get 24 = 4x. Then, divide both sides by 4 to find x = 6.

15. Use "cross-products" for the proportion $\frac{3}{8} = \frac{9}{x}$ to get 3x = 72. Then, divide both sides by 3 to find x = 24.

17. $\frac{n}{40} = \frac{9}{2}$

19. $3 \div \frac{1}{2} = 3 \times 2 = 6$

21. Let x inches be the length. Then, you must have the proportion
$$\frac{10}{19} = \frac{35}{x}$$
Now, use "cross-products" to get 10x = 19 × 35 = 665. Then divide both sides by 10 to find x = 66.5. Thus, the flag should be 66.5 in. long.

23. The given rate is $\frac{10}{32}$, so if x is the number of runs the pitcher would allow in 9 innings, you have the proportion $\frac{10}{32} = \frac{x}{9}$. Use "cross-products" to get 90 = 32x. Divide by 32 to find $x = \frac{90}{32} = \frac{45}{16} = 2.81$ (to two decimal places).

25. Let x be the number of fish in the lake. Assuming that the ratio of the total number of fish to the total number tagged is the same as the corresponding ratio for the sample of 53 fish, we have the proportion
$$\frac{x}{250} = \frac{53}{5}$$
Now, multiply both sides by 250 to get
$$x = \frac{250 \times 53}{5} = 50 \times 53 = 2650$$
Thus, there are about 2650 fish in the lake.

27. (a) R=kt
(b) Substitute R = 112.5 and t = 2.5 into the equation in part (a) to get 112.5 = 2.5k. Then divide both sides by 2.5 to find k = 45.
(c) Use k = 45 and R = 108 to get the equation 108 = 45t. Then divide both sides by 45 to find t = 2.4 min.

SECTION 5.6   Ratio, Proportion and Variation   131

29. (a) $T = kh^3$

 (b) Substitute $T = 196$ and $h = 70$ into the equation in part (a) to get $196 = k(70)^3$. Then divide both sides by $(70)^3$ to find k:
$$k = \frac{196}{343000} = \frac{1}{1750} = 0.0005714.$$

 (c) $T = \dfrac{75^3}{1750} = \dfrac{75 \times 75 \times 75}{1750} = \dfrac{3 \times 15 \times 75}{14} = 241.07$

 Thus, the threshold weight is 241 lb.

31. (a) $f = \dfrac{k}{d}$

 (b) Substitute $f = 8$ and $d = \dfrac{1}{2}$ into the equation in part (a) to get $8 = \dfrac{k}{1/2} = 2k$. Then divide both sides by 2 to find $k = 4$.

 (c) Put $d = \dfrac{1}{4}$ and $k = 4$ into the equation in part (a). This gives $f = \dfrac{4}{1/4} = 16$. So the f number is 16.

33. Since the pressure P varies inversely as the volume V, we write $P = \dfrac{k}{V}$ or $PV = k$. To find k, substitute $P = 24$ and $V = 18$ to find $k = 18 \times 24 = 432$. This gives the equation $PV = 432$. For $P = 40$, we get $40V = 432$. Divide by 40 to find $V = \dfrac{432}{40} = 10.8$ in.^{3}.

35. A ratio is simply a fraction such as $\dfrac{a}{b}$. A proportion is an equation between two ratios such as $\dfrac{a}{b} = \dfrac{c}{d}$.

CHAPTER 5 EQUATIONS, INEQUALITIES, PROBLEM SOLVING

PRACTICE TEST 5

STUDY TIPS You won't have "mental blocks" on a test if you are thoroughly prepared and confident you will get an A. Mental blocks usually occur when you are under prepared! To be really prepared:
- (1) Begin studying at least a week in advance of your test.
- (2) Ask questions about material you do not understand
- (3) Do the practice test at the end of this manual and go over the answers, especially the ones you missed.

1. (a) $x + 7 = 2$ Subtract 7 from both sides to get $x = -5$.
   (b) $x - 4 = 9$ Add 4 to both sides to get $x = 13$.

2. (a) $x + 5 > 4$ Subtract 5 from both sides to get $x > -1$. The set of integers satisfying this inequality is $\{0, 1, 2, 3, \ldots\}$.

   (b) $2 + x \geq -x - 1$ Add x to both sides to get the inequality:
   $2 + 2x \geq -1$ Subtract 2 from both sides to get:
   $2x \geq -3$ Now, divide both sides by 2 to obtain:
   $x \geq -\frac{3}{2}$ Thus, the solution set is $\{-1, 0, 1, 2, 3, \ldots\}$.

3. $2x + 2 = 3x - 2$
   $2 = x - 2$ (Subtract 2x from both sides.)
   $4 = x$ (Add 2 to both sides.)
   The solution is $x = 4$.

4. $2x + 8 \geq -x - 1$
   $3x + 8 \geq -1$ (Add x to both sides.)
   $3x \geq -9$ (Subtract 8 from both sides.)
   $x \geq -3$ (Divide both sides by 3.)
   The solution set consists of all real numbers greater than or equal to -3.

5. (a) $x - 3 \leq 0$ Add 3 to both sides to get $x \leq 3$. This inequality is satisfied by all real numbers less than or equal to 3. The solution set is shown on the graph.

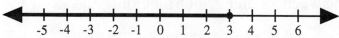

(b) $-2x + 4 > x + 1$
$\quad\quad 4 > 3x + 1$ \quad (Add 2x to both sides.)
$\quad\quad 3 > 3x$ \quad\quad (Subtract 1 from both sides.)
$\quad\quad 1 > x$ \quad\quad\quad (Divide both sides by 3.)

The last inequality is satisfied by all real numbers that are less than 1. This solution set is shown on the graph.

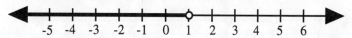

6. (a) $x + 2 \geq 3$ **and** $x \leq 4$
$\quad\quad x \geq 1$ **and** $x \leq 4$ \quad (Subtract 2 from both sides of the first inequality.)

The solution set of this system of inequalities is the set of all real numbers between 1 and 4, inclusive. This set is shown on the graph.

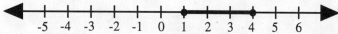

(b) $x - 3 \geq 1$ **and** $x < 0$
$\quad\quad x \geq 4$ **and** $x < 0$ \quad (Add 3 to both sides of the first inequality.)

There are no numbers that are greater than or equal to 4 and that are less than zero, so the solution set is empty.

7. (a) $x < 0$ **or** $x - 2 < 1$
$\quad\quad x < 0$ **or** $x < 3$ \quad (Add 2 to both sides of the second inequality.)

The solution set of this system of inequalities is the set of all real numbers that are less than 3. This set is shown on the graph.

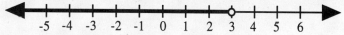

(b) $x + 2 < 3$ **or** $x - 1 > 2$
$\quad\quad x < 1$ **or** $x > 3$ \quad (Subtract 2 from both sides of the first inequality and add 1 to both sides of the second inequality.)

The solution set of this system of inequalities is the set of all real numbers that are less than 1 or that are greater than 3. This set is shown on the graph.

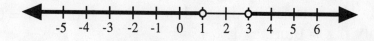

8. Since |3| = 3 and |-3| = 3, the solutions of |x| = 3 are x = ± 3.

9. |x| < 2 means that -2 < x < 2, so the solution set consists of all the real numbers between -2 and 2, not including the -2 or the 2. The graph shows this set.

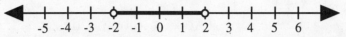

10. |x + 2| < 1 is equivalent to -1 < x + 2 < 1. By subtracting 2 from the members of this inequality, we get -3 < x < -1. The solution set is the set of all real numbers between -3 and -1, not including the -3 or the -1. This set is shown on the graph.

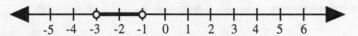

11. |x - 2| < 2 is equivalent to -2 < x - 2 < 2.
By adding 2 to each side, we get 0 < x < 4.
The solution set of this system is the set of all real numbers that are between 0 and 4. The endpoints are not included. The graph shows this solution set.

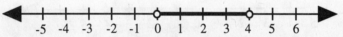

12. |x| > 2 is equivalent to the system x < -2 **or** x > 2. The solution set of this system is the set of all real numbers that are less than -2 or that are greater than +2. The graph shows this set.

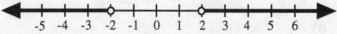

13. |x - 2| > 3 is equivalent to the system x - 2 < -3 or x - 2 > 3. By adding 2 to both sides of these inequalities, we get the system x < -1 or x > 5. The solution set of this system is the set of all real numbers that are either less than -1 or that are greater than 5. The graph shows this set.

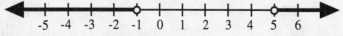

14. (a) Since $1 \times 2 = 2$ and $1 + 2 = 3$, $x^2 + 3x + 2 = (x + 1)(x + 2)$.
    (b) Since $1 \times (-4) = -4$ and $1 + (-4) = -3$, $x^2 - 3x - 4 = (x + 1)(x - 4)$.

15. (a) $(x - 1)(x + 2) = 0$ is satisfied if and only if $x - 1 = 0$ or $x + 2 = 0$, that is, if and only if $x = 1$ or $x = -2$.
    (b) $x(x - 1) = 0$ is satisfied if and only if $x = 0$ or $x - 1 = 0$, that is, if and only if $x = 0$ or $x = 1$.

16. Since $5 \times 2 = 10$ and $5 + 2 = 7$, $x^2 + 7x + 10 = (x + 5)(x + 2)$. So we may replace $x^2 + 7x + 10 = 0$ by $(x + 5)(x + 2) = 0$. This equation is satisfied if and only if $x + 5 = 0$ or $x + 2 = 0$, that is, if and only if $x = -5$ or $x = -2$.

17. Since $2 \times (-5) = -10$ and $2 + (-5) = -3$, $x^2 - 3x - 10 = (x + 2)(x - 5)$. So we may replace $x^2 - 3x - 10 = 0$ by $(x + 2)(x - 5) = 0$. This equation is satisfied if and only if $x + 2 = 0$ or $x - 5 = 0$, that is, if and only if $x = -2$ or $x = 5$.

18. In the equation $2x^2 + 3x - 5 = 0$, $a = 2$, $b = 3$, and $c = -5$. Substitute these values in the quadratic formula
$$x = \frac{-b \pm \sqrt{b^2 - 4ac}}{2a}$$
to get $x = \frac{-3 \pm \sqrt{9 + 40}}{4} = \frac{-3 \pm \sqrt{49}}{4} = \frac{-3 \pm 7}{4} = \frac{4}{4}$ or $\frac{-10}{4}$. Thus, the solutions are $x = 1$ and $x = -\frac{5}{2}$.

19. In the equation $3x^2 + 5x - 2 = 0$, $a = 3$, $b = 5$, and $c = -2$. Substitute these values in the quadratic formula (See Problem 18 above) to get $x = \frac{-5 \pm \sqrt{25 + 24}}{6} = \frac{-5 \pm \sqrt{49}}{6} = \frac{-5 \pm 7}{6} = \frac{2}{6}$ or $\frac{-12}{6}$. Thus, the solutions are $x = \frac{1}{3}$ and $x = -2$.

20. (a) $9x^2 - 16 = 0$
$\quad\quad 9x^2 = 16$ (Add 16 to both sides.)
$\quad\quad x^2 = \frac{16}{9}$ (Divide both sides by 16.)
$\quad\quad x = \pm \frac{4}{3}$ (Take square roots.)

(b) $25x^2 - 4 = 0$
$\quad\quad 25x^2 = 4$ (Add 4 to both sides.)
$\quad\quad x^2 = \frac{4}{25}$ (Divide both sides by 25.)
$\quad\quad x = \pm \frac{2}{5}$ (Take square roots.)

21. Let x cm be the length of the shorter leg. Then the hypotenuse is of length (x + 9) cm and the other leg is of length (x + 7) cm. Thus,
$$(x + 9)^2 = x^2 + (x + 7)^2$$
or $\quad x^2 + 18x + 81 = x^2 + x^2 + 14x + 49$

$\qquad\qquad 0 = x^2 - 4x - 32 \qquad$ (Subtract $x^2 + 18x + 81$.)
$\qquad\qquad 0 = (x - 8)(x + 4) \qquad$ (Factor.)

This means that x = 8 or x = -4. Since x must be positive, we discard the -4. If the shorter leg is 8 cm long, then the hypotenuse is 17 cm long and the other leg is 15 cm long. You can check that
$$17^2 = 8^2 + 15^2.$$

22. The number is 10x + y. So the equation is 10x + y = 5(x + y).

23. Let m be the number of miles you could drive for a charge of $63. Then the required equation is
$\quad 21 + 0.21m = 63$
$\qquad 0.21m = 42 \qquad$ (Subtract 21 from both sides.)
$\qquad 21m = 4200 \qquad$ (Multiply both sides by 100.)
$\qquad m = 200 \qquad$ (Divide both sides by 21.)

Thus, you could drive 200 miles for the $63 charge.

24. Let x and x + 1 be the integers. Then the required equation is
$\quad 3(x + x + 1) = 45$
or $\quad 3(2x + 1) = 45$
$\qquad 2x + 1 = 15 \qquad$ (Divide both sides by 3.)
$\qquad 2x = 14 \qquad$ (Subtract 1 from both sides.)
$\qquad x = 7 \qquad$ (Divide both sides by 2.)

Thus, the two integers are x = 7 and x + 1 = 8.
Check: 7 + 8 = 15 and 3 × 15 = 45.

25. Let x and x + 1 be the integers.
Then $12x > 9(x + 1)$
or $\quad 12x > 9x + 9 \qquad$ (Simplify.)
$\qquad 3x > 9 \qquad$ (Subtract 9 from both sides.)
$\qquad x > 3 \qquad$ (Divide both sides by 3.)

Thus, the smaller integer must be greater than 3. Since the smallest such integer is 4, the two integers must be 4 and 5.
Check: 12 × 4 = 48, 9 × 5 = 45 and 48 > 45. That x = 4 is the smallest integer that satisfies the given condition is shown by trying x = 3. This gives 12 × 3 = 36 and 9 × 4 = 36 so that 12 × 3 equals, but is not greater than, 9 × 4.

26. (a) 801 to 688;   801:688; $\dfrac{801}{688}$

   (b) 688 + 801 + 501 = 1990 = total number of stocks. So the ratio of losers to total number is $\dfrac{801}{1990}$.

27. Let x be the required number of years. Then the older ring will be (35 + x) years old and the younger ring will be (9 + x) years old. For the older ring to be twice as old as the younger, we must have
   35 + x = 2(x + 9)
   35 + x = 2x + 18   (Simplify.)
   35 = x + 18   (Subtract x from both sides.)
   17 = x   (Subtract 18 from both sides.)
   Check: In 17 years, the older ring will be 35 + 17 = 52 years old and the younger ring will be 9 + 17 = 26 years old. Since 52 = 2 × 26, the answer 17 years is correct.

28. (a) $\dfrac{50}{8} = 6\dfrac{1}{4}$ ¢ per oz for the 8 oz box

   $\dfrac{76}{12} = 6\dfrac{1}{3}$ ¢ per oz for the 12 oz box

   (b) The 8 oz box is the better buy.

29. Let n be the number of nickels and d be the number of dimes. Then n + d = 20, so that n = 20 - d, and
   5n + 10d > 120
   5(20 - d) + 10 d > 120   (Substitute 20 - d for n.)
   100 + 5d > 120   (Simplify.)
   5d > 20   (Subtract 100 on both sides.)
   d > 4   (Divide both sides by 5.)
   Thus, the least number of dimes is 5.
   Check: If there are 5 dimes (50 ¢), there must be 15 nickels (75 ¢) to give a sum of $1.25 > $1.20. If there were only 4 dimes (40 ¢), there would have to be 16 nickels (80 ¢) giving a sum of $1.20 which is not greater than $1.20.

30. (a) $\dfrac{x}{8} = \dfrac{9}{5}$ (b) Multiply both sides by 8 to get x = $\dfrac{72}{5}$ = 14.4. So the second side is 14.4 ft long.

138 CHAPTER 5 EQUATIONS, INEQUALITIES, PROBLEM SOLVING

31. (a) $C = kx^2$
    (b) Put $C = 100$ and $x = 150$ in the equation of part (a). This gives $100 = k(150)^2 = 22{,}500k$. Now divide both sides by 22,500 to get $k = \dfrac{1}{225}$.
    (c) Put $k = \dfrac{1}{225}$ and $x = 180$ in the equation of part a to get $C = \dfrac{(180)^2}{225} = \dfrac{32400}{225} = 144$, so the cost is \$144 per hr.

32. (a) $t = \dfrac{k}{I}$
    (b) Put $t = \dfrac{1}{30}$ and $I = 300$ in the equation of part a. This gives $\dfrac{1}{30} = \dfrac{k}{300}$. Now multiply both sides by 300 to find $k = 10$.
    (c) Put $k = 10$ and $I = 600$ in the equation of part a to get $t = \dfrac{10}{600} = \dfrac{1}{60}$. Thus, the correct exposure time is $\dfrac{1}{60}$ sec.

CHAPTER 6
FUNCTIONS AND GRAPHS

EXERCISE 6.1

STUDY TIPS The best way of learning about relations and functions is to study some examples that you understand. Say you are making $5 per hour. The amount A of money you earn is a **function** of the number of hours H that you work. We write $A = f(H) = 5H$. If you work 1, 2 or 3 hours, you make 5, 10 or 15 dollars, respectively. The set of ordered pairs $\{(1, 5), (2, 10), (3, 15)\}$ gives the **relationship** between the number of hours worked and your earnings. The set $\{1, 2, 3\}$ is the **domain** and the set $\{5, 10, 15\}$ is the **range**. As a matter of fact, any set of ordered pairs is a relation. In general, if $y = f(x)$ the x's are in the domain, the y's are in the range. A relation is a **function** if for every x in the domain there is only **one** y. (If you look at the ordered pairs, there are never two distinct pairs with the same first coordinate. [(3, 4) and (3, 5), for example.] Geometrically, this means that there are never two points on the same vertical line. A picture is worth a thousand words, so if you want to understand relations and functions even better look at their pictures, called **graphs**. Since both functions and relations are sets of ordered pairs, graphs are points of ordered pairs satisfying the function or relation. Keep in mind that $f(x)$ and y are always the same. If $f(x) = 2x + 4$, do not panic because there is no y, this means that $y = f(x) = 2x + 4$. The ordered pairs you need to find the graph are found by giving different values to x and finding the corresponding y values. If the function is of the form $f(x) = \mathbf{ax + b}$, the result is a **line.** If $f(x) = \mathbf{ax^2 + bx + c}$, it is a **parabola** opening up (like a soup bowl) when a is positive, and opening down (invert the soup bowl) when a is negative.

1. The domain is the set of first numbers in the pairs of the relation: $\{1, 2, 3\}$. The range is the set of second numbers in these pairs: $\{2, 3, 4\}$.

3. The domain is the set of first numbers in the pairs of the relation: $\{1, 2, 3\}$. The range is the set of second numbers in these pairs: $\{1, 2, 3\}$.

5. The domain is the set of x values: {x | x is a real number}. The range is the set of y values: {(y | y is a real number}.

7. The domain is the set of x values: {x | x is a real number}. The range is the set of y values: {y | y is a real number}.

9. The domain is the set of x values: {x | x is a real number} The range is the set of y values: {y | y ≥ 0}.

11. The domain is the set of x values: {x | x ≥ 0}.
The range is the set of y values: {y | y is a real number}.

13. The domain is the set of x values: {x | x ≠ 0}.
The range is the set of y values: {y | y ≠ 0}.

15. This **is** a function because one real value of y corresponds to each real value of x.

17. This is **not** a function, because two values of y correspond to each positive value of x. For instance, (4, 2) and (4, -2) both belong to this relation.

19. This **is** a function. The domain is (x | x ≥ 0), and to each value of x in this domain there corresponds one y value, the nonnegative square root of the x value.

21. This **is** a function. The domain consists of the set of real numbers, and to each number in this domain there corresponds one y value, the cube root of x.

23. (a) $f(0) = 3 \times 0 + 1 = 1$ (b) $f(2) = 3 \times 2 + 1 = 7$
    (c) $f(-2) = 3 \times (-2) + 1 = -5$

25. (a) $F(1) = \sqrt{1 - 1} = \sqrt{0} = 0$ (b) $F(5) = \sqrt{5 - 1} = \sqrt{4} = 2$
    (c) $F(26) = \sqrt{26 - 1} = \sqrt{25} = 5$

27. (a) $f(x + h) = 3(x + h) + 1 = 3x + 3h + 1$
    (b) $f(x + h) - f(x) = 3x + 3h + 1 - (3x + 1) = 3h$
    (c) $\dfrac{f(x + h) - f(x)}{h} = \dfrac{3h}{h} = 3$ for $h \neq 0$.

SECTION 6.1 Graphing Relations and Functions

29. $g(x) = x^2$. The missing numbers are $\frac{1}{16}$, 4.41, and ± 8.

31. (a) $g(0) = 2(0)^3 + (0)^2 - 3(0) + 1 = 1$
    (b) $g(-2) = 2(-2)^3 + (-2)^2 - 3(-2) + 1 = -16 + 4 + 6 + 1 = -5$
    (c) $g(2) = 2(2)^3 + 2^2 - 3(2) + 1 = 16 + 4 - 6 + 1 = 15$

33. (a) Put $a = 50$ in the function $U(a) = -a + 190$. This gives $U(50) = -50 + 190 = 140$. Thus, the upper limit is 140 beats per min.
    (b) Put $a = 60$ in the same function as in part (a). This gives $U(60) = -60 + 190 = 130$. So the upper limit is 130 beats per min.

35. (a) Put $h = 70$ in the function $w(h) = 5h - 190$. This gives $w(70) = 5 \times 70 - 190 = 350 - 190 = 160$, so his weight should be 160 lb.
    (b) Put $w(h) = 200$ to get the equation $200 = 5h - 190$. Then add 190 to both sides to get $390 = 5h$. Now divide both sides by 5 to find $h = 78$. Thus, his height should be 78 in.

37. (a) Put $d = 10$ in the function $P(d) = 63.9d$ to get $P(10) = 63.9 \times 10 = 639$. Thus, the pressure is 639 lb per sq ft.
    (b) Put $d = 100$ in the same function to get $P(100) = 63.9 \times 100 = 6390$. Thus, the pressure is 6390 lb per sq ft.

39. $S = f(t) = \frac{1}{2}gt^2 = 16t^2$
    (a) $S = f(3) = 16(3^2) = 16(9) = 144$ ft
    (b) $S = f(5) = 16(5^2) = 16(25) = 400$ ft

41. The table of values and the graph for the relation $\{(x, y) \mid y = x, x$ an integer between -1 and 4, inclusive$\}$ appear below.

x	y
-1	-1
0	0
1	1
2	2
3	3
4	4

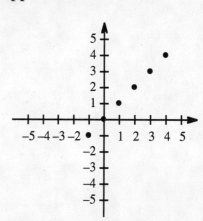

142 CHAPTER 6 FUNCTIONS AND GRAPHS

43. The table of values and the graph for the relation $\{(x, y) \mid y = 2x + 1,$ x an integer between 0 and 5, inclusive$\}$ appear below.

x	y = 2x + 1
0	1
1	3
2	5
3	7
4	9
5	11

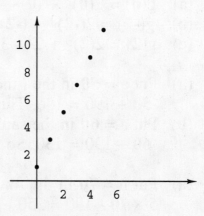

45. The table of values and the graph for the relation $\{(x, y) \mid 2x - y = 4,$ x an integer between -2 and 2, inclusive$\}$ appear below.

x	y = 2x - 4
-2	-8
-1	-6
0	-4
1	-2
2	0

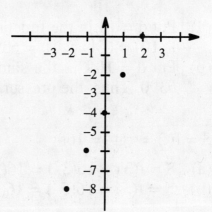

SECTION 6.1   Graphing Relations and Functions   143

47. The table of values and the graph for the relation $\{(x, y) \mid y = \sqrt{x},$ $x = 0, 1, 4, 9, 16, 25,$ or $36\}$, appear below.

x	$\sqrt{x}$
0	0
1	1
4	2
9	3
16	4
25	5
36	6

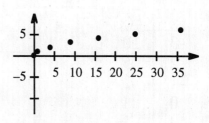

49. The table of values and the graph for the relation $\{(x, y) \mid x + y < 5,$ x, y nonnegative integers$\}$ appear below.

x	$0 \leq y < 5 - x$
0	0, 1, 2, 3, 4
1	0, 1, 2, 3
2	0, 1, 2
3	0, 1
4	0

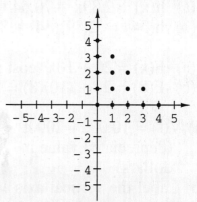

51. The table of values and the graph for the function given by $f(x) = x + 1$, x an integer between -3 and 3, inclusive, appear below.

x	f(x)
-3	-2
-2	-1
-1	0
0	1
1	2
2	3
3	4

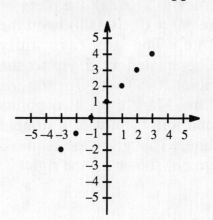

53. The table of values and the graph for the function given by
$g(x) = x^2 + 1$, x an integer between -3 and 3, inclusive, appear below.

x	g(x)
-3	10
-2	5
-1	2
0	1
1	2
2	5
3	10

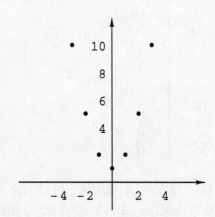

55. (a) $h(x) = 2.89x + 70.64$
    (b) $h(34) = (2.89)(34) + 70.64 = 168.9$, so the height was about 169 cm.

57. (a) $F(x) = 20 + 10x$ (cost in dollars)
    (b) $F(8) = 20 + (10)(8) = 20 + 80 = 100$, so the total cost is $100.

59. (a) $V(t) = 10,000 - 2000t$
    (depreciated value in dollars)
    (b) Label the vertical axis V and the horizontal axis t.
    Since the initial value of the truck is $10,000, the point (t, V) = (0, 10,000) is on the line. The truck is fully depreciated in 5 yr, so the point (t, V) = (5,0) is on the line. Mark these two points and join them with a straight line. This gives the required graph, shown on the right.

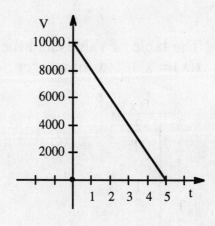

61. (a) The table of values and the graph of C(t) appear below.

t	C(t)
$0 < t \leq 1$	0.55
$1 < t \leq 2$	0.78
$2 < t \leq 3$	1.01
$3 < t \leq 4$	1.24
$4 < t \leq 5$	1.47

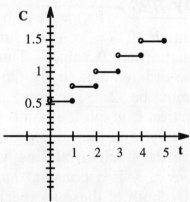

(b) Since the cost of the call was $3.31, you have to solve the equation

$\quad 0.55 + 0.23(t - 1) = 3.31$
$\quad 0.55 + 0.23t - 0.23 = 3.31 \quad$ (Simplify.)
$\quad\quad\quad 0.32 + 0.23t = 3.31 \quad$ (Combine terms on the left.)
$\quad\quad\quad\quad\quad 0.23t = 2.99 \quad$ (Subtract 0.32 from both sides.)
$\quad\quad\quad\quad\quad\quad\quad t = 13 \quad$ (Divide both sides by 0.23.)

The charge was for 13 minutes, so they talked for more than 12, but not more than 13 minutes.

63. $g(x) = \sqrt{x - 1}$ is real if and only if $x - 1 \geq 0$, that is, if $x \geq 1$. Thus, we would exclude all values of x less than 1 if g is to have real values.

65. The graph of a function f(x) is a picture of the set of points $\{(x, y) \mid y = f(x), x \text{ an element of the domain of } f\}$

67. Yes. It is a relation where there is exactly one value of y for each value of x in the domain.

69. Notice that if the units digit in the temperature reading is multiplied by 4, the result is the number of chirps per minute. Thus, the desired function is $f(x) = 4(x - 40)$.

71. The table shows that the distance is given by multiplying the square of the time (in seconds) by 16. Thus, the desired function is $f(t) = 16t^2$.

73. For each table entry, the number of seconds is the square root of the number of units of length, so the suggested rule is $f(x) = \sqrt{x}$.

75. If $f(x) = \sqrt{x} = 100$, then, squaring both sides of $\sqrt{x} = 100$, we obtain $x = 100^2 = 10,000$. Thus, the length would be 10,000 units.

EXERCISE 6.2

STUDY TIPS $f(x) = ax + b$ is a linear function (its graph is a **line**.) For example, $f(x) = 2x + 4$ is linear. The easiest values to use in graphing this equation are **0** values. Thus, if $x = 0$, $y = f(0) = 2 \cdot 0 + 4 = 4$ and we have the ordered pair (**0**, 4). On the other hand, if $y = 0$, $f(x) = 2x + 4 = 0$ and x must be -2. The corresponding ordered pair is (-2, **0**). To graph the function f, graph the points (-2, 0) and (0, 4), the **x-** and **y-intercepts** respectively, and then draw a line through them. Two special cases of lines are the **horizontal line** $y = f(x) = c$, where no matter what value you give x, y is a constant and the **vertical line** $x = c$. (You should recognize both of these as special lines because they involve one variable only.) We also study the distance between the points (x_1, y_1) and (x_2, y_2). You need Formula (3) on page 314 to work this type of problem.

1. If $y = ax$ and $y = 2.25$ for $x = 3$, then $2.25 = 3a$, so that $a = 0.75$ or $\frac{3}{4}$.
   Thus, $y = \frac{3}{4} x$. If $y = 3$, then $3 = \frac{3}{4} x$, so that $x = \mathbf{4}$.

3. Since $f(x) = 3x + 6$,
   then $f(0) = 6$ and $f(-2) = 0$.
   Thus, (0, 6) and (-2, 0) are two points on the line. Mark these points and draw the line through them, as shown on the graph.

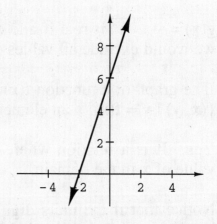

5. Since $f(x) = 3$ for all values of x, the graph is a straight line parallel to the x-axis and 3 units above this axis as shown on the graph.

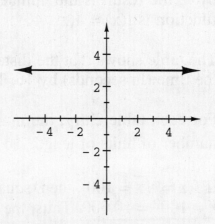

SECTION 6.2 Linear Functions and Relations   147

7. Here we are given x = -1 for all values of y. Thus, the graph is a straight line parallel to the y-axis and 1 unit to the left of this axis as shown in the figure.

9. Since f(x) = -x + 2, then f(0) = 2 and f(2) = 0. Thus, (0, 2) and (2, 0) are two points on the line. Mark these points and draw the straight line through them as in the figure.

11. With g(x) = -3x - 6, we get g(0) = -6 and g(-2) = 0. So (0, -6) and (-2, 0) are two points on the line. Mark these points and draw the straight line through them as in the figure.

13. The equation 3x + 2y = 6 is satisfied by the pairs (0, 3) and (2, 0), so these are two points on the line. Mark these points and draw the straight line through them as in the figure.

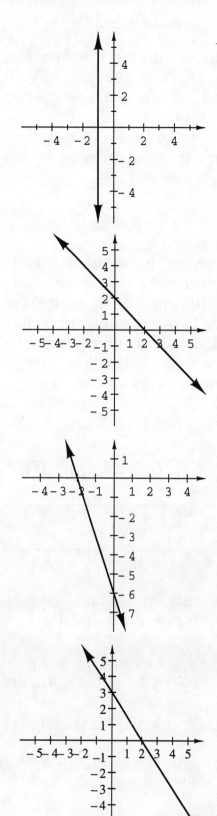

148 CHAPTER 6 FUNCTIONS AND GRAPHS

15. The equation -2x + 3y = 6 is satisfied by the pairs (0, 2) and (-3, 0), so these are two points on the line. Mark these points and draw the straight line through them as in the figure.

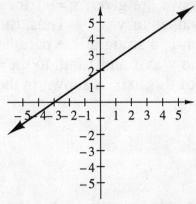

17. Two of the solutions of the equation 4x - 3y = 12 are (3, 0) and (0, -4), and these are two points on the graph. Mark these two points and draw the straight line through them as shown in the figure.

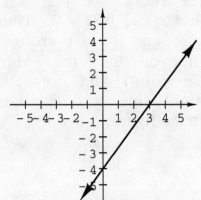

19. $d = \sqrt{(-1-2)^2 + (0-4)^2} = \sqrt{3^2 + 4^2} = \sqrt{25} = $ **5** units

21. $d = \sqrt{(-1+4)^2 + (3+5)^2} = \sqrt{3^2 + 8^2} = \sqrt{73} \approx $ **8.54** units

23. $d = \sqrt{(1-4)^2 + (-1+8)^2} = \sqrt{3^2 + 7^2} = \sqrt{58} \approx $ **7.62** units

25. These points are on a line parallel to the y-axis, so the distance is $d = |y_2 - y_1| = |0 - (-2)| = $ **2** units.

27. These points are on a line parallel to the y-axis, so the distance is $d = |y_2 - y_1| = |7 - 3| = $ **4** units.

29. (a) E(x) = 500 + 25x and S(x) = 1000 + 20x

    (b) E(0) = 500, E(200) = 5500.

SECTION 6.2   Linear Functions and Relations

Mark the points (0, 500) and (200, 5500) on the graph and draw the straight line through these points to get the graph of $S(0) = 1000$, $S(200) = 5000$. Mark the points (0, 1000) and (200, 5000) on the graph and draw the straight line through these points to get the graph of $S(x)$.

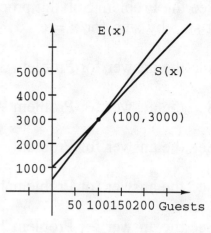

(c) The two lines in the graph intersect at (100, 3000). This point checks in the two equations. Thus, the cost is the same for 100 persons.

31. (a) Since the cost is $1.16 for the first minute or fraction thereof, plus $0.65 for each additional minute or fraction thereof, the table of values is as follows:

t	$C(t)$
$0 < t \le 1$	$1.16
$1 < t \le 2$	$1.81
$2 < t \le 3$	$2.46
$3 < t \le 4$	$3.11
$4 < t \le 5$	$3.76

(b) The graph is shown below

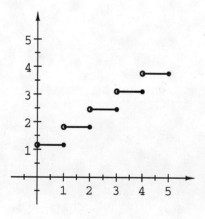

33. Use the distance formula to find the square of the length of each side of the triangle:
$$(AB)^2 = (a_1 - b_1)^2 + (a_2 - b_2)^2$$
$$(AC)^2 = (a_1 - c_1)^2 + (a_2 - c_2)^2$$
$$(BC)^2 = (b_1 - c_1)^2 + (b_2 - c_2)^2$$
The triangle is a right triangle if and only if one of these squares equals the sum of the other two squares. So we can check this.

35. See the Problem Solving procedure following Example 1 and follow it step by step for **x = c**

37. See the answer for Problem 3.

39. See the answer for Problem 9.

41. See the answer for Problem 11.

43. See the answer for Problem 13.

45. See the answer for Problem 15.

47. See the answer for Problem 17.

SECTION 6.3   Equations of a Line   151

EXERCISE 6.3

STUDY TIPS We shift emphasis from graphs to equations, that is, we want to explore the different ways in which the equation of a line can be written, as well as the information necessary to write these equations. We start by looking at the **inclination** of a line, called its **slope** (Definition 6.3). Do not worry about which formula you have to use to find the equation of a line because the name of the formula will tell you when to apply it. For example, if you are asked for the slope of a line, you must certainly use Definition 6.3 and two points must be given so you can find this slope. Now, if you are given two points and are asked to find the equation of the line through them, find the slope and then use the **point-slope** form. (Of course, if you are given a point and the slope of a line, through that point you can use the **point-slope** formula directly.) What if you have the slope and the y-intercept? Use the **slope-intercept** formula. For your convenience, all this information is in Table 6.8. Keep in mind that parallel lines have the **same** slope and perpendicular lines have slopes that are **negative reciprocals**. If a line has slope 4, a line perpendicular to it must have slope $-\frac{1}{4}$, the negative reciprocal of 4.

1. $m = \frac{y_2 - y_1}{x_2 - x_1} = \frac{4 - 2}{3 - 1} = \frac{2}{2} = 1$

3. $m = \frac{y_2 - y_1}{x_2 - x_1} = \frac{0 - 5}{5 - 0} = -1$

5. $m = \frac{y_2 - y_1}{x_2 - x_1} = \frac{-4 - (-3)}{7 - (-1)} = -\frac{1}{8}$

7. $m = \frac{y_2 - y_1}{x_2 - x_1} = \frac{3 - 0}{12 - 0} = \frac{1}{4}$

9. $m = \frac{y_2 - y_1}{x_2 - x_1} = \frac{5 - 5}{2 - 3} = 0$

11. Use the equation $y - y_1 = m(x - x_1)$ to get $y - 2 = \frac{1}{2}(x - 1)$. Then solve for y: $y = \frac{1}{2}x + \frac{3}{2}$

13. Use the equation $y - y_1 = m(x - x_1)$ to get $y - 4 = -1(x - 2)$. Then solve for y: $y = -x + 6.$

15. $m = 0$ means that the line is parallel to the x-axis. Since the line goes through (4, 5), the equation must be $y = 5$.

17. By comparing $y = x + 2$ with the equation $y = mx + b$, we find that
    (a) **m = 1**   (b) **b = 2**

19. First solve $3y = 4x$ for y to get $y = \frac{4}{3}x$; then compare with $y = mx + b$ to find: (a) $m = \frac{4}{3}$ (b) **b = 0**

21. First solve $x + y = 14$ for y to get $y = -x + 14$; then compare with $y = mx + b$ to find: (a) **m = -1**   (b) **b = 14**

23. Compare the equation $y = 6$ with $y = mx + b$ to find:
    (a) **m = 0**   (b) **b = 6**

25. The line whose equation is $x = 3$ is parallel to the y-axis. Thus,
    (a) The slope is **not** defined.
    (b) The line does **not** intersect the y-axis.

27. For the line through the points (1, -1) and (2, 2), find the slope:
$$m = \frac{2 - (-1)}{2 - 1} = 3$$
    Then use the point-slope form to get:
$$y - (-1) = 3(x - 1)$$
    or $y + 1 = 3x - 3$.
    Then write the equation in the form $Ax + By = C$: **3x - y = 4**

29. For the line through the points (3, 2) and (2, 3), find the slope:
$$m = \frac{3 - 2}{2 - 3} = -1$$
    Then use the point-slope form to get:
$$y - 2 = -1(x - 3)$$
    or $y - 2 = -x + 3$.
    Now write the equation in the form $Ax + By = C$: **x + y = 5**

31. For the line through the points (0, 0) and (1, 10), find the slope
$$m = \frac{10 - 0}{1 - 0} = 10$$
    Then use the point-slope form to get:
$$y - 0 = 10(x - 0)$$
    or $y = 10x$
    Now write the equation in the form $Ax + By = C$: **10x - y = 0**

SECTION 6.3 Equations of a Line

33. Use the first two entries of the table: (62, 134) and (63, 139). With h in place of x and w in place of y, substitute in the point slope form of the line:

$$m = \frac{139 - 134}{63 - 62} = 5$$
$$w - 134 = 5(h - 62)$$
or $\quad w - 134 = 5h - 310$
and $\quad$ **w = 5h - 176**

35. Use the first two entries of the table: (62, 123) and (63, 128). With h in place of x and w in place of y, substitute in the point slope form of the line:

$$m = \frac{128 - 123}{63 - 62} = 5$$
$$w - 123 = 5(h - 62)$$
or $\quad w - 123 = 5h - 310$
and $\quad$ **w = 5h - 187**

37. To find the slopes, compare with the slope-intercept form $y = mx + b$: For $y = 2x + 5$, $m = 2$. Solve $4x - 2y = 7$ for y to get $y = 2x - \frac{7}{2}$, which shows that $m = 2$ for the second line. Since the slopes are equal, the lines **are** parallel.

39. To find the slopes, compare with the slope-intercept form $y = mx + b$. Solve the first equation, $2x + 5y = 8$, for y to get $y = -\frac{2}{5}x + \frac{8}{5}$. This result shows that $m_1 = -\frac{2}{5}$. Solve the second equation, $5x - 2y = -9$, for y to get $y = \frac{5}{2}x + \frac{9}{2}$, which shows that $m_2 = \frac{5}{2}$. Since the slopes are not equal, the lines **are not** parallel.

41. To find the slopes, compare with the slope-intercept form $y = mx + b$. Solve the first equation, $x + 7y = 7$, for y to get $y = -\frac{1}{7}x + 1$, which gives $m_1 = -\frac{1}{7}$. Solve the second equation, $2x + 14y = 21$, for y to get $y = -\frac{1}{7}x + \frac{3}{2}$, which gives $m_2 = -\frac{1}{7}$. Since $m_1 = m_2$, the lines **are** parallel.

43. Since the required line is to be parallel to the line $4x - y = 7$, it must have an equation of the form $4x - y = C$, with C determined so that the line passes through the point $(1, -2)$. By substituting $(1, -2)$ into the equation, we find that $(4)(1) - (-2) = C$, so $C = 6$. Thus, an equation of the desired line is **$4x - y = 6$.**

45. (a) In order to be perpendicular to the line $2x + 5y = 7$, the desired line must have an equation of the form $5x - 2y = C$, with C chosen so that the line goes through $(2, 0)$. By substituting $(2, 0)$ into the equation, we find that $(5)(2) - (2)(0) = C$, so $C = 10$. Thus, an equation of the desired line is **$5x - 2y = 10$.**

    (b) The equation $y = 2x - 3$ can be written in the form $2x - y = 3$. In order to be perpendicular to the line $2x - y = 3$, the desired line must have an equation of the form $x + 2y = C$, with C chosen so that the line goes through $(1, 1)$. By substituting $(1, 1)$ into the equation, we find that $1 + (2)(1) = C$, so $C = 3$. Thus, an equation of the desired line is **$x + 2y = 3$.**

    (c) In order to be perpendicular to the line $x - 2y = 3$, the desired line must have an equation of the form $2x + y = C$, with C chosen so that the line goes through $(2, -2)$. By substituting $(2, -2)$ into the equation, we find that $(2)(2) + (-2) = C$, so $C = 2$. Thus, an equation of the desired line is **$2x + y = 2$.**

    (d) In order to be perpendicular to the line $4x + 5y = 9$, the desired line must have an equation of the form $5x - 4y = C$, with C chosen so that the line goes through $(1, 1)$. By substituting $(1, 1)$ into the equation, we find that $(5)(1) - (4)(1) = C$, so $C = 1$. Thus, an equation of the desired line is **$5x - 4y = 1$.**

47. For the first line, $m_1 = \dfrac{200 - 0}{100 - 0} = 2$, and for the second line $m_2 = \dfrac{-405 - 10}{790 - 0} = \dfrac{-415}{790} \neq -\dfrac{1}{2}$. Since $m_2 \neq -\dfrac{1}{m_1}$, the two lines are **not** perpendicular.

49. If (x_1, y_1) and (x_2, y_2) are any two distinct points on a horizontal line, $y_1 = y_2$ and $x_1 \neq x_2$. Thus, the slope is $m = \dfrac{y_2 - y_1}{x_2 - x_1} = \dfrac{0}{x_2 - x_1} = 0$.

51. (a) $m = \dfrac{1}{1.266} = \mathbf{0.79}$ (b) No

SECTION 6.3 Equations of a Line 155

53. (a) $m = \dfrac{0.4}{1.6} = \mathbf{0.25}$ (b) Yes

55. $m = \dfrac{20 - 16}{124 - 108} = \dfrac{4}{16} = 0.25$, which is not safe for parking.
    $x = \mathbf{106}$ will give the maximum allowable slope.

57. Let r (dollars) be the hourly rate and f (dollars) the fixed fee. Then the table gives the following:
    (1) $3r + f = 110$
    (2) $r + f = 60$
    (3) $5r + f = 170$
    Subtract Equation (2) from Equation (1) to get
    $2r = 50$
    $r = 25$
    Then Equation (2) becomes
    $25 + f = 60$
    which gives $f = 35$.
    Thus, the fixed fee is **$35** and the hourly rate is **$25.** To check that this is true, substitute 25 for r and 35 for f in Equation 3 to obtain:
    (3) $5(25) + 35 = 170$, a true statement.

156 CHAPTER 6 FUNCTIONS AND GRAPHS

EXERCISE 6.4

STUDY TIPS If you can graph one equation, you can certainly graph two of them. What does it mean if they do intersect? It means that the point of intersection satisfies **both** equations and thus, is a solution of both equations. Unfortunately, graphs sometimes are hard to read, especially if large numbers are involved. Because of this, we also use the system of **elimination** to solve systems of equations. The main idea is to multiply one or both equations by numbers that cause the resulting coefficients for one of the variables to be of equal magnitude and opposite in sign then add the two equations to eliminate that variable As before, you can have **one** solution (an ordered pair of numbers satisfying **both** equations), **no** solution (a **contradiction**, like $0 = 7$ or $-3 = 5$) or **infinitely many** solutions (a **true** statement like $7 = 7$ or $-4 = -4$).

1. Two points on the line $x + y = 3$ are $(0, 3)$ and $(3, 0)$. Mark these points and draw the line through them. The line $2x - y = 0$ passes through the origin, $(0, 0)$, and through the point $(2, 4)$. Mark these points and draw the line through them. The graph shows that the lines intersect at $(1, 2)$. Check that this point satisfies both equations:
   $1 + 2 = 3$ and $2 - 2 = 0$.
   Thus, the solution is **(1, 2)**.

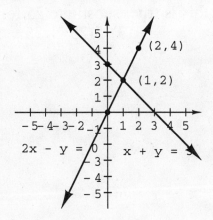

SECTION 6.4 Two Linear Equations in Two Variables

3. Two points on the line $2x - y = 10$ are $(0, -10)$ and $(5, 0)$. Mark these points and draw the line through them. Two points on the line $3x + 2y = 1$ are $(1, -1)$ and $(5, -7)$. Mark these two points and draw the line through them. The graph shows that the lines intersect at $(3, -4)$. Check that this point satisfies both equations:
$(2)(3) - (-4) \quad = 6 + 4 = 10$
$(3)(3) + (2)(-4) = 9 - 8 = 1$
Thus, the solution is **(3, -4)**.

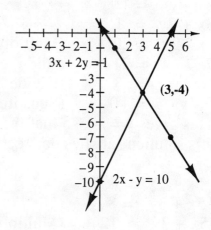

5. Two points on the line $3x + 4y = 4$ are $(0, 1)$ and $(-4, 4)$. Mark these points and draw the line through them. Two points on the line $2x - 6y = 7$ are $(7/2, 0)$ and $(-1, -3/2)$. Mark these two points and draw the line through them. The graph shows that the two lines intersect at the point $(2, -1/2)$. Check that this point satisfies both equations:
$(3)(2) + (4)(-1/2) = 6 - 2 = 4$ and
$(2)(2) - (6)(-1/2) = 4 + 3 = 7$.
Thus, the solution is **(2, -1/2)**.

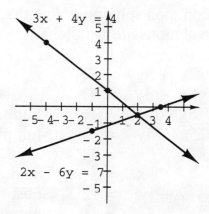

7. $x + y = 3$
    $2x - y = 0$
    $3x \quad = 3$ (Add the equations, term by term.)
    $\quad x = 1$ (Divide both sides by 3.)
    $2 - y = 0$ (Substitute $x = 1$ into the second equation.)
    This gives $y = 2$, so that the solution is **(1, 2)**. You can check this.

9. $x + y = 6$
    $3x - 2y = 8$
    $2x + 2y = 12$ (Multiply both sides of the first equation by 2.)
    $5x \quad = 20$ (Add the last two equations, term by term.)
    $\quad x = 4$ (Divide both sides by 5.)
    $4 + y = 6$ (Substitute $x = 4$ into the first equation.)
    Thus, $y = 2$, and the solution is **(4, 2)**, which checks in both equations.

11. $2x - y = 10$
    $3x + 2y = 1$
    $4x - 2y = 20$ (Multiply both sides of the first equation by 2.)
    $7x = 21$ (Add the last two equations, term by term.)
    $x = 3$ (Divide both sides by 7.)
    $6 - y = 10$ (Substitute $x = 3$ into the first equation.)

This gives $y = -4$, so that the solution is **(3, -4)**. You can check that this solution satisfies both equations.

13. $5x + y = 4$
    $15x + 3y = 8$
    $15x + 3y = 12$ (Multiply both sides of the first equation by 3.)
    $0 = -4$ (Subtract the last equation, term by term, from the second equation.)

Since the last result is impossible, the two given equations have **no** common solution. (Note that the two lines corresponding to these equations are parallel.)

15. $2x + 5y = 12$
    $5x - 3y = -1$
    $6x + 15y = 36$ (Multiply both sides of the first equation by 3.)
    $25x - 15y = -5$ (Multiply both sides of the second equation by 5.)
    $31x = 31$ (Add the last two equations, term by term.)
    $x = 1$ (Divide both sides by 31.)
    $2 + 5y = 12$ (Substitute $x = 1$ into the first equation.)
    $5y = 10$ (Subtract 2 from both sides.)
    $y = 2$ (Divide both sides by 5.)

Thus, the solution is **(1, 2)**, which you can check by substituting into the given equations.

17. $3x + 4y = 4$
    $2x - 6y = 7$
    $9x + 12y = 12$ (Multiply both sides of the first equation by 3.)
    $4x - 12y = 14$ (Multiply both sides of the second equation by 2.)
    $13x = 26$ (Add the last two equations term by term.)
    $x = 2$ (Divide both sides by 13.)
    $6 + 4y = 4$ (Substitute $x = 2$ into the first equation.)
    $4y = -2$ (Subtract 6 from both sides.)
    $y = -1/2$ (Divide both sides by 4.)

This gives the solution **(2, -1/2)**, which you can check in the given equations.

SECTION 6.4 Two Linear Equations in Two Variables

19. $11x + 3 = -3y$
    $5x + 2y = 5$
    $11x + 3y = -3$ (Add 3y and subtract 3 on both sides of the first equation.)
    $-15x - 6y = -15$ (Multiply both sides of the second equation by -3.)
    $22x + 6y = -6$ (Multiply both sides of the third equation by 2.)
    $7x = -21$ (Add the last two equations term by term.)
    $x = -3$ (Divide both sides by 7.)
    $-33 + 3 = -3y$ (Substitute x = -3 into the first equation.)
    $-30 = -3y$ (Simplify.)
    $10 = y$ (Divide both sides by -3.)
    This gives the solution **(-3, 10)**, which you can check in the two given equations.

21. $x = 2y - 3$
    $x = -2y - 1$
    $2x = -4$ (Add the two equations term by term.)
    $x = -2$ (Divide both sides by 2.)
    $-2 = 2y - 3$ (Substitute x = -2 in the first equation.)
    $1 = 2y$ (Add 3 to both sides.)
    $y = 1/2$ (Divide both sides by 2.)
    This gives the solution **(-2, 1/2)**, which checks in both equations.

23. $2x + 3y + 11 = 0$
    $5x + 6y + 20 = 0$
    $-4x - 6y - 22 = 0$ (Multiply both sides of the first equation by -2.)
    $x - 2 = 0$ (Add the last two equations term by term.)
    $x = 2$ (Solve for x.)
    $4 + 3y + 11 = 0$ (Substitute x = 2 into the first equation.)
    $3y + 15 = 0$ (Simplify.)
    $y = -5$ (Divide by 3 and solve for y.)
    This gives the solution **(2, -5)**, which checks in both given equations.

25. $3x - 12y = -8$
    $2x + 2y = 3$
    $12x + 12y = 18$ (Multiply both sides of the second equation by 6.)
    $15x = 10$ (Add the first and third equations term by term.)
    $x = 2/3$ (Divide both sides by 15.)
    $2 - 12y = -8$ (Substitute x = 2/3 into the first equation.)
    $-12y = -10$ (Subtract 2 from both sides.)
    $y = 5/6$ (Divide both sides by -12.)
    This gives the solution **(2/3, 5/6)**, which checks in both equations.

160 CHAPTER 6 FUNCTIONS AND GRAPHS

27. $r - 4s = -10$
$2r - 8s = 13$
$-2r + 8s = 20$ (Multiply both sides of the first equation by -2.)
$0 = 33$ (Add the last two equations term by term.)

Since the last result is an impossible equation, the given system has **no** solution. (Note that the graphs of the two equations are parallel lines.)

29. $6u - 2v = -27$
$4u + 3v = 8$
$18u - 6v = -81$ (Multiply both sides of the first equation by 3.)
$8u + 6v = 16$ (Multiply both sides of the second equation by 2.)
$26u = -65$ (Add the last two equations term by term.)
$u = -5/2$ (Divide both sides by 26.)
$-15 - 2v = -27$ (Substitute $u = -5/2$ into the first equation.)
$-2v = -12$ (Add 15 to both sides.)
$v = 6$ (Divide both sides by -2.)

This gives the solution $u = -5/2$, $v = 6$, or **(-5/2, 6)** which checks in both equations.

31. (a) A: $y = 8x + 1000$
B: $y = 10x + 800$
$-y = -8x - 1000$ (Multiply both sides of the first equation by -1)
$0 = 2x - 200$ (Add the last two equations term by term.)
$200 = 2x$ (Add 200 to both sides.)
$100 = x$ (Divide both sides by 2.)

This shows that the costs are equal ($1800) for **100** guests.

(b) The first equation is satisfied by (0,1000) and (200, 2600) The line through these points is the graph of the first equation. The second equation is satisfied by (0, 800) and by (200, 2800). The line through these points is the graph of the second equation. These lines are shown in the figure.

(c) The graph shows that company A is cheaper for more than 100 guests.

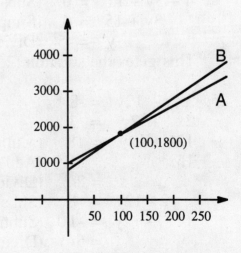

SECTION 6.4 Two Linear Equations in Two Variables

33. Put $4x = 2x + 8$.
    $2x = 8$ (Subtract 2x from both sides.)
    $x = 4$ (Divide both sides by 2.)
    The answer is **4** days.

35. With x as the number of days, the supply will be $y = 10 + 5x$.
    Put $7x = 10 + 5x$
    $2x = 10$ (Subtract 5x from both sides.)
    $x = 5$ (Divide both sides by 2.)
    The answer is **5** days.

37. Step 1. Graph the two equations. The graphs will be straight lines.
    Step 2. If the lines are not parallel or coincident, read the coordinates of the point of intersection. This is the solution.
    Step 3. Check the solution in the given equations. If the lines are parallel, then there is no solution. If the lines are coincident then all points on the line satisfy both equations.

39. In order to break even, we must have income = expense. Thus,
    Income = Expense
    $150x = 50x + 10{,}000$
    $100x = 10{,}000$ (Subtract 50x.)
    $x = 100$ (Divide by 100.)

    This shows that they must sell **100** pairs of skates to break even.

41. **Loss** =   **Expense** - **Income**
    $= (50)(80) + 10{,}000 - (150)(80)$
    $= 4000 + 10{,}000 - 12{,}000$
    $= 2000$

    Thus, the loss is **$2000**.

EXERCISE 6.5

STUDY TIPS Linear inequalities are graphed by using the graphs of the corresponding linear equations, but there are two things to keep in mind:
(1) If the inequality has the < or > sign, graph a **dotted** line instead of a solid line.
(2) To determine which region to shade, select a test point [(0,0) is the easiest], and see if the test point satisfies the inequality. If it does, select all points on that side of the line. If it does not, select the points on the other side of the line.

1. $x + 2y \geq 2$ (0, 1) and (2, 0) are two points on the line $x + 2y = 2$. Draw the line through these points. The point (0, 2) makes $x + 2y = 4 > 2$, so the required region is above the line as shown on the graph. Note that the line is drawn full to show that all the points on it satisfy the given inequality.

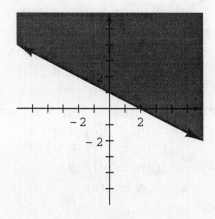

3. $x \leq 4$ Draw the line $x = 4$. Since (0, 0) satisfies the inequality, all the points in the region to the left of the line $x = 4$ also satisfy the inequality. Thus, the region to the left of this line is shaded. The solution set consists of all the points on the line and in the shaded region

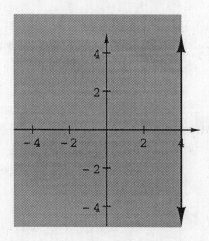

5. $3x - y < 6$

Two points on the line $3x - y = 6$ are $(2, 0)$ and $(0, -6)$. Since the points on this line are not in the solution set of the given inequality, draw a dashed line through these points. The point $(0, 0)$ satisfies the inequality, so all points on the same side of the line as the origin form the required solution set. The shaded region in the figure is the graph of this set.

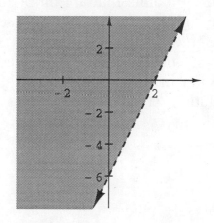

7. $2x + y \leq 4$

$(2, 0)$ and $(0, 4)$ are two points on the line $2x + y = 4$. Draw a full line through these points. (All the points on this line are part of the solution set.) The point $(0, 0)$ satisfies the inequality $2x + y < 4$, so all points on the same side of the line as the origin also satisfy this inequality. The shaded region in the figure is the graph of the required solution set.

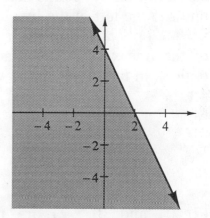

9. $4x + y > 8$

$(2, 0)$ and $(0, 8)$ are two points on the line $4x + y = 8$. Draw a dashed line through these two points, as the points on this line do not satisfy the given inequality. The point $(0, 0)$ does not satisfy the inequality, so the points on the side of the line opposite the origin do satisfy. Shade the region to the right of the line as shown in the figure.

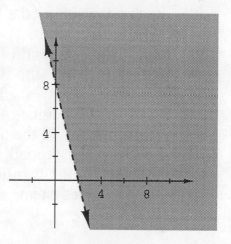

164 CHAPTER 6 FUNCTIONS AND GRAPHS

11. All the points that are on or to the right of the line x = -4 and are to the left of the line x = 3 satisfy this system. The shaded region in the graph shows the solution set

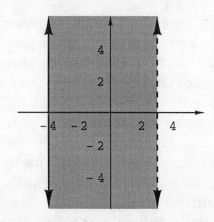

13. x ≤ -1 or x > 1
    The solution set of this system consist of all the points that are on or to the left of the line x = -1 or that are to the right of the line x = 1. The shaded regions in the graph shows this set.

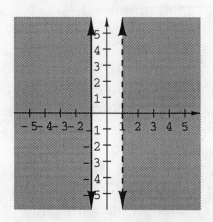

15. x - y ≥ 2, x + y ≤ 6
    (2, 0) and (0, -2) are two points on the line x - y = 2. Draw a line through these two points. For (0, 0), x - y < 2, so all points on the side of the line opposite the origin satisfy the first inequality. (3, 0) and (0, 3) are two points on the line x + y = 6. Draw a line through these two points. For

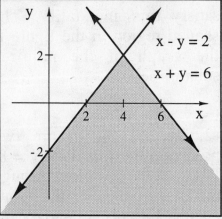

(0, 0), x + y < 6 so all the points on the same sides of this line as the origin satisfy the second inequality. Thus, the shaded region along with all points on the two lines below their point of intersection is the graph of the given system.

17. $2x - 3y \leq 6$, $4x - 3y \geq 12$

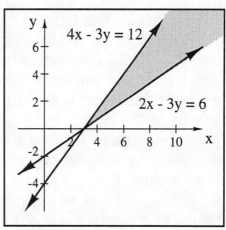

(3, 0) and (0, -2) are points on the line $2x - 3y = 6$. Draw a solid line through these points. (3, 0) and (0, -4) are points on the line $4x - 3y = 12$. Draw a solid line through these points. Now, the point (0, 0) makes $2x - 3y = 0 < 6$, so all points on the same side of this line as the origin make $2x - 3y < 6$.
The point (0, 0) makes $4x - 3y = 0$, which is less than 12, so all points on the opposite side of this line from the origin make $4x - 3y > 12$. The shaded region, along with its boundaries, is the graph of the solution set of the given system.

19. $2x - 3y \leq 5$, $x \geq y$, $y \geq 0$

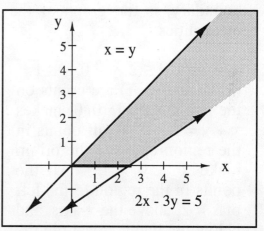

(4, 1) and (-2.5, 0) are points on the line $2x - 3y = 5$. Draw a solid line through these points. (0, 0) and (2, 2) are points on the line $x = y$. Draw a solid line a through these points. The line $y = 0$ is the x-axis. (0, 0) makes $2x - 3y = 0 < 5$, so the origin side of the line $2x - 3y = 5$ is the region where $2x - 3y < 5$. At the point (1, 0), $x > y$, so the region below the line $x = y$ is where $x > y$. The region above the x-axis is where $y > 0$. The shaded region, along with its boundaries, is the graph of the solution set of this system.

21. $x + 3y \leq 6$, $x \geq 0$, $y \geq 0$

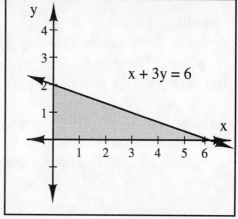

(0, 2) and (6, 0) are points on the line $x + 3y = 6$. Draw a solid line through these points. (0, 0) makes $x + 3y = 0 < 6$, so the origin side of this line is the region where $x + 3y < 6$. The system $x \geq 0$, $y \geq 0$ describes the first quadrant along with its boundaries.
The shaded region, along with its boundaries, is the graph of the given system of inequalities.

23. $x \geq 1$, $y \geq 1$, $x - y \leq 1$, $3y - x < 3$
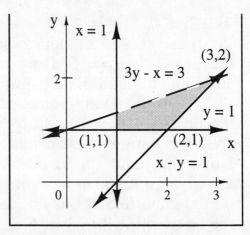
The region to the right of the line $x = 1$ and above the line $y = 1$ and its boundaries is the graph of the solution set of the system $x \geq 1$ **and** $y \geq 1$. The points $(1, 0)$ and $(2, 1)$ are on the line $x - y = 1$. Draw a solid line through these points. $(0, 0)$ makes $x - y = 0 < 1$, so all points on the origin side of this line make $x - y < 1$. The points $(0, 1)$ and $(3, 2)$ are on the line $3y - x = 3$. Draw a dashed line through these points. $(0, 0)$ makes $3y - x = 0 < 3$, so all points on the origin side of this line make $3y - x < 3$. The shaded region, along with the solid line portions of the boundary, is the graph of the solution set of the given system of inequalities.

25. $x + y \geq 1$, $x \leq 2$, $y \geq 0$, $y \leq 1$
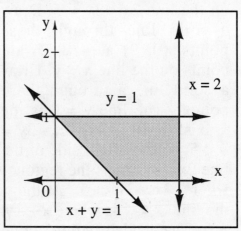
$(1, 0)$ and $(0, 1)$ are points on the line $x + y = 1$. $(0, 0)$ makes $x + y = 0 < 1$, so all points in the region opposite the origin make $x + y > 1$. All the points in the region to the left of $x = 2$, above the x-axis, and below the line $y = 1$ are in the solution set of the system $x \leq 2$, $y \geq 0$, $y \leq 1$. The shaded region, including its boundaries is the graph of the solution set of the given system of inequalities.

27. Conditions **(b)** are the correct ones. x and y are to be greater than or equal to zero means that the points must be restricted to the first quadrant. The conditions in (a) would allow all the points to the left of the line $3x + 2y = = 6$, that are to the right of $x = 2$ and below $y = 3$.

29. Conditions **(a)** are the correct ones. Conditions (b) would allow only the points that are both above the line $y = 1$ and to the right of the line $x = 2$, as well as the boundaries of this region.

SECTION 6.5 Linear Inequalities

31. The line $x + y = 1$ goes through the points $(1, 0)$ and $(0, 1)$. The condition $x + y > 1$ is true only for the points to the right of this line. The line $x - y = 1$ goes through the points $(1, 0)$ and $(0, -1)$. The condition $x - y < 1$ is true only for the points to the left of this line. The common region is that shown in (b).

33. The solution set of the inequality $62 \leq h \leq 76$ is the set of all points between and on the lines $h = 62$ and $h = 76$. The points $(62, 108)$ and $(76, 164)$ are on the line $w = 4h - 140$. In the figure a line is drawn joining these points. The points $(62, 134)$ and $(76, 204)$ are on the line $w = 5h - 176$. A solid line is drawn joining these two points. The point $(70, 150)$, which is inside the region enclosed by the four lines that have been drawn makes

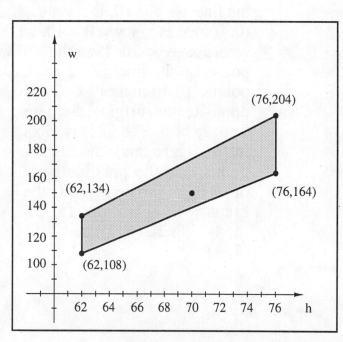

$w > 4h - 140$, since $150 > (4)(70) - 140 = 280 - 140 = 140$
and $w < 5h - 176$, since $150 < (5)(70) - 176 = 350 - 176 = 174$. This shows that the desired region is that enclosed by the four lines drawn in the figure. Note that the boundary lines are also included in the graph of the solution set of the given inequalities.

35. Suppose $c \neq 0$.
    Step 1. Find the intercepts of the line $ax + by = c$.
    Step 2. Draw a dashed line through these intercepts.
    Step 3. Substitute $(0, 0)$ into the equation. This gives zero for the left side.
    Step 4. If $c > 0$, shade the region opposite the origin. If $c < 0$, shade the region on the origin side of the line.
    Suppose $c = 0$.
    Step 1. Draw the line $ax + by = 0$. This line goes through the origin and the point $(b, -a)$.
    Step 2. Substitute (a, b) into the left side. This gives $a^2 + b^2 > 0$, so shade the region that is on the same side of the line as the point (a, b).

37. The graph would show x = k as a solid line with the region to the right of this line as the shaded region.

39. y > 1, x + y < 10, 3x - 2y > 6   The region above the line y = 1 corresponds to the inequality y > 1. (3, 7) and (10, 0) are points on the line x + y = 10. Draw a dashed line through these two points. (0, 0) makes x + y = 0 < 10, so the origin side of the line is the region where x + y < 10. (See the arrow in the figure.) (2, 0) and (4, 3) are points on the line 3x - 2y = 6. Draw a dashed line through these two points. (0, 0) makes 3x - 2y = 0, which is less than 6, so the region opposite the origin is where 3x - 2y > 6. The heavy dots in the figure mark the integer points, and the possible pairs of integers can be read off the graph: (4, 2), (5, 2), (5, 3), (5, 4), (6, 2), (6, 3), and (7, 2).

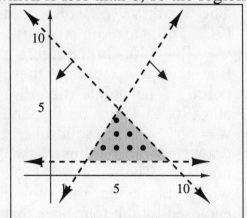

41. Cutting up the 2-inch square by connecting the midpoints of the opposite sides gives four smaller squares, each 1 inch on a side. Since there are five points, at least two of them must be inside or on the perimeter of one of the four small squares. The diagonal of this small square is $\sqrt{1^2 + 1^2} = \sqrt{2}$, so the two points cannot be more than $\sqrt{2}$ inches apart.

43. Leave everything as in the given flowchart except for interchanging the shading instructions at the end, so that if the answer to the question, "Is c > 0 ?" is "No", the half-plane not containing the origin will be shaded, and if the answer is "Yes", then the half-plane containing the origin will be shaded. You can check this by substituting (0, 0) in the inequality.

SECTION 6.6 Linear Programming

EXERCISE 6.6

STUDY TIPS The idea of linear programming is to maximize or minimize a given quantity. The restrictions given are called the **constraints** and the procedure is to graph those constraints on the same set of axes to produce the **feasible region**. The maximum or minimum for the function to be maximized or minimized must occur at one of the corners of the feasible region.

1. $x \geq 1, x \leq 4, y \leq 4, x - 3y \leq -2$

   To minimize $C = 2x + y$ subject to the given constraints, you must find the feasible region and check the vertices. First, draw the four lines: $x = 1$, $x = 4$, $y = 4$, and $x - 3y = -2$, as in the figure. Then, check the point $(0, 4)$ in the inequality $x - 3y \leq -2$. For $x = 0, y = 4, x - 3y = -12 < -2$, so the feasible region is above the line $x - 3y = -2$. The diagram shows this region and the vertices to be checked. The following table shows this check.

Vertex	$C = 2x + y$	
(1, 1)	3	← **minimum**
(4, 2)	10	
(4, 4)	12	
(1, 4)	6	

   The minimum value of C is 3 at the point (1, 1).

3. $x + y \geq 1, 2y - x \leq 1, x \leq 1$

   To find the minimum value of $W = 4x + y$ subject to the given constraints, you must find the feasible region and check the vertices.

First, draw the three lines
$x + y = 1$, $2y - x = 1$, and $x = 1$
as in the figure. By substituting
the point $(0, 0)$ into $x + y$ and
into $2y - x$, you can verify that
the feasible region is above the
line $x + y = 1$ and below the
line $2y - x = 1$, so that this
region is as shown in the
figure. There are three vertices
to be checked to find the
minimum value of W. This is
done in the following table:

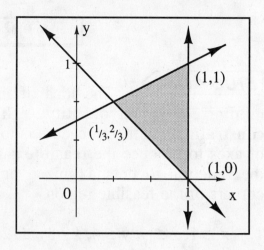

Vertex	$W = 4x + y$	
$(\frac{1}{3}, \frac{2}{3})$	2	← **minimum**
$(1, 0)$	4	
$(1, 1)$	5	

Thus, the minimum value of W is 2 at the point $(1/3, 2/3)$.

5. $8 \le 3x + y \le 10$, $x \ge 1$, $y \ge 2$
    To find the minimum value of
    $C = x + 2y$ subject to the
    given constraints, you must
    find the feasible region and
    check the vertices. First, draw
    the four lines $3x + y = 8$,
    $3x + y = 10$, $x = 1$, $y = 2$. as
    in the figure. By substituting
    $(2, 3)$ into $3x + y$, you can
    verify that the feasible region
    is the shaded region in the
    figure. The vertices are the
    points $(1, 5)$, $(2, 2)$, $(2, 8/3)$,
    and $(1, 7)$.

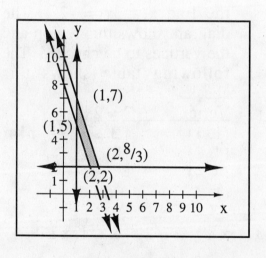

The following table shows the value of C at each of these points.

Vertex	$C = x + 2y$	
$(1, 5)$	11	
$(2, 2)$	6	← **minimum**
$(8/3, 2)$	20/3	
$(1, 7)$	15	

Thus, the minimum value of C is 6 at the point $(2, 2)$.

7. $2x + y \geq 6$, $0 \leq y \leq 4$, $0 \leq x \leq 2$

To find the maximum value of $P = x + 2y$ subject to the given constraints, you must find the feasible region and check the vertices.

First, draw the three lines $2x + y = 6$, $y = 4$, and $x = 2$. The lines $y = 0$ and $x = 0$ are the axes. By substituting $(2, 4)$ into $2x + y$ to get 8, you verify that the feasible region is above the line $2x + y = 6$. Hence, this region is shaded as shown in the figure. The vertices are the points $(2, 2)$, $(2, 4)$, and $((1, 4)$. The quantity P is evaluated at these points in the following table.

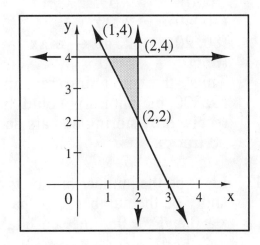

Vertex	$P = x + 2y$	
(2, 2)	6	
(2, 4)	**10**	← maximum
(1, 4)	9	

Thus, the maximum value of $P = x + 2y$ is 10 at the point $(2, 4)$.

9. Let c be the number of cars and t be the number of trucks to be parked in the lot. Then, since the total number of cars and trucks cannot exceed 100, $c + t \leq 100$. A car uses 100 sq ft and a truck uses 200 sq ft, and the total usable area is 12,000 sq ft, so that $100c + 200t \leq 12{,}000$, or, dividing by 100, $c + 2t \leq 120$. Of course, $c \geq 0$ and $t \geq 0$. If R is the monthly revenue, then $R = 20c + 35t$. Hence, the problem is to find the maximum value of R subject to the constraints: $c \geq 0$, $t \geq 0$, $c + t \leq 100$, and $c + 2t \leq 120$. The feasible region is found by drawing the lines $c + t = 100$, and $c + 2t = 120$. The lines $c = 0$ and $t = 0$ are the two axes. At $(0, 0)$, $c + t = 0$ (< 100) and $c + 2t = 0$ (< 120), so the feasible region is as in the figure on the next page. The vertices are $(0, 0)$, $(100, 0)$, $(80, 20)$, $(0, 60)$.

The table shows the value of R at each vertex.

Vertex	R = 20c + 35t
(0, 0)	0
(100, 0)	2000
(80, 20)	**2300** ← max
(0, 60)	2100

Thus, the maximum revenue ($2300 per month) would be received by storing 80 cars and 20 trucks.

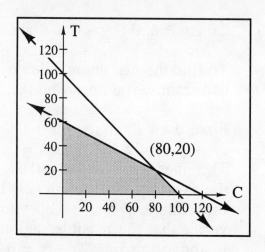

11. Let x be the number of tablets to be packed in a bottle of type 1 tablets, and y be the number to be packed in a bottle of type 2 tablets. Then, $x \geq 0$ and $y \geq 0$. Since each tablet contains 1 mg of vitamin B_1 and the two bottles are to contain at most 100 mg of B_1, $x + y \leq 100$. Similarly, each type 1 tablet contains 1 mg of vitamin B_2, and each type 2 tablet contains 2 mg of B_2, and the two bottles are to contain at most 150 mg of vitamin B_2, so $x + 2y \leq 150$. The total profit P(cents) on the two bottles is $P = 2x + 3y$. Thus, the problem is to maximize P subject to the constraints $x \geq 0$, $y \geq 0$, $x + y \leq 100$, and $x + 2y \leq 150$. The feasible region is found by drawing the lines $x + y = 100$ and $x + 2y = 150$. The lines $x = 0$ and $y = 0$ are the two axes. Since (0, 0) makes $x + y = 0$ (< 100) and makes $x + 2y = 0$ (< 150), the feasible region is as shown in the figure.

The vertices to be checked are (100, 0), (50, 50), and (0, 75). The table below shows that the maximum profit is obtained by packing 50 tablets in each of the two bottles.

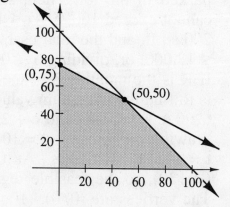

Vertex	P = 2x + 3y
(100, 0)	200
(50, 50)	**250** ← max.
(0, 75)	225

13. Let x be the number of ounces of additive X and y be the number of ounces of additive Y to be put into each bottle. Then, to have at least 32 units of vitamin A, $16x + 4y \geq 32$, or $4x + y \geq 8$. (Divide by 4.) To have at least 10 units of vitamin B, $2x + 2y \geq 10$, or $x + y \geq 5$. (Divide by 2.) To have 40 units of vitamin C, $4x + 14y \geq 40$, or $2x + 7y \geq 20$. (Divide by 2.) Since the total amount of additives is not to exceed 10 ounces, $x + y \leq 10$. If C cents is the total cost of the additives per bottle, then $C = 20x + 40y$. Thus, the problem is to minimize C under the constraints: $4x + y \geq 8$, $x + y \geq 5$, $2x + 7y \geq 20$, and $x + y \leq 10$. The feasible region is found by drawing the lines $4x + y = 8$, $x + y = 5$, $2x + 7y = 20$, and $x + y = 10$. The point (4, 4), which is inside the region enclosed by these four lines, satisfies all four of the inequalities, so this region is the feasible region shown in the figure. The vertices to be checked are (3, 2), (10, 0), (0, 10) (0, 8), and (1,4).

The table below shows that the minimum cost is obtained by using 3 oz of additive X and 2 oz of additive Y in each bottle.

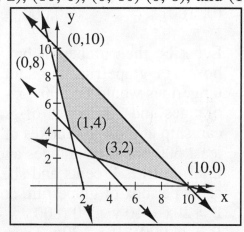

Vertex	$C = 20x + 40y$	
(3, 2)	**140**	← min.
(10, 0)	200	
(0, 10)	400	
(0, 8)	320	
(1, 4)	180	

15. Let x be the number of batches to be bought from Refinery 1 and y be the number to be bought from Refinery II. The cost will then be $C = 300x + 500y$ dollars. Since each of the batches has 1 unit of A, $x + y \geq 100$. Each of the x batches has 3 units of B, and each of the y batches has 4 units of B, so $3x + 4y \geq 340$. Each of the x batches has 1 unit of C and each of the y batches has 5 units of C Therefore, $x + 5y \geq 150$. Furthermore, $x \leq 100$ and $y \leq 100$. Hence, the problem is to minimize the cost subject to the constraints: $x + y \geq 100$, $3x + 4y \geq 340$, $x + 5y \geq 150$, $x \leq 100$, and $y \leq 100$. The feasible region can be found by drawing the lines $x + y = 100$, $3x + 4y = 340$, $x + 5y = 150$, $x = 100$, $y = 100$, as in the figure. You can verify that the point (80, 40) satisfies all of the constraints, so the feasible region

is that shaded in the figure. The vertices to be checked are (60,40), (100,10), (100,100), and (0,100) The following table gives the values of C at these vertices and shows that the dealer should buy 100 batches from Refinery I and 10 batches from Refinery II to minimize the cost.

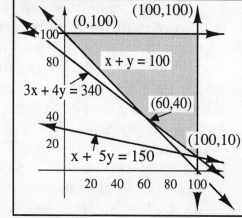

Vertex	C = 300x + 500y
(60, 40)	38,000
(100, 10)	**35,000** ← **min.**
(100, 100)	80,000
(0, 100)	50,000

17. Let x be the number of boxes of oranges, and y be the number of boxes of grapefruit she should load. Then the number of boxes of tangerines would be 800 - x - y. She must ship at least 200 boxes of oranges and 100 boxes of grapefruit, so $x \geq 200$ and $y \geq 100$. She can ship at most 200 boxes of tangerines, so she must ship a total of at least 600 boxes of oranges and grapefruit, that is, $x + y \geq 600$. If the total profit is P cents and she makes 20 cents per box of oranges, 10 cents per box of grapefruit, and 30 cents per box of tangerines, then
P = 20x + 10 y + 30 (800 - x - y), which simplifies to
P = 24,000 - 10x - 20y. Thus, the problem is to maximize P subject to the constraints: $x \geq 200, y \geq 100$ and $x + y \geq 600$. The feasible region can be found by drawing the lines x = 200, y = 100, and x + y = 600. Since (0, 0) does not satisfy any of the three inequalities, the feasible region is the one shaded in the figure. The vertices to be checked are (200, 400) and (500,100). The value of P at each of these points is given in the following table, which shows that the maximum profit is obtained by shipping 500 boxes of oranges, 100 boxes of grapefruit, and 200 boxes of tangerines.

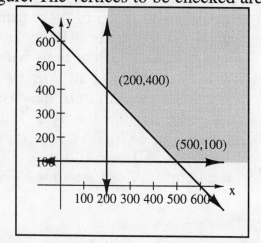

Vertex	P = 24,000 - 10x - 20y
(200, 400)	14,000
(500, 100)	**17,000** ← **max.**

19. Let s be the number of shrubs and t the number of trees to be displayed. The profit, P dollars, is given by $P = 7s + 6t$. The space restriction is $3s + 2t \leq 12$, and the time restriction is $s + 2t \leq 8$. Neither s nor t can be negative, so the problem is to maximize P subject to the constraints $3s + 2t \leq 12$, $s + 2t \leq 8$, $s \geq 0$, and $t \geq 0$. The feasible region can be found by drawing the lines $3s + 2t = 12$, $s + 2t = 8$, $s = 0$ and $t = 0$, as in the figure. The point $(2, 2)$ is inside the region bounded by these four lines and this point satisfies all four of the constraints. Thus, the feasible region is the shaded region in the figure. The vertices to be checked are $(4, 0)$, $(2, 3)$, and $(0, 4)$.

The following table gives the value of P at each of the vertices and shows that the maximum profit is obtained by displaying 2 shrubs and 3 trees.

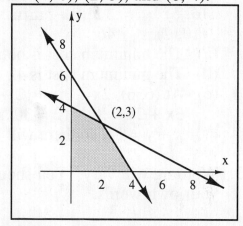

Vertex	$P = 7s + 6t$	
(4, 0)	28	
(2, 3)	**32**	← max.
(0, 4)	24	

21. Let x be the number of ounces of orange juice and y be the number of ounces of grapefruit juice to be put into each can. Then, since a can is to contain not more than 15 ounces, $x + y \leq 15$. The orange juice has 2 units of vitamin A per ounce and the grapefruit juice has 3 units of A per ounce, so $2x + 3y \geq 26$. The orange juice has 3 units of vitamin C per ounce and the grapefruit juice has 2 units of C per ounce, so $3x + 2y \geq 30$. Each of the juices has 1 unit of vitamin D per ounce, so $x + y \geq 12$. The cost C cents per can is given by $C = 4x + 3y$. Thus, the problem is to minimize C under the constraints:

$x + y \leq 15$, $2x + 3y \geq 26$, $3x + 2y \geq 30$, and $x + y \geq 12$.

The feasible region can be found by drawing the lines $x + y = 15$, $2x + 3y = 26$, $3x + 2y = 30$, and $x + y = 12$, as in the figure. The point $(10, 4)$ satisfies all of the constraints, so the feasible region is

176 CHAPTER 6 FUNCTIONS AND GRAPHS

that shaded in the figure. The vertices to be checked are (13, 0), (15, 0), (0,15), (6, 6), and (10, 2). The table below gives the value of C for each of these points.

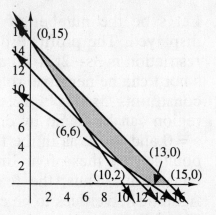

Vertex	C = 4x + 3y
(13, 0)	52
(15, 0)	60
(0, 15)	45
(6, 6)	**42** ← **min.**
(10, 2)	46

(a) The minimum cost is obtained by using 6 oz of each juice.
(b) The minimum cost is 42 cents.
(c) At (6, 6), $2x + 3y = 12 + 18 = 30$, so there will be 30 units of A, $3x + 2y = 18 + 12 = 30$, so there will be 30 units of C, and $x + y = 12$, so there will be 12 units of D.

23. Wording will vary. You should refer to the text and then restate in your own words.

25. (a) The odds are 5 to 3 in favor of the champion, so if the champion wins, Gary wins $3 for each $5 that he bet on the champion to win, and he loses whatever he bet on the challenger to win. If he bet x dollars on the champion and y dollars on the challenger, and if the champion wins, then Gary has a net gain of $(\frac{3}{5}x - y)$ dollars.

(b) The odds are 2 to 1 in favor of the challenger, so if the challenger wins, Gary wins $2 for each $1 that he bet on the challenger to win and loses whatever he bet on the champion to win. Thus, if the challenger wins, Gary's net gain is **(-x + 2y)** dollars.

(c) Since the total amount that Gary is betting is (x + y) dollars, the problem is to minimize x + y subject to the constraints:

$x \geq 0$, $y \geq 0$, $(\frac{3}{5}x - y) \geq 100$, which simplifies to $3x - 5y \geq 500$, and $-x + 2y \geq 100$.

To solve this problem, you must find the feasible region, so draw the lines $3x - 5y = 500$ and $-x + 2y = 100$ as in the figure. These lines intersect at the point (1500, 800). At the point (2000, 0),

$3x - 5y = 6000$, which is greater than 500, so the feasible region lies below the line $3x - 5y = 500$. Also, at the point $(2000, 0)$, $-x + 2y = -2000$, which is less than 100, so the feasible region lies above the line $-x + 2y = 100$. The feasible region is shaded in the figure. There is only one vertex,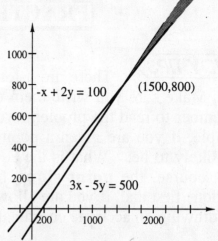
(1500, 800), so the minimum must occur here. This shows that the least amount that Gary can bet for a net gain of at least $100 is $1500 on the champion to win and $800 on the challenger to win, for a total of $2300.

Note: The graph shows that if you stay in the feasible region, but move away from the point (1500, 800) both x and y increase, so that the total $x + y$ will be greater than 2300.

178 CHAPTER 6 FUNCTIONS AND GRAPHS

PRACTICE TEST 6

STUDY TIPS There are a lot of formulas to be remembered for this test. Make sure you write them on your test paper as soon as you get it. Remember to read the problems very carefully and see what is given. For example, if you are given a point and the slope of a line, the question is very likely to be: What is the equation of this line? The formula to use is, of course, the **point-slope** formula. Moreover, answer the easy questions first and, if you are allowed, use graph paper so you do not waste time drawing an accurate coordinate system.

1. The domain is the set of x values:{**0, 2, 3, 5**}; the range is the set of y values: {**-1, 2, 3, 4**}.

2. The domain is the set of x values, which for this relation is **the set of all real numbers**; the range is also **the set of all real numbers**.

3. The domain is the set of x values: {**1, 2, 3, 4**}; the range is the set of y values: {**1, 2, 3, 4**}.

4. The relations in (**b**) and (**c**) are both functions. In each of these there is just one y value for each value of x in the domain.

5. Since $f(x) = x^2 - x$,
   (a) $f(0) = 0^2 - 0 = 0$ (b) $f(1) = 1^2 - 1 = 0$
   (c) $f(-2) = (-2)^2 - (-2) = 4 + 2 = 6$

6. We must solve the equation $\quad 35.30 = 15 + 0.10m$ for m.
   Multiply both sides by 10: $\quad 353 = 150 + m$
   Subtract 150 from both sides: $\quad 203 = m$
   Thus, the person drove **203 mi**.

7. Since x is an integer between -1 and 3 inclusive, the table of values is as follows:

x	y
-1	-3
0	0
1	3
2	6
3	9

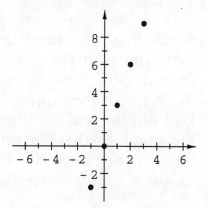

The corresponding points are the heavy dots shown in the graph

8. Since x and y are nonnegative integers, the table of values is as follows:

x	y
0	0
0	1
0	2
1	0
1	1
2	0

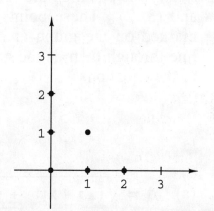

The corresponding points are the heavy dots shown in the graph.

9. Here, x is an integer between -2 and 2, inclusive, so the table of values is as follows:

x	$g(x) = 2x^2 - 1$
-2	7
-1	1
0	-1
1	1
2	7

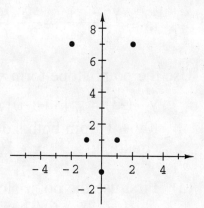

The corresponding points are the heavy dots shown in the graph.

180 CHAPTER 6 FUNCTIONS AND GRAPHS

10. The function defined by $f(x) = 2x - 6$ is a linear function, so we need to find two points on the graph and draw the line through them for the graph of the function. We find $f(0) = -6$ and $f(3) = 0$, so the graph is the straight line through the points $(0, -6)$ and $(3, 0)$

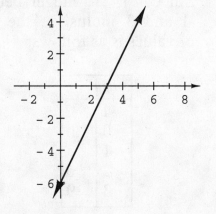

11. $3x - 2y = 5$ is the equation of a straight line. Two points that satisfy the equation are $(1, -1)$ and $(3, 2)$. These points are marked on the graph and the line through them is the graph of the equation.

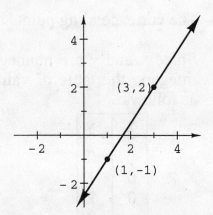

12. (a) $d = \sqrt{(x_2 - x_1)^2 + (y_2 - y_1)^2} = \sqrt{(7 - 4)^2 + (3 - 7)^2}$
    $= \sqrt{9 + 16} = \sqrt{25} = $ **5**

    (b) The two points have the same x coordinate, so are on the same vertical line. Thus, $d = |y_2 - y_1| = |-2 - 8| = $ **10.**

13. $m = \dfrac{y_2 - y_1}{x_2 - x_1} = \dfrac{-2 - (-3)}{9 - (-1)} = \dfrac{1}{10}$

14. Use the point-slope form $y - y_1 = m(x - x_1)$ to get
    $y - (-3) = \dfrac{1}{10}[x - (-1)]$ or $10y + 30 = x + 1$. Then, by subtracting $10y + 1$ from both sides, you can get the general form of the equation: **x - 10 y = 29**.

15. (a) First use the point-slope form to get $y - (-1) = -2(x - 3)$ or $y + 1 = -2x + 6$. Then, subtract 1 from both sides to get the answer **y = -2x + 5**.

    (b) Solve the equation $2y = 4 - 8x$ for y to get $y = -4x + 2$ as the slope-intercept form of the equation. Then read off m = **-4** and b = **2**.

16. (a) Solve each equation for y: $y = -2x + 1$, $y = -4x + \frac{4}{3}$. This shows that the slope of the first line is -2 and the slope of the second line is -4, so the lines are not parallel. By equating the two expressions for y, we get the equation:

    $-2x + 1 = -4x + \frac{4}{3}$

    $2x = \frac{1}{3}$ (Add 4x and subtract 1 on both sides.)

    $x = \frac{1}{6}$ (Divide both sides by 2.)

    $y = -\frac{1}{3} + 1 = \frac{2}{3}$ (Put $x = \frac{1}{6}$ in the first equation.)

    Thus, the lines intersect at $(\frac{1}{6}, \frac{2}{3})$.

    (b) The slope of the line $y = 2x - 5$ is $m = 2$. Solving the second equation for y, we get $y = 2x - \frac{7}{2}$, so the slope of the second line is also $m = 2$. Thus, the two lines **are** parallel.

17. Since the required line is to be parallel to the line $2x - 3y = 5$, its equation must be of the form $2x - 3y = C$, where C is to be determined so that the line passes through the point (1, -2). So, we substitute into the equation to get $(2)(1) - (3)(-2) = C$, which gives $C = 8$. Thus, the answer is **$2x - 3y = 8$.**

18. To find the point of intersection of the lines, we solve the system

    $3x + 2y = 9$
    $2x - 3y = 19$
    $9x + 6y = 27$ (Multiply both sides of the first equation by 3.)
    $4x - 6y = 38$ (Multiply both sides of the second equation by 2.)
    $13x = 65$ (Add the equations term by term.)
    $x = 5$ (Divide both sides by 13.)
    $15 + 2y = 9$ (Substitute $x = 5$ into the first equation.)
    $2y = -6$ (Subtract 15 from both sides.)
    $y = -3$ (Divide both sides by 2.)

    Thus, the solution is **(5, -3)**. (You should check this answer in the second equation.)

19. To graph the solution set of the given inequality, draw the line $4x - 3y = 12$, as in the figure. The point $(0, 0)$ makes the left side, $4x - 3y = 0$, which satisfies the given inequality. Thus, all points on the origin side of the line are in the solution set. This region is shaded in the figure. Note that the line is drawn solid to show that it belongs in the graph of the solution set.

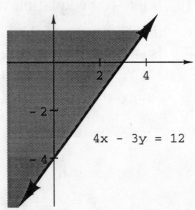

20. To graph the solution set of the system $x + 3y \leq 6$ and $x - y \geq 2$ draw the lines $x + 3y = 6$ and $x - y = 2$, as in the figure The point $(0, 0)$ satisfies the first inequality, so the desired region is on the origin side of the line $x + 3y = 6$. $(0, 0)$ does not satisfy the second inequality, so the desired region is on the side of the line $x - y = 2$ opposite the origin. The shaded region along with the heavily drawn lines is the desired graph.

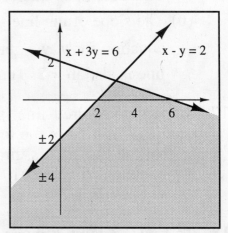

21. To graph the solution set of the system $x + 2y \leq 3$, $x \leq y$, $x \geq 0$, draw the lines $x + 2y = 3$, $x = y$, and $x = 0$. The point $(0, 0)$ satisfies the first inequality, so the desired region is on the origin side of the line $x + 2y = 3$. The point $(1,2)$ satisfies the second inequality, so the region is on the side above the line $x = y$. All points to the right of the y axis satisfy the third inequality. The shaded region along with its boundaries is the graph of the solution set of the given system.

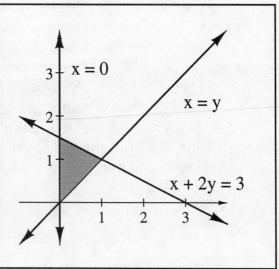

22. The first equation can be written in the form $2x - y = 3$. If both sides of this equation are multiplied by 3, the result is $6x - 3y = 9$, which is exactly the second equation. This means that the two lines corresponding to these equations are coincident, so that any solution of one equation is also a solution of the other equation. A general solution can be obtained by letting $x = a$, where a is any real number. Then, the first equation gives $y = 2a - 3$. Thus, the solution set can be written as **{(a, 2a - 3)| a is any real number}**.

23. To find the maximum value of $C = x + 2y$ subject to the constraints: $3x + y \leq 8$, $x \leq 1$, $y \geq 2$, $x \geq 0$, first draw the lines $3x + y = 8$, $x = 1$, $y = 2$, and $x = 0$ as in the figure. The point (0, 0) satisfies the first inequality, so the feasible region lies below the line $3x + y = 8$. Because of the other constraints, this region must lie to the right of the y axis, to the left of the line $x = 1$ and above the line $y = 2$. This region is shaded in the figure.

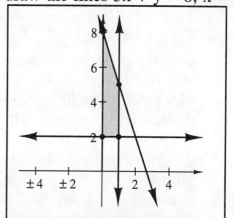

The vertices to be checked are (0, 2), (1, 2), (1, 5), and (0, 8). The value of C at each vertex is shown below. The maximum value of C is 16 at the point (0, 8)

Vertex	C = x + 2y
(0, 2)	4
(1, 2)	5
(1, 5)	11
(0, 8)	**16** ← max.

24. To minimize $P = 3y - 2x$ subject to the constraints: $y - x \leq 2$, $x + y \leq 4$, $x \leq 3$, $x \geq 0$, $y \geq 0$, draw the lines $y - x = 2$, $x + y = 4$, $x = 3$, $x = 0$, and $y = 0$, as in the figure on the next page. The point (0, 0) satisfies the first three of the constraints, so the feasible region is on the origin side of the corresponding lines. The remaining two inequalities are satisfied by all first quadrant points. Thus, the feasible region is the shaded region in the graph. The vertices to be checked are (0, 0), (3, 0), (3, 1), (1, 3), and (0, 2).

The table below gives the value of P at each of these points and shows that the minimum value of P is -6 at the point (3,0).

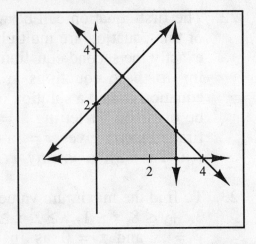

Vertex	P = 3y - 2x	
(0,0)	0	
(3,0)	**-6**	← min.
(3,1)	-3	
(1,3)	7	
(0,2)	6	

25. Let x be the number of hr that Machine A operates per week, and let y be the number of hr that Machine B operates per week. Then, since this is a 40-hr week, $x + y \leq 40$. Machine A produces 10 items per hr and Machine B produces 12 items per hr, and at least 420 items must be produced each week. Thus, $10x + 12y \geq 420$ or, dividing by 2, $5x + 6y \geq 210$. Also, $x \geq 0$ and $y \geq 0$. It costs $20/hr to operate Machine A and $25/hr to operate Machine B, so the weekly machine cost is C dollars, with $C = 20x + 25y$. Thus, the problem is to minimize C subject to the constraints: $x + y \leq 40$, $5x + 6y \geq 210$, $x \geq 0$, $y \geq 0$.

To find the feasible region, first draw the lines $x + y = 40$ and $5x + 6y = 210$ as in the figure. The point (0,0) satisfies the first constraint but not the second, so the feasible region is on the origin side of the line $x + y = 40$ and on the side of $5x + 6y = 210$ opposite the origin. This region is shown shaded in the figure. Note that all points in this region satisfy

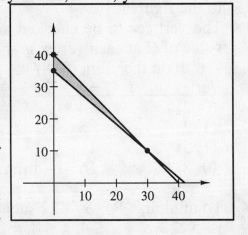

the last two inequalities. The vertices to be checked are (0, 40), (0, 35), and (30, 10). The table gives the value of C for each of these points and shows that C is a minimum at the point (30, 10). Thus, Machine A should be operated 30 hr/wk and Machine B 10 hr/wk for minimum machine cost.

Vertex	C = 20x + 25y	
(0, 40)	1000	
(0, 35)	875	
(30,10)	**450**	← min

CHAPTER 7
GEOMETRY

EXERCISE 7.1

STUDY TIPS The study of geometry contains many undefined terms such as point, line, and plane. You should learn the meaning of these concepts by looking at concrete examples and studying Figures 7.2-7.5 7.12-7.15, and 7.17 in the text. Since the material also relates to operations with sets, review the concept of intersection and union studied in Chapter 1. Lines, rays, and points can be combined to form angles. An **angle** is the figure formed by two rays with a common endpoint called the **vertex**. Example 2 will show you three different ways of naming an angle. Angles can be measured by using a tool called a protractor and then can be classified according to their measure. Remember that a **straight** angle measures **180°**, a **right** angle **90°**, an **acute** angle **less than 90°**, and an **obtuse** angle **more than 90° but less than 180°**. **Supplementary** angles are angles whose measures **add up to 180°**, while **complementary** angles are angles whose **sum is 90°** You should master Example 5, since the concepts in the example are used to prove the important fact that the sum of the measures of the angles of any triangle is 180°

1. (a) P●——●→Q (b) P←●——○Q (c) P●——●Q

3. The segment $\overline{BC}$ 5. The ray $\overrightarrow{AD}$ 7. The segment $\overline{AD}$

9. The segment $\overline{AD}$ 11. ∅   13. The point C

15. (a) AB, AC, AD, BC, BD, CD
    (b) AB and CD; AC and BD; AD and BC
    (c) None of the edges determine parallel lines.

17. True   19. True   21. True   23. False (They can be skew lines.)

25. If a plane, say EFG, contains line m, it must contain the point where line m intersects plane ABC. So plane EFG must intersect plane ABC, and the given statement is false.

27. (a) ∠BAC is the same as ∠α.   (b) ∠β is the same as ∠EAF.

29. ∠BAC, ∠CAD, ∠DAE, ∠EAF   31. ∠BAE, ∠CAE, ∠CAF

33. (a) ∠DAE (b) ∠CAD   35. (a) ∠DAF   (b) ∠BAE

37. m∠DAE = 90° - 55° = 35°   39. m∠α = 90° - 75° = 15°

41. ∠B is vertical to the 70° angle.   43. m∠A = 180° - 70° = 110°

45. 90° - 70° = 20°

47. m∠A = m∠C = 180° - 70° = 110°, so m∠A + m∠C = 220°.

49. (a) m∠A = 180° - 30° = 150°   51. ∠A, ∠C, ∠E, ∠G
    (b) m∠B = 30°
    (c) m∠C = m∠A = 150°

53. (a) m∠B = 90° - 41° = 49°   (b) m∠B = 180° - 41° = 139°

55. Since the two angles are complementary, the sum of their measures is 90°. Thus, $(3x + 15) + (2x - 5) = 90$
    Simplify: $5x + 10 = 90$
    Subtract 10: $5x = 80$
    Divide by 5: $x = 16$

57. The angles are supplementary, so their measures must add to 180°. Thus, we have the equation: $10x - 5 + 2x + 5 = 180$
    Simplify: $12x = 180$
    Divide by 12 $x = 15$
                            $2x + 5 = 35$
                            $10x - 5 = 145$

    This shows that x = 15 and one angle measures 35°; the other measures 145°.

59. The two angles are complementary, so their measures must add to 90°. Thus we have the equation: $2x + 10 + 5x + 10 = 90$
    Simplify: $7x + 20 = 90$
    Subtract 20: $7x = 70$
    Divide by 7: $x = 10$
    Thus, 2x + 10 = 30, 5x + 10 = 60, so that one angle measures 30° and the other measures 60°.

SECTION 7.1 Points, Lines, Planes and Angles

61. (a) It moves through 1/12th of 360°, that is, through 30°.
    (b) It moves through 6/12ths of 360°, that is, through 180°.

63. $m\angle A + m\angle B + m\angle C = 180°$. Thus, if the number of degrees in $\angle B$ is x, then $37 + x + 53 = 180$. This gives $x = 90$, so $m\angle B = 90°$.

65. The total number of degrees in the three angles is 180, so we have
    the equation: $(x + 10) + (2x + 10) + (3x + 10) = 180$
    Simplify: $6x + 30 = 180$
    Subtract 30: $6x = 150$
    Divide by 6: $x = 25$

67. (a) $m\angle A + m\angle B = 180°$ Angles A and B form a straight angle.
    (b) $m\angle C + m\angle B = 180°$ Angles C and B form a straight angle.
    (c) $m\angle A + m\angle B = m\angle C + m\angle B$ Both sides equal 180°.
    (d) $m\angle A = m\angle C$ Subtract $m\angle B$ from both sides.

69. Point, line, and plane are all undefined.

71. The ray may have its endpoint only in common with the plane, or it may have some other single point in common with the plane, or it may lie entirely in the plane.

73. (a) One of the usual meanings of <u>acute</u> is "sharp" or "intense". Thus "an acute pain" means "a sharp or intense pain."
    (b) One of the ordinary meanings of <u>obtuse</u> is "dull". Thus, "obtuse intelligence" means "dull intelligence" or "stupidity."

75. $48.2 + 75.9 + 56.1 = 180.2$, so the sum of the angles is 0.2° too large.

77. $360 - 310 = 50$, so the surveyor's bearing would be N 50° W.

79. $360 - 40 = 320$, so the navigator's bearing would be 320°.

188 CHAPTER 7 GEOMETRY

EXERCISE 7.2

STUDY TIPS Since we have studied lines, we now take a sequence of connected straight line segments, called **broken lines**, and construct polygons which are classified according to the number of sides they have. (See the table on page 368). Practice by looking at traffic signs and determining how many sides they have. If the angles of a polygon are all of equal size, try naming the polygon. The emphasis here, however, is on **triangles** and their classification according to their angles (Table 7.3) or the number of equal sides (Table 7.4). Study the illustrations in these tables to become familiar with the classification of triangles. Other quadrilaterals are shown and named in Table 7.5. If you have worked with model airplanes or looked at house plans you are familiar with the idea of **similar** figures. The same idea applies to triangles (See Definition 7.1.). The formula on page 374 gives the sum of the measures of the interior angles of any convex polygon.

1. (a) 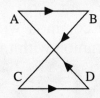 (b)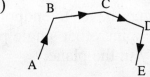

3. (a) C, D, I, J, L, M, N, O, P, S, U, V, W, Z   (b) B, D, O

5. (a) D, O   (b) A, E, F, H, K, Q, R, T, X, Y

7. Convex

9. Parallelogram

11. Rectangle

13. Trapezoid

15. Parallelogram

17. A scalene, right triangle

19. A scalene, acute triangle

21. An isosceles, acute triangle

23. A scalene, obtuse triangle

25. Triangles (a) and (c) are similar because $\dfrac{4/3}{4} = \dfrac{1}{3} = \dfrac{2/3}{2}$.

SECTION 7.2 Triangles and Other Polygons

proportional. Thus,
$$\frac{x}{4} = \frac{y}{5} = \frac{4}{3}$$
so that $x = \frac{16}{3} = 5\frac{1}{3}$ and $y = \frac{20}{3} = 6\frac{2}{3}$.

29. PQ is parallel to AB, so triangle PQC is similar to triangle ABC. Thus, corresponding sides of these triangles are proportional. This gives the proportion $\frac{AC}{PC} = \frac{BC}{QC}$. Let x be the length of BC, so that x - 6 is the length of QC. Since AP is of length 3 and PC is of length 4, AC is of length 7. With these lengths, our proportion gives the equation:

$$\frac{7}{4} = \frac{x}{x - 6}$$

By "cross-multiplication," we get

$$7x - 42 = 4x$$
so that $\quad\quad\quad\quad 3x = 42 \quad$ (Subtract 4x and add 42.)
Thus, $\quad\quad\quad\quad\quad x = 14$

31. Let x be the length of PC. Then, since the length of AP is 2, the length of AC is x + 2. Using the same proportion as in Problem 29, we get the equation:

$$\frac{x + 2}{x} = \frac{5}{2}$$
$$2x + 4 = 5x \quad \text{(Use "cross-multiplication".)}$$
so that $\quad\quad\quad\quad 4 = 3x$
and $\quad\quad\quad\quad x = \frac{4}{3} = 1\frac{1}{3}$.

33. Let x be the length of the side of the smaller triangle that corresponds to the 6 cm side of the larger triangle, and let y be the length of the side of the smaller triangle that corresponds to the 9 cm side of the larger triangle. Then, since the triangles are similar,

$$\frac{x}{6} = \frac{y}{9} = \frac{7}{12}$$

This proportion gives $x = \frac{6 \times 7}{12} = \frac{7}{2} = 3\frac{1}{2}$ cm

and $\quad\quad\quad\quad y = \frac{9 \times 7}{12} = \frac{21}{4} = 5\frac{1}{4}$ cm.

35. Let the sides of the second triangle be x, y, and z in. long, respectively.

35. Let the sides of the second triangle be x, y, and z in. long, respectively. Then $x + y + z = 36$.
    Since the triangles are similar,
    $$\frac{x}{2} = \frac{y}{3} = \frac{z}{4}$$
    This gives $x = \frac{1}{2} z$ and $y = \frac{3}{4} z$.
    Substituting into the first equation, we get
    $$\frac{1}{2} z + \frac{3}{4} z + z = 36$$
    (Multiply by 4.) $\quad 2z + 3z + 4z = 144$
    Thus, $\quad 9z = 144, \; z = 16$
    $$x = \frac{1}{2} z = 8 \text{ and } y = \frac{3}{4} z = 12.$$
    The respective sides of the second triangle are 8 in., 12 in., and 16 in. long.

37. Let the telephone pole be x ft high. Then the similar right triangles in the figure give the proportion
    $$\frac{x}{5} = \frac{30}{8}$$
    Thus, $x = \frac{5 \times 30}{8} = \frac{75}{4} = 18\frac{3}{4}$, so the pole is $18\frac{3}{4}$ ft high.

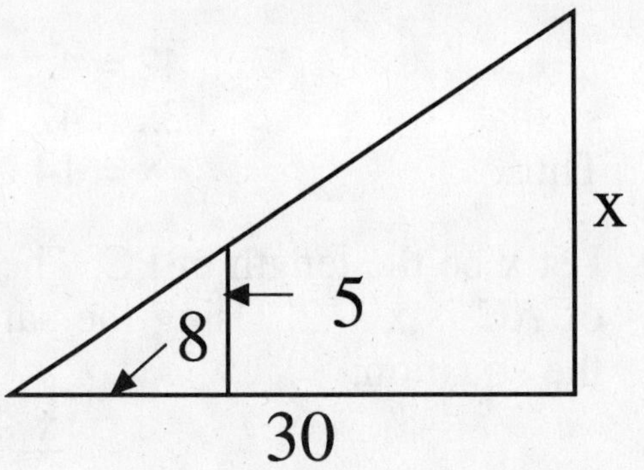

39. Let the height of the tree be **h** ft. Then the two similar right triangles give the proportion
    $$\frac{h}{5} = \frac{40}{8}$$
    Thus, $h = 5 \times 5 = 25$, so the tree is 25 ft high.

41. The marked angles are equal and the vertical angles in the figure are equal, so the two triangles are similar. If the tunnel is **t** meters long, then we have the proportion:
    $$\frac{t}{20} = \frac{540}{18}$$
    Since $\frac{540}{18} = 30$, we have $t = 20 \times 30 = 600$, that is, the tunnel is 600 m long.

SECTION 7.2   Triangles and Other Polygons

BC means that $\angle A = \angle B$. Since $m\angle A + m\angle B = 180° - m\angle C$, we have that $2m\angle A = 100°$. Thus, $m\angle A = m\angle B = 50°$.

(b) $\angle B$ is supplementary to the 140° angle, so $m\angle B = 40°$. As in part (a), $\angle A = \angle B$. Hence, $m\angle A = 40°$. $m\angle A + m\angle B + m\angle C = 180°$. Therefore, $80° + m\angle C = 180°$. Consequently, $m\angle C = 100°$.

45. AE and BD are parallel, so $\angle ABD$ equals the 120° angle. $\angle CBD$ is supplementary to $\angle ABD$, which makes $m\angle CBD = 60°$. Because CE is perpendicular to BD, $m\angle CDB = 90°$. Since the sum of the measures of the angles of triangle BCD is 180°, $m\angle BCD = 180° - 60° - 90°$. Thus, $m\angle BCD = 30°$

47. The sum of the measures of the angles of a regular 14-sided polygon is
$(14 - 2)180° = 12 \times 180° = 2160°$. (See the formula on page 374.)

49. The sum of the measures of the angles of a regular pentagon is
$(5 - 2)180° = 3 \times 180° = 540°$. (See the formula on page 374.) Since there are five angles, the measure of each must be 108°.

51. The sum of the measures of the angles of a regular octagon is
$(8 - 2)180° = 6 \times 180° = 1080°$. (See the formula on page 374.) Since there are eight angles, the measure of each must be 135°.

53. The sum of the measures of the angles of a regular decagon is
$(10 - 2)180° = 8 \times 180° = 1440°$. (See the formula on page 374.) Since there are ten angles, the measure of each must be 144°.

55. (a) $(E \cup I \cup S) = T \subset P$

57.

(b)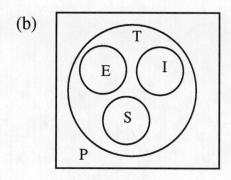

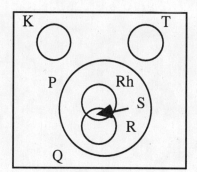

59. Answers will vary.

192 CHAPTER 7 GEOMETRY

61. (a) 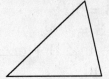 (b)   (c)

63. (a)   (b)   (c) Impossible

SECTION 7.3 Perimeter and Circumference

EXERCISE 7.3

STUDY TIPS The **perimeter** of a geometric figure is simply the distance around it. In the case of a circle, the perimeter is called the **circumference** and is given by the formula $C = \pi d$ or $C = 2\pi r$. Be careful with your answers, since 3.14 is just an approximation for π. Table 7.8 gives the perimeter of several geometric figures. All of these perimeters are measured using linear units (Table 7.6).

1. $10 + 25 + 30 = 65$, so the perimeter is **65 cm.**

3. $2 + 5 + 1.6 + 4 = 12.6$, so the perimeter is **12.6 yd.**

5. $2(61.2 + 31.1) = 184.6$, so the perimeter is **184.6 m.**

7. $2(19.4 + 9.2) = 57.2$, so the perimeter is **57.2 in.**

9. $5 \times 6 = 30$, so the perimeter is **30 cm.**

11. If the width was x ft, then $2x + 2 \times 30 = 80$, so that $2x + 60 = 80$. by subtracting 60 from both sides and dividing both sides by 2, we obtain $x = 10$. Therefore, the width was **10 ft.**

13. This distance is 1/4th of the perimeter, that is, $360/4 =$ **90 ft.**

15. Let x ft be the width. Then, the length is $(x + 198)$ ft, and, in terms of x, the perimeter is $[2x + 2(x + 198)]$ ft. Thus, we have the equation:
$$2x + 2(x + 198) = 2468$$
Simplify: $\qquad 4x + 396 = 2468$
Subtract 396: $\qquad 4x = 2072$
Divide by 4: $\qquad x = 518$
$\qquad\qquad\qquad x + 198 = 716$
Thus, the dimensions are **716** ft by **518** ft.

17. $C = \pi d = 7\pi \approx (7)(3.14) \approx$ **22.0 m**.

19. $C = 2\pi r = 2\pi(10) = 20\pi \approx (20)(3.14) \approx$ **62.8 ft**.

21. $C = \pi d = 3\pi \approx (3)(3.14) \approx$ **9.42 cm**.

23. $C = 2\pi r = 2\pi(4.5) = 9\pi \approx (9)(3.14) \approx$ **28.3 yd**.

25. $C = \pi d = 61\pi \approx (61)(3.14) \approx$ **192 cm.**

27. $C = 2\pi r = 2\pi(8) = 16\pi \approx (16)(3.14) \approx$ **50.2 cm.**

29. $C = 2\pi r = 2\pi(4) = 8\pi \approx (8)(3.14) \approx$ **25.1 ft.**

31. $\pi d = 15\pi$, so d = **15 cm** and r = **7.5 cm.**

33. $\pi d = 7$, so $d = \dfrac{7}{\pi} \approx \dfrac{7}{3.14} \approx$ **2.23 cm.**

35. $\pi d = 4\dfrac{1}{8}$, so $d = \dfrac{4.125}{\pi} \approx$ **1.31 in.**

37. The answer to Problem 36 is approximately 776 revolutions. Thus, 20,000 mi would be approximately (20,000)(776) or **15,500,000** (rounded from 15,520,000) revolutions.

39. The diameter of the pool is 20 - 4 = 16 yd. Thus, the circumference of the pool is $\pi d = 16\pi \approx (16)(3.14) \approx$ **50.2 yd.**

41. A bicycle tire.

43. The worn tire has a smaller circumference, so will turn more times per mile.

45. This is one fourth of the large circumference: $\dfrac{3}{4}$ in.

47. Point b travels $\dfrac{3}{4}$ in.

SECTION 7.4 Area Measure and Pythagorean Theorem

EXERCISE 7.4

STUDY TIPS After studying **perimeters** (measured in **linear** units), the next logical step is to study **areas**, measured in **square** units. (See the Table on page 386.) Familiarize yourself with the formulas for the area of a rectangle, a parallelogram (same as a rectangle!), a triangle and a circle. Remember to write all these formulas on 3 by 5 cards and practice by reading the examples and doing the problems. We end the section with a theorem that applies to right triangles, the Pythagorean theorem. In essence, if you know the length of two of the sides of a right triangle, you can always find the length of the other side by using this theorem. Keep in mind that the longest side of a right triangle has a special name, the **hypotenuse.**

1. $A = \frac{1}{2}bh = \frac{1}{2}(5)(6) =$ **15 in.^{2}**   3. $A = bh = (5)(3) =$ **15 cm.^{2}**

5. The required area can be calculated as the area of the outside rectangle minus the area of the triangle to be taken out. Thus,
$$A = 10 \times 3 - \frac{1}{2} \times 2 \times 6 = 30 - 6 = \mathbf{24\ ft^2}$$

7. This area can be calculated as the sum of the area of the rectangle and the triangle on the left end. Thus,
$$A = 6 \times 4 + \frac{1}{2}(4 \times 3) = 24 + 6 = \mathbf{30\ ft^2}$$

9. This area can be calculated as the sum of the area of the rectangle and the semicircle on the left end. Thus,
$$A = 40 \times 20 + \frac{1}{2}\pi(10^2) \approx 800 + (50)(3.14) \approx \mathbf{957\ cm.^2}$$

11. This area can be calculated as the area of the square plus the area of the four semicircles. Thus,
$$A = 8^2 + 2\pi(4^2) = 64 + 32\pi \approx 64 + (32)(3.14) \approx \mathbf{164\ cm^2}$$

13. The shaded area is the area of the rectangle minus the area of the inscribed semicircle. Thus,
$$A = 6 \times 3 - \frac{1}{2}\pi(3^2) \approx 18 - (3.14)(\tfrac{9}{2}) \approx \mathbf{3.87\ cm^2}$$

15. The shaded area is the area of the large semicircle minus the area of the two smaller semicircles. Thus,

    $A = \dfrac{1}{2}\pi (5^2) - \pi(\dfrac{5}{2})^2 = \dfrac{25}{2}\pi - \dfrac{25}{4}\pi = \dfrac{25}{4}\pi \approx (6.25)(3.14) \approx$ **19.6 ft²**

17. The diagonal is the hypotenuse of a right triangle with two adjacent sides of the rectangle as the sides of the triangle. Thus, if x ft is the length of the unknown side of the rectangle, then $d^2 = x^2 + 15^2$, that is, $17^2 = x^2 + 15^2$. This equation gives $x^2 = 17^2 - 15^2 = 289 - 225 = 64$, so that x = 8, so the other side is **8 ft** long.

19. Since the triangle is isosceles, the perpendicular from the base to the opposite vertex divides the triangle into two congruent right triangles with hypotenuse 17 in. long and base side 8 in. long. Thus, we have $h^2 + 8^2 = 17^2$ or $h^2 = 17^2 - 8^2 = 289 - 64 = 225$. Thus, h = 15 so the height of the triangle is **15 in.**

21. If the length of the wall is x ft, then the right triangle shows that $x^2 + 120^2 = 130^2$, so that $x^2 = 130^2 - 120^2 = 16{,}900 - 14{,}400 = 2{,}500$. Thus, x = 50. Therefore, the cost is (50)($12) = **$600.**

23. In the figure, H is the foot of the perpendicular from the vertex D to the base AB. Let x ft be the distance from H to B. The diagonal BD is the hypotenuse of a right triangle with sides HB and HD. Thus, we 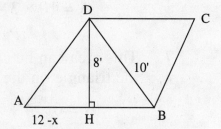 have the equation $x^2 + 8^2 = 10^2$ or $x^2 + 64 = 100$. This gives $x^2 = 36$, so x = 6, and 12 - x = 6. Thus, the point H is the midpoint of the base AB, so triangle AHD is congruent to triangle BHD. Therefore, AD = BD. This shows that the length of AD and of BC is 10 ft. The length of DC is the length of AB, **12 ft.**

25. $A = 120 \times 53\dfrac{1}{3} = 120 \times \dfrac{160}{3} = 40 \times 160 = 6400$. Thus, **6400 yd²** of turf are needed.

27. Let w ft be the width. Then, 16w = 288, so that $w = \dfrac{288}{16} = 18$. The room must be **18 ft** wide.

SECTION 7.4 Area Measure and Pythagorean Theorem

29. Let w be the width. Then, 70w = 6,720, so that $w = \frac{6720}{70} = 96$. The screen is **96 ft** wide.

31. $A = \pi r^2 = \pi 4^2 = \mathbf{16\pi \text{ cm}^2}$

33. Since the cost of the glass is $4.50 at $3/ft^2, the area of the glass is 4.50 ÷ 3 = 1.5 ft^2. Let x ft be the width of the frame, so that 1.5x ft is the length. The area of the glass in terms of x is (1.5x)(x) = 1.5x^2. Hence, we have the equation
$$1.5x^2 = 1.5$$
Thus, $\qquad x^2 = 1$ and $x = 1$.

Thus, the dimensions are 1 ft by 1.5 ft, that is, **12 in. by 18 in.**

35. Let x ft be the amount to be cut from the length and the width. Then the new dimensions will be (4 - x) ft and (5 - x) ft; the final area will be (4 - x:)(5 - x) ft^2. The original area, 20 ft^2, is to be reduced by $4\frac{1}{4}$ ft^2, so the final area is $15\frac{3}{4}$ or $\frac{63}{4}$ ft^2. Thus, we have the equation

$$(4-x)(5-x) = \frac{63}{4}$$

or $\qquad\qquad\qquad x^2 - 9x + 20 = \frac{63}{4}$

Multiply by 4: $\qquad\quad 4x^2 - 36x + 80 = 63$
Subtract 63: $\qquad\qquad 4x^2 - 36x + 17 = 0$
Factor the left side: $\quad (2x - 17)(2x - 1) = 0$

The first factor gives 2x - 17 = 0 or x = 8.5, which is impossible. The second factor gives x = 1/2, which means John would cut 1/2 ft off the length and the width, giving the new dimensions $3\frac{1}{2}$ **ft by** $4\frac{1}{2}$ **ft.** (To check this, note that the new area is $\frac{7}{2} \times \frac{9}{2} = \frac{63}{4}$ ft^2.)

37. Let triangle ABC be an equilateral triangle of side s. Let D be the midpoint of the base AB. Draw the line CD. Triangles ADC and BDC are congruent (by the SSS statement), so ∠ADC = ∠BDC (corresponding angles of congruent triangles). Since the sum of the two angles is 180°, the angles are right angles. Thus, triangle ADC is a right triangle. Now, let the length of DC be h. The length of AC is s, and the length of AD is s/2. By the Pythagorean theorem,

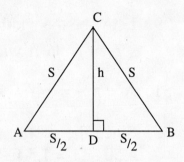

$$h^2 + (s/2)^2 = s^2$$

so that

$$h^2 + \frac{s^2}{4} = s^2$$

and

$$h^2 = \frac{3s^2}{4}$$

Thus, $h = \frac{s\sqrt{3}}{2}$ and the area of triangle ABC is

$$A = \frac{1}{2}bh = \frac{1}{2}s\left(\frac{s\sqrt{3}}{2}\right) = \frac{s^2\sqrt{3}}{4}$$

39. (a) The original area is $A = bh$. The new area is $A_1 = (2b)(2h) = 4bh$, so the **original area is multiplied by 4.**
    (b) The new area is $A_2 = (3b)(3h) = 9bh$, so the **original area is multiplied by 9.**
    (c) The new area is $A_3 = (kb)(kh) = k^2bh$, so the **original area is multiplied by k^2.**

41. For the circle, $C = 2\pi r$ and $A = \pi r^2$. Thus, we have $2\pi r = 20$, so that $r = \frac{10}{\pi}$ and $A = \pi\left(\frac{10}{\pi}\right)^2 = \frac{100}{\pi} \approx 31.8$ cm². For the square, $P = 4s$ and $A = s^2$. Thus, we have $4s = 20$, so that $s = 5$ and $A = 25$. Consequently, the area of the circle is larger by about **6.8 cm².**

SECTION 7.4 Area Measure and Pythagorean Theorem 199

43. As the diagram shows, the side of the octagon is of length a - 2s, and this is the length of the hypotenuse of each of the four right triangles cut from the square. Hence, $s^2 + s^2 = (a - 2s)^2$ or $2s^2 = (a - 2s)^2$ Taking positive square roots, we find that

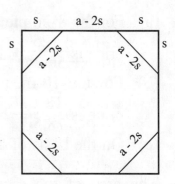

$$\sqrt{2}\, s = a - 2s$$

or $2s + \sqrt{2}\, s = a$ (Add 2s to both sides.)
Thus, $(2 + \sqrt{2})s = a$ (Factor the left side.)
$(2 - \sqrt{2})(2 + \sqrt{2})s = a(2 - \sqrt{2})$ (Multiply both sides by $2 - \sqrt{2}$.)
$2s = a(2 - \sqrt{2})$ (Simplify the left side.)
$s = \dfrac{2 - \sqrt{2}}{2} a$ (Divide both sides by 2.)

45. (a) **in.² and cm²**   (b) **mi² and km²**   (c) and (d) **yd² and m²**

47. The perimeter of the first room is 20 + 24 = 44 ft, of the second room is 28 + 30 = 58 ft, and of the third room is 24 + 24 = 48 ft. Hence, the sum of the perimeters is 44 + 58 + 48 = 150 ft. The ceiling is 8 ft high, so the total wall area to be painted is 8 × 150 or 1200 ft².
(a) Since a gallon of paint covers 450 ft², the number of gallons needed is $\dfrac{1200}{450} = \dfrac{8}{3} = 2\dfrac{2}{3}$.
(b) The paint is sold only by the gallon, so 3 gallons will have to be bought at a cost of 3 × 14 = **$42.**

49. The entire lot area is 100 × 200 = 20,000 ft². The area occupied by the shed is 10 × 10 = 100 ft²; the area occupied by the house is 30 × 50 = 1500 ft²; the area occupied by the drive is 10 × 50 = 500 ft². Thus, the lawn area is
20,000 - 100 - 1500 - 500 = **17,900 ft²**
(a) A bag of fertilizer covers 1200 ft², so it will take
$\dfrac{17900}{1200} = 14.9+$ bags
Thus, **15** bags are needed.
(b) The cost will be 4 × 15 = **$60.**

51. (a) For the 8-in. pie, $A = \pi r^2 \approx (3.14)(4^2) = 50.24$ in.^{2} Thus, the unit price is $\frac{125}{50.24} = $ **2.49 cents/in.^{2}**.

(b) For the 10-in. pie, $A \approx (3.14)(5^2) = 78.5$ in.^{2} Thus, the unit price is $\frac{185}{78.5} = $ **2.36 cents/in.^{2}**.

(c) On the basis of the unit price, the **10-in. pie** is the better buy.

53. The area of the rectangle taken away is Wx, and the area of the rectangle added on is hy. Since

$$\frac{y}{W} = \frac{x}{h}, \text{ it follows that hy = Wx}$$

Thus, the area of the new rectangle is equal to the area of the original rectangle.

EXERCISE 7.5

STUDY TIPS After the study of perimeter (*linear* units), area (*square* units) comes the study of **volume** (measured in **cubic** units. (See the table on page 597.) We find the volumes of three dimensional objects such as cubes and cylinders. The formulas are given in Table 7.11. As before, memorize these formulas and practice with them. One last word of warning: do not confuse *surface area* with *volume*. The *surface area* of a shoe box is, roughly, the area of the paper it will take to cover the box. The volume is the amount (volume) of sand it would take to fill the box. **Surface area** is measured in **square units**, **volume** in **cubic** units.

1. (a) A, B, C, D, E   (b) AB, AC, AD, AE, BC, BE, CD, CE, DE

3. ABCD   5. 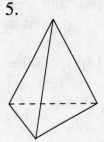   7.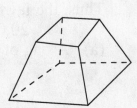

SECTION 7.5 Volume and Surface Area

9. (a) If the original edge is of length a, then $V = a^3$. The new edge is of length 2a, so the volume is $(2a)^3 = 8a^3$. Thus, the **volume is multiplied by 8.**

   (b) If the length of the edge is tripled, the new volume is $(3a)^3 = 27a^3$. Thus, the **volume is multiplied by 27.**

11. (a) $V = lwh = 20 \times 10 \times 8 =$ **1600 in.^{3}**

    (b) $S = 2lw + 2lh + 2wh = 2 \times 20 \times 10 + 2 \times 20 \times 8 + 2 \times 10 \times 8 =$ **880 in.^{2}**

13. (a) $V = (6x)^3 + \frac{1}{3}(6x)^2(4x) = 216x^3 + 48x^3 =$ **264x^3**

    (b) Since the height of the pyramid is 4x and the distance from the center of the base to the center of an edge of the base is 3x, the slant height of the pyramid is 5x. (Use the Pythagorean theorem.) Thus, the external surface area of the pyramidal part of the solid is $(4)(\frac{1}{2})(6x)(5x) = 60x^2$. The external surface area of the cubical part is $(5)(6x)^2 = 180x^2$. Consequently, the total external surface area of the solid is the sum $60x^2 + 180x^2 =$ **240x^2**.

15. $V = Bh = 10 \times 5 =$ **50 in.^{3}**   17. $V = \frac{1}{3} Bh = \frac{1}{3}(4^2)(6) =$ **32 in.^{3}**

19. The volume of the cube is $V = 10^3 = 1000$ cm^3; the volume of the pyramid is $V_1 = \frac{1}{3}Bh = \frac{1}{3}(10^2)(15) = 500$ cm^3. Thus, the total volume is 1500 cm^3, which is **1.5 liters.**

21. (a) $V = \pi r^2 h = \pi(5^2)(9) = 225\pi \approx (225)(3.14) \approx$ **707 in.^{3}**
    $S = 2\pi rh + 2\pi r^2 = 2\pi(5)(9) + 2\pi 5^2 = 140\pi \approx (140)(3.14) \approx$ **440 in.^{2}**

    (b) $V = \frac{1}{3}\pi r^2 h = \frac{1}{3}\pi(5^2)(9) = 75\pi \approx (75)(3.14) \approx$ **236 in.^{3}**
    $S = \pi r^2 + \pi rs = \pi(5^2) + \pi(5)\sqrt{106} = (25 + 5\sqrt{106})\pi \approx (76.5)(3.14) \approx$ **240 in.^{2}**

    (Use the Pythagorean theorem to find the slant height. $s^2 = 5^2 + 9^2$.)

23. (a) $V = \pi r^2 h = \pi(3^2)(4) = 36\pi \approx (36)(3.14) \approx$ **113 ft^3**
    $S = 2\pi rh + 2\pi r^2 = 2\pi(3)(4) + 2\pi(3^2) = 42\pi \approx (42)(3.14) \approx$ **132 ft^2**

    (b) $V = \frac{1}{3}\pi r^2 h = \frac{1}{3}\pi(3^2)(4) = 12\pi \approx (12)(3.14) \approx$ **37.7 ft^3**
    $S = \pi r^2 + \pi rs = \pi(3^2) + \pi(3)(5) = 24\pi \approx (24)(3.14) \approx$ **75.4 ft^2**

202 CHAPTER 7 GEOMETRY

25. $V = \frac{4}{3}\pi r^3 = \frac{4}{3}\pi(6^3) = 288\pi \approx (288)(3.14) \approx$ **904 in.^{3}**

27. $V = \pi r^2 h = \pi(58^2)(754) = 2{,}536{,}456\pi \approx (2{,}536{,}456)(3.14) \approx$ **7,960,000 ft^3**

29. $V = \frac{1}{3}\pi r^2 h = \frac{1}{3}\pi(16^2)(12) = 1024\pi \approx (1024)(3.14) \approx$ **3220 m^3**

31. $r = (1.5)(2.54)$ cm and $h = (4)(2.54)$ cm, so
$V = \pi r^2 h = \pi[(1.5)(2.54)]^2(4)(2.54) = \pi(9)(2.54)^3 \approx (3.14)(147.5)$
≈ 463 cm^3
Thus, the can will hold about **463 gm**.

33. (a) Baseball: $r = \frac{9}{2\pi}$, so $S_1 = 4\pi r^2 = 4\pi(\frac{9}{2\pi})^2 = \frac{81}{\pi} \approx$ **25.8 in.^{2}**

and $V_1 = \frac{4}{3}\pi r^3 = \frac{4}{3}\pi\left(\frac{9}{2\pi}\right)^3 = \frac{243}{2\pi^2} \approx$ **12.3 in.^{3}**

Soccer Ball: $r = \frac{27}{2\pi}$, so $S_2 = 4\pi r^2 = 4\pi\left(\frac{27}{2\pi}\right)^2 = \frac{729}{\pi} \approx$ **232 in.^{2}**

and $V_2 = \frac{4}{3}\pi r^3 = \frac{4}{3}\pi\left(\frac{27}{2\pi}\right)^3 = \frac{6561}{2\pi^2} \approx$ **333 in.^{3}**

Basketball: $r = \frac{30}{2\pi}$, so $S_3 = 4\pi r^2 = 4\pi\left(\frac{30}{2\pi}\right)^2 = \frac{900}{\pi} \approx$ **287 in.^{2}**

and $V_3 = \frac{4}{3}\pi r^3 = \frac{4}{3}\pi\left(\frac{30}{2\pi}\right)^3 = \frac{4500}{\pi^2} \approx$ **456 in.^{3}**

(b) $\frac{S_1}{S_3} = \frac{81/\pi}{900/\pi} = \frac{9}{100}$ (c) $\frac{V_1}{V_2} = \frac{243/(2\pi^2)}{6561/(2\pi^2)} = \frac{1}{27}$

35. (a) $V = \frac{4}{3}\pi r^3 = \frac{4}{3}\pi\left(\frac{27}{2\pi}\right)^3 = \frac{6561}{2\pi^2} \approx$ **333 in.^{3}**

(b) The volumes drilled out are approximately cylinders of radii 5/8 and 1/2 in., respectively, and height 5/2 in. This volume is thus
$\pi(5/8)^2(5/2) + \pi(1/2)^2(5/2) + \pi(1/2)^2(5/2) = (\frac{125}{128} + \frac{5}{8} + \frac{5}{8})\pi = \frac{285}{128}\pi$
in.^{3} or about 6.99 in.^{3}. Thus, the volume remaining is about 326 in.^{3} so that the final weight is about $\frac{326}{333}(16) \approx$ **15.66 lb**.

SECTION 7.5 Volume and Surface Area

37. (a) $V = Bh = (7 \times 12) \times 17 = \mathbf{1428}$ **in.**$^\mathbf{3}$
    (b) S = area of base + area of four sides
       $= 7 \times 12 + 2(7 \times 17) + 2(12 \times 17) = \mathbf{730}$ **in.**$^\mathbf{2}$
    (c) Two tons of recycled paper are needed. This means that
       $\dfrac{2 \times 2000 \times 16}{2} = \mathbf{32{,}000}$ bags have to be recycled to save 34 trees.

39. Answers will vary.

41. No. This is true only if the radius is more than three units long. (Compare $\dfrac{4}{3}\pi r^3$ and $4\pi r^2$.)

43. Since the sum of the face angles at a vertex must be less than 360°, three, four or five equilateral triangles, three squares or three pentagons can be put together at any vertex. There are no other possibilities, so only the five regular polyhedrons listed are possible.

45.
Figure	F	V	E		
7.59B	6	8	12	F + V = 14	= E + 2
7.59C	6	5	9	F + V = 11	= E + 2
7.59D	7	10	15	F + V = 17	= E + 2
7.60	5	6	9	F + V = 11	= E + 2
7.61	6	6	10	F + V = 12	= E + 2

Thus, Euler's formula is **F + V = E + 2.**

EXERCISE 7.6

STUDY TIPS Have you ever tried to trace or redraw a figure without lifting your pencil and without tracing any line twice? Some figures are impossible to trace this way! In geometry, we say these figures are **not traversable.** If you study the traversability rules on page 406, you will know which figures can be traversed and which cannot. The key is to know whether the vertices (the endpoints of the arcs forming the figure) are odd or even.

1. (a) 3   (b) 0   (c) This network is traversable; all three vertices are possible starting points.

3. (a) 3 (B, D, and E)   (b) 2 (A and C)
   (c) Traversable; start at either A or C.

5. (a) 1 (A only)   (b) 4 (B, C, D, E)
   (c) Not traversable; it has more than 2 odd vertices.

7. (a) 5 (A, C, D, E, G)   (b) 2 (B, F)
   (c) Traversable; start at B or F.

9. (a) 1 (The vertex of the pyramid)   (b) 4 (The vertices of the base)
   (c) Not traversable; it has more than 2 odd vertices.

11. Think of each region as a vertex with the individual line segments in its boundary as the number of paths to the vertex. The boundary of region A has four segments, so the corresponding network point would be even. The boundary of region B has five segments, so the corresponding network point would be odd. The boundary of region C has four segments, so the corresponding network point would be even. The boundary of region D has five segments, so the corresponding network point would be odd. The boundary of region E has 10 segments, so the corresponding network point would be even. Thus, the network would have two odd vertices (B and D). By starting in region B or D, it is possible to draw a simple connected broken line that crosses each line segment exactly once.

SECTION 7.6 Non-Euclidean Geometry, Topology, Networks   205

13. Region A has three doorways, so the corresponding vertex would be odd. Regions B, C, D and E each has two doorways, so the corresponding vertices would be even. Region F has three doorways, so the corresponding vertex would be odd. Thus, the network would have two odd vertices, A and F. By starting in A and ending in F (or vice versa), it is possible to take a walk and pass through each doorway exactly once. It is not possible to start and end outside, because the walk must start in one of the regions corresponding to one odd vertex and end in the region corresponding to the other odd vertex.

15. A and D each has three doorways and all the other rooms have an even number of doorways. Thus, the corresponding network would have two odd vertices, A and D. By starting in either A or D and ending in the other, it is possible for a walk to pass through each doorway exactly once. It is not possible to start and end outside.

17. All the rooms and the outside have an even number of doorways, so the corresponding network would have no odd vertices. The walk can start in any room or outside and end in the same place and pass through each doorway exactly once.

19. Rooms B and D each has three doorways and all the other rooms have an even number of doorways, so the corresponding network would have two odd vertices. The walk can start in either B or D and end in the other, passing through each doorway exactly once. It is not possible to start and end outside.

21. Room A and the outside D each has three doorways and B and C each has an even number of doorways, so the corresponding network would have two odd vertices. By starting in A and ending in D (or vice versa), a walk could pass through each doorway exactly once. It is not possible to start and end outside.

23. Given a line and any point not on that line, there is one and only one line through that point and parallel to the given line.

25. Given a line and any point not on that line, there is no line through that point that is parallel to the given line.

27. In hyperbolic geometry   29. The surface of a rectangular box

31. The surface of a sphere

206 CHAPTER 7 GEOMETRY

33. (a), (c) and (f) are topologically equivalent and
    (b), (d) and (e) are topologically equivalent.

35. (a), (c) and (f) are of genus 1.
    (b), (d) and (e) are of genus 2.

37. Topo is correct. If you cut through a loop of the left-hand figure, you can unwind it into a single strip as you can with the circular cord.

39. Since each arc has two endpoints, the total number of endpoints must be even. An odd vertex accounts for an odd number of endpoints, while an even vertex accounts for an even number of endpoints. Thus, there must be an even number of odd vertices.

41. V: 3, R: 2, A: 3   43. V: 5, R: 2, A: 5   45. V: 2, R: 3, A: 3

47. V: 6, R: 3, A: 7   Nos. 41-48 all fit Euler's formula, $V + R = E + 2$.

EXERCISE 7.7

STUDY TIPS If you have seen or read *Jurassic Park* you are ahead of the game when studying this section. What happened in the book (and the movie) can be characterized by a controlled situation that developed into **chaos**, random happenings that do not yield easily to scientific study. This is one of the subjects in this section. A calculator will help in doing the problems there! Section B is devoted to a new type of geometry called **fractal geometry.** At this level, we are content with having you enjoy the drawing that result in this geometry.

1. | x | 0.2 | 0.862 | 0.850 | 0.865 | 0.846 | 0.868 | 0.841 | 0.871 |
    |---|-----|-------|-------|-------|-------|-------|-------|-------|
    | y | 0.56 | 0.416 | 0.446 | 0.409 | 0.456 | 0.401 | 0.468 | 0.393 |

x	0.835	0.874	0.829	**0.875**	**0.827**	**0.875**	**0.827**
y	0.482	0.385	0.496	**0.383**	**0.501**	**0.383**	**0.501**

    There are four attractors: **0.383, 0.501, 0.827 and 0.875.**

3.
x	0.4	0.605	0.609	0.613	0.615	0.618	0.620	0.621	0.623
y	0.72	0.717	0.714	0.712	0.710	0.708	0.707	0.706	0.705

x	0.624	0.625	0.626	0.628	0.629	0.630	**0.631**	**0.631**
y	0.704	0.703	0.702	0.701	0.700	**0.699**	**0.699**	**0.699**

There are two attractors: **0.631 and 0.699.**

5. Take s as the length of one side of the original triangle.
   (a) 4s   (b) $\frac{16}{3}s$

7. On completing the second stage, the fractal is as shown

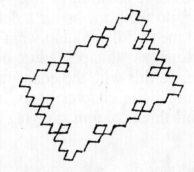

9. $\frac{1}{1024}$ sq. in.

11. On completing the second stage, the fractal is as shown

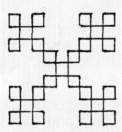

13. $\frac{25}{81}s^2$

15. The remaining area approaches zero.

17. $\frac{\sqrt{3}}{6}s^2$

19. The perimeter increases without limits.

208 CHAPTER 7 GEOMETRY

PRACTICE TEST 7

STUDY TIPS Hopefully, you have been taking notes both during your individual study and in class. This chapter contains so many concepts that we shall offer a slightly different test-preparation technique. Get a ruled writing tablet or pad with a one inch margin (usually it has two red, parallel, vertical lines for margin. By the way, see how geometry terms come up in everyday life?) Now, divide the writing area into two columns. To the left of the red line, write the concept, key idea or word you are studying. In the middle column, write the definition, theorem, procedure or formula that corresponds to the idea. If it is a definition, think of examples that illustrate it and write them in the third column. If it is a theorem, do the same as for a definition. If it is a procedure or a formula, write in the middle column what the procedure or formula does. What kind of problems use this procedure or formula? Write examples. To get some practice, do this with the concepts of line, angle, the Pythagorean Theorem. Try reading the chapter again and jotting down the important concepts using this three column studying technique.

1. (a) $\overline{XY}$ (b) Point Y (c) $\overrightarrow{WZ}$

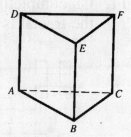

2. (a) $\overleftrightarrow{AB}$ and $\overleftrightarrow{DE}$; $\overleftrightarrow{BC}$ and $\overleftrightarrow{EF}$;

   $\overleftrightarrow{AC}$ and $\overleftrightarrow{DF}$, $\overleftrightarrow{AD}$, $\overleftrightarrow{BE}$ and $\overleftrightarrow{CF}$

   (b) $\overleftrightarrow{AB}$ and $\overleftrightarrow{CF}$; $\overleftrightarrow{AB}$ and $\overleftrightarrow{DF}$; $\overleftrightarrow{AB}$ and $\overleftrightarrow{EF}$; $\overleftrightarrow{BC}$ and $\overleftrightarrow{AD}$

   $\overleftrightarrow{BC}$ and $\overleftrightarrow{DE}$; $\overleftrightarrow{BC}$ and $\overleftrightarrow{DF}$; $\overleftrightarrow{AC}$ and $\overleftrightarrow{BE}$; $\overleftrightarrow{AC}$ and $\overleftrightarrow{DE}$;

   $\overleftrightarrow{AC}$ and $\overleftrightarrow{EF}$; $\overleftrightarrow{DE}$ and $\overleftrightarrow{CF}$; $\overleftrightarrow{EF}$ and $\overleftrightarrow{AD}$; $\overleftrightarrow{DF}$ and $\overleftrightarrow{BE}$.

   (c) $\overleftrightarrow{AD}$, $\overleftrightarrow{AB}$ and $\overleftrightarrow{AC}$; $\overleftrightarrow{AD}$, $\overleftrightarrow{DE}$ and $\overleftrightarrow{DF}$; $\overleftrightarrow{BE}$, $\overleftrightarrow{AB}$ and $\overleftrightarrow{BC}$; $\overleftrightarrow{BE}$, $\overleftrightarrow{EF}$

   and $\overleftrightarrow{ED}$; $\overleftrightarrow{CF}$, $\overleftrightarrow{AC}$ and $\overleftrightarrow{BC}$; $\overleftrightarrow{CF}$, $\overleftrightarrow{EF}$ and $\overleftrightarrow{DF}$.

PRACTICE TEST 7   209

3. (a) $\frac{4}{12} \times 360° = \mathbf{120°}$ (b) $\frac{2}{12} \times 360° = \mathbf{60°}$

4. Let $m\angle B = x°$. Then $m\angle A = 3x°$. Since the two angles are supplementary, $x + 3x: = 180$, that is, $4x = 180$. so that $x = 45$. Therefore, m $\angle A = \mathbf{135°}$ and m $\angle B = \mathbf{45°}$.

5. (a) $\angle C$ and the 50° angle are supplementary, so $m\angle C = \mathbf{130°}$.
   (b) $\angle E$ and the 50° angle are equal, so $m\angle E = \mathbf{50°}$.
   (c) $\angle D$ and $\angle C$ are equal, so m $\angle D = \mathbf{130°}$.

6. (a) m $\angle C = 180° - 38° - 43° = \mathbf{99°}$
   (b) Let m $\angle C = x°$, so that m $\angle A = m \angle B = 2x°$. Then,
   $$x + 2x + 2x = 180$$
   or $\qquad 5x = 180$ and $x = 36$.
   Thus, m $\angle A = m \angle B = \mathbf{72°}$ and m $\angle C = \mathbf{36°}$.

7. (a) 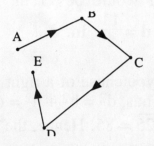   (b)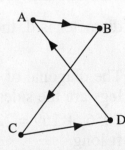

8.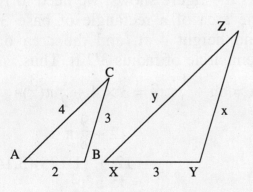

   Since the triangles are similar, the corresponding sides are proportional. Thus, we have (see the figure)

   $$\frac{x}{3} = \frac{y}{4} = \frac{3}{2}$$

   Therefore, $x = \frac{9}{2} = \mathbf{4\frac{1}{2}}$ in.

   and $\qquad y = \frac{12}{2} = \mathbf{6}$ in.

9. (a) The sum of the measures of the 9 angles is $(9 - 2)180°$, so the measure of one angle is $\frac{(9 - 2)180°}{9} = \mathbf{140°}$.
   (b) The sum of the measures of the 10 angles is $(10 - 2)180°$, so the measure of one angle is $\frac{(10 - 2)180°}{10} = \mathbf{144°}$.

10. If w is the width, then the length is 2w. Hence, the perimeter is 6w, and this is to be 120 yd. Thus, 6w = 120, so that w = 20. This means that the width is **20 yd** and the length is **40 yd.**

11. 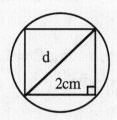 The figure shows that d is the hypotenuse of a right triangle whose legs are each 2 cm long. Thus, $d^2 = 2^2 + 2^2 = 8$, so $d = \sqrt{8} = 2\sqrt{2}$ cm, and the circumference of the circle is C= πd = **2√2π cm**

12. The area that is inside the circle and outside the square is $\pi r^2 - s^2$, where r is the radius of the circle and s is the side of the square. Thus, this area is $\pi(\sqrt{2})^2 - 2^2 =$ **(2π - 4) cm²**.

13. The circumference of the cylinder would be 11 in. Hence, if the diameter is d, then πd = 11, so that $\mathbf{d = \dfrac{11}{\pi}}$ **in.**

14. The diagonal of a rectangle is the hypotenuse of a right triangle whose legs are the sides of the rectangle. Thus, $d^2 = b^2 + h^2 = (84)^2 + (13)^2 = 7056 + 169 = 7225$ so that $d = \sqrt{7225} = 85$. Hence, the diagonal is **85** ft long.

15. 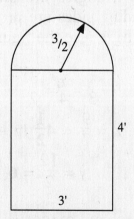 As the figure shows, we need to find the area of a rectangle of base 3 ft and height 4 ft, and the area of a semicircle of radius 3/2 ft. Thus,

$$A = bh + \frac{1}{2}\pi r^2 = 3 \times 4 + \frac{1}{2}\pi(\frac{3}{2})^2$$
$$= 12 + \frac{9}{8}\pi$$
$$\approx 12 + (1.125)(3.14)$$
$$\approx \mathbf{15.5 ft^2}$$

16. The area of the square is s^2, where s is the length of the side, and the area of the circle is πr^2, where r = s/2. Thus, the area inside the square and outside the circle is
$A = 2^2 - \pi(1^2) = 4 - \pi \approx 4 - 3.14 =$ **0.86 in.²**

17. S = 2(lw + lh + wh) = 2(5 × 3 + 5 × 2 + 3 × 2) = **62 ft²**

18. The volume of the hemisphere is $\frac{1}{2}(\frac{4}{3}\pi r^3) = \frac{2}{3}\pi(2^3) = \frac{16}{3}\pi$ in.³r The volume of the cone is $\frac{1}{3}\pi r^2 h = \frac{1}{3}\pi(2^2)h = \frac{4}{3}\pi h$ in.³, where h is the number of inches in the height of the cone Since the volumes are equal, we have the equation
$$\frac{4}{3}\pi h = \frac{16}{3}\pi$$
Dividing both sides by $\frac{4}{3}\pi$, we get h = 4. Thus, the cone is **4** in. high.

19. The surface area of the sphere is $4\pi r^2$ and of the cylinder is $2\pi rh + 2\pi r^2$. Since these areas are to be equal, we have
$$2\pi rh + 2\pi r^2 = 4\pi r^2$$
or $\qquad\qquad 2\pi rh = 2\pi r^2$ (Subtract $2\pi r^2$ from both sides.)
By dividing both sides by $2\pi r$, we find that
$$h = r$$
so the height of the cylinder is equal to the given radius, **10 in.**

20. The volume of a pyramid is given by the formula
$$V = \frac{1}{3}BH$$
where B is the area of the base and H is the height of the pyramid. The given triangle is the base of the pyramid, and its area is $\frac{1}{2}$bh, where b is the length of the base and h is the height of the triangle. The figure shows that b = 5 ft and h = 3 ft, so the area of the triangle is $\frac{1}{2}(5)(3) = \frac{15}{2}$ ft². The volume of the pyramid is $V = \frac{1}{3}(\frac{15}{2})(4) =$ **10 ft³**.

21. (a) The figure shows a traversable network; it has no odd vertices.

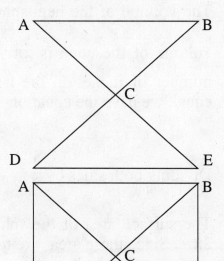

(b) The figure shows a nontraversable network; it has four odd vertices: A, B, D, and E.

22. Given a line and any point not on that line, there is one and only one line through that point that is parallel to the given line.

23. (a) The button. It can be cut four times and still be in one piece. This cannot be done with any of the other objects.
    (b) Zero (No holes)
    (c) Zero (No holes)
    (d) Four (Four holes)

24. Step 1. $y = 2(0.1)((0.1 - 1) = (0.2)(-0.9) = -0.18$
    Step 2. $y = 2(-0.18)(-0.18 - 1) = (-0.36)(-1.18) = 0.425$
    Step 3. $y = 2(0.425)(0.425 - 1) = (0.850)(-0.575) = -0.489$
    Step 4 $y = 2(-0.489)(-0.489 - 1) = (-0.978)(-1.489) = 1.456$

25. The sketch is shown

CHAPTER 8
MATHEMATICAL SYSTEMS AND MATRICES

EXERCISE 8.1

STUDY TIPS Make sure you study Definition 8.1 and keep in mind how the elements of a matrix are numbered. For example, a_{32} denotes the element in the **3rd** row and **2nd** column. Remember that two matrices are **equal** if and only if they have the **same dimensions** and **corresponding elements are equal**. Part A deals with some simple operations on matrices. Addition or subtraction of matrices can be performed only for matrices of the same dimensions. To multiply a matrix by any number, you multiply each element by this number. Examples 1-3 illustrate these operations. Read Part B very carefully. This explains what matrices can be multiplied (if the first matrix is an **m × n** and the second is a **p × q**, **n = p** for multiplication to be possible and the product is an **m × q** matrix). At the bottom of page 435 we define the product of two matrices using row-column multiplication (RC for short). Understand Definitions 8.6 and 8.7 and their application in the examples. The identity for matrix multiplication is defined in Part C.

1. (a) To calculate 4A, multiply each element of A by 4.

    $$4A = \begin{bmatrix} 8 & 4 \\ 0 & -4 \end{bmatrix}$$

   (b) To calculate -3B, multiply each element of B by -3.

    $$-3B = \begin{bmatrix} 6 & -12 \\ -9 & -3 \end{bmatrix}$$

   (c) To calculate C - B, subtract each element of C from the corresponding element of B.

    $$C - B = \begin{bmatrix} 5 & 1 \\ -1 & -1 \end{bmatrix}$$

3. (a) To calculate A + B + C, add the corresponding elements of the three matrices.

$$A + B + C = \begin{bmatrix} 3 & 10 \\ 5 & 0 \end{bmatrix}$$

(b) To calculate A + B - C, add the elements of B to the corresponding elements of A, and subtract the corresponding elements of C.

$$A + B - C = \begin{bmatrix} -3 & 0 \\ 1 & 0 \end{bmatrix}$$

5. (a) To calculate 7A + 4B - 2C, add 4 times each element of B to 7 times the corresponding element of A, and subtract 2 times the corresponding element of C.

$$7A + 4B - 2C = \begin{bmatrix} 0 & 13 \\ 8 & -3 \end{bmatrix}$$

(b) To calculate 2A - 2B - 3C, subtract 2 times each element of B and 3 times each element of C from 2 times the corresponding element of A.

$$2A - 2B - 3C = \begin{bmatrix} -1 & -21 \\ -12 & -4 \end{bmatrix}$$

7. (a) To find A + B, add the corresponding elements of A and B

$$A + B = \begin{bmatrix} 0 & 1 & 3 \\ 7 & 3 & -3 \\ 4 & 3 & 0 \end{bmatrix}$$

(b) To find A - B, subtract the elements of B from the corresponding elements of A.

$$A - B = \begin{bmatrix} 2 & -3 & 1 \\ -1 & -3 & -1 \\ 4 & 1 & 2 \end{bmatrix}$$

(c) To find A - C, subtract the elements of C from the corresponding elements of A.

$$A - C = \begin{bmatrix} 1 & 0 & -1 \\ 2 & 2 & -6 \\ 1 & 5 & 1 \end{bmatrix}$$

9. (a) To find 2A + 3B, add 3 times the elements of B to 2 times the corresponding elements of A.

$$2A + 3B = \begin{bmatrix} -1 & 4 & 7 \\ 18 & 9 & -7 \\ 8 & 7 & -1 \end{bmatrix}$$

(b) To find -2A + 3B, subtract 2 times the elements of A from 3 times the corresponding elements of B.

$$-2A + 3B = \begin{bmatrix} -5 & 8 & -1 \\ 6 & 9 & 1 \\ -8 & -1 & -5 \end{bmatrix}$$

11. (a) To find 3A - 2C, subtract
    2 times the elements of C
    from 3 times the corre-
    sponding elements of A.

$$3A - 2C = \begin{bmatrix} 3 & -1 & 0 \\ 7 & 4 & -14 \\ 6 & 12 & 3 \end{bmatrix}$$

(b) To find B + C, add the
elements of C to the
corresponding elements of B.

$$B + C = \begin{bmatrix} -1 & 1 & 4 \\ 5 & 1 & 3 \\ 3 & -2 & -1 \end{bmatrix}$$

13. To find AB, do a "row-column" multiplication of A × B.

$$AB = \begin{bmatrix} 1\times3 + (-2)\times1 + 1\times2 & 1\times2 + (-2)\times1 + 1\times0 & 1\times0 + (-2)\times(-1) + 1\times1 \\ 2\times3 + 0\times1 + 2\times2 & 2\times2 + 0\times1 + 2\times0 & 2\times0 + 0\times(-1) + 2\times1 \\ (-1)\times3 + 1\times1 + 3\times2 & (-1)\times2 + 1\times1 + 3\times0 & (-1)\times0 + 1\times(-1) + 3\times1 \end{bmatrix}$$

$$= \begin{bmatrix} 3 & 0 & 3 \\ 10 & 4 & 2 \\ 4 & -1 & 2 \end{bmatrix}$$

15. To find BA, do a "row-column" multiplication of B × A.

$$BA = \begin{bmatrix} 3\times1 + 2\times2 + 0\times(-1) & 3\times(-2) + 2\times0 + 0\times1 & 3\times1 + 2\times2 + 0\times3 \\ 1\times1 + 1\times2 + (-1)\times(-1) & 1\times(-2) + 1\times0 + (-1)\times1 & 1\times1 + 1\times2 + (-1)\times3 \\ 2\times1 + 0\times2 + 1\times(-1) & 2\times(-2) + 0\times0 + 1\times1 & 2\times1 + 0\times2 + 1\times3 \end{bmatrix}$$

$$= \begin{bmatrix} 7 & -6 & 7 \\ 4 & -3 & 0 \\ 1 & -3 & 5 \end{bmatrix}$$

17. First find A - B and A + B.

$$A - B = \begin{bmatrix} -2 & -4 & 1 \\ 1 & -1 & 3 \\ -3 & 1 & 2 \end{bmatrix} \qquad A + B = \begin{bmatrix} 4 & 0 & 1 \\ 3 & 1 & 1 \\ 1 & 1 & 4 \end{bmatrix}$$

To find (A - B)(A + B), do a "row-column" multiplication of
(A - B) × (A + B). This gives the matrix

$$\begin{bmatrix} -2\times4 + (-4)\times3 + 1\times1 & -2\times0 + (-4)\times1 + 1\times1 & -2\times1 + (-4)\times1 + 1\times4 \\ 1\times4 + (-1)\times3 + 3\times1 & 1\times0 + (-1)\times1 + 3\times1 & 1\times1 + (-1)\times1 + 3\times4 \\ -3\times4 + 1\times3 + 2\times1 & -3\times0 + 1\times1 + 2\times1 & -3\times1 + 1\times1 + 2\times4 \end{bmatrix}$$

Thus,

$$(A - B)(A + B) = \begin{bmatrix} -19 & -3 & -2 \\ 4 & 2 & 12 \\ -7 & 3 & 6 \end{bmatrix}$$

19. First find A^2 and B^2 by using "row-column" multiplication.

$$A^2 = \begin{bmatrix} 1\times1 + (-2)\times2 + 1\times(-1) & 1\times(-2) + (-2)\times0 + 1\times1 & 1\times1 + (-2)\times2 + 1\times3 \\ 2\times1 + 0\times2 + 2\times(-1) & 2\times(-2) + 0\times0 + 2\times1 & 2\times1 + 0\times2 + 2\times3 \\ -1\times1 + 1\times2 + 3\times(-1) & -1\times(-2) + 1\times0 + 3\times1 & -1\times1 + 1\times2 + 3\times3 \end{bmatrix}$$

$$B^2 = \begin{bmatrix} 3\times3 + 2\times1 + 0\times2 & 3\times2 + 2\times1 + 0\times0 & 3\times0 + 2\times(-1) + 0\times1 \\ 1\times3 + 1\times1 + (-1)\times2 & 1\times2 + 1\times1 + (-1)\times0 & 1\times0 + 1\times(-1) + (-1)\times1 \\ 2\times3 + 0\times1 + 1\times2 & 2\times2 + 0\times1 + 1\times0 & 2\times0 + 0\times(-1) + 1\times1 \end{bmatrix}$$

Thus,

$$A^2 = \begin{bmatrix} -4 & -1 & 0 \\ 0 & -2 & 8 \\ -2 & 5 & 10 \end{bmatrix} \qquad B^2 = \begin{bmatrix} 11 & 8 & -2 \\ 2 & 3 & -2 \\ 8 & 4 & 1 \end{bmatrix}$$

Then find $A^2 - B^2$ by subtracting each element of B^2 from the corresponding element of A^2. The result is

$$A^2 - B^2 = \begin{bmatrix} -15 & -9 & 2 \\ -2 & -5 & 10 \\ -10 & 1 & 9 \end{bmatrix}$$

21. Use "row-column" multiplication to find I^2.

$$I^2 = \begin{bmatrix} 1\times1 + 0\times0 & 1\times0 + 0\times1 \\ 0\times1 + 1\times0 & 0\times0 + 1\times1 \end{bmatrix} = \begin{bmatrix} 1 & 0 \\ 0 & 1 \end{bmatrix} = I$$

23. Use "row-column" multiplication to find $A \times B$ and $B \times A$.

$$AB = \begin{bmatrix} 2\times1 + 1\times(-1) & 2\times(-1) + 1\times2 \\ 1\times1 + 1\times(-1) & 1\times(-1) + 1\times2 \end{bmatrix} = \begin{bmatrix} 1 & 0 \\ 0 & 1 \end{bmatrix} = I$$

$$BA = \begin{bmatrix} 1\times2 + (-1)\times1 & 1\times1 + (-1)\times1 \\ -1\times2 + 2\times1 & -1\times1 + 2\times1 \end{bmatrix} = \begin{bmatrix} 1 & 0 \\ 0 & 1 \end{bmatrix} = I$$

SECTION 8.1 Matrix Operations

25. Use "row-column" multiplication to find A × B and B × A.

$$AB = \begin{bmatrix} 2\times 3 + 5\times(-1) & 2\times(-5) + 5\times 2 \\ 1\times 3 + 3\times(-1) & 1\times(-5) + 3\times 2 \end{bmatrix} = \begin{bmatrix} 1 & 0 \\ 0 & 1 \end{bmatrix} = I$$

$$BA = \begin{bmatrix} 3\times 2 + (-5)\times 1 & 3\times 5 + (-5)\times 3 \\ -1\times 2 + 2\times 1 & -1\times 5 + 2\times 3 \end{bmatrix} = \begin{bmatrix} 1 & 0 \\ 0 & 1 \end{bmatrix} = I$$

27. Use "row-column" multiplication to find A × B and B × A.

$$AB = \begin{bmatrix} 1\times 5 + 0\times(-2) + 1\times(-4) & 1\times 1 + 0\times 0 + 1\times(-1) & 1\times(-2) + 0\times 1 + 1\times 2 \\ 0\times 5 + 2\times(-2) + (-1)\times(-4) & 0\times 1 + 2\times 0 + (-1)\times(-1) & 0\times(-2) + 2\times 1 + (-1)\times 2 \\ 2\times 5 + 1\times(-2) + 2\times(-4) & 2\times 1 + 1\times 0 + 2\times(-1) & 2\times(-2) + 1\times 1 + 2\times 2 \end{bmatrix}$$

$$BA = \begin{bmatrix} 5\times 1 + 1\times 0 + (-2)\times 2 & 5\times 0 + 1\times 2 + (-2)\times 1 & 5\times 1 + 1\times(-1) + (-2)\times 2 \\ -2\times 1 + 0\times 0 + 1\times 2 & -2\times 0 + 0\times 2 + 1\times 1 & -2\times 1 + 0\times(-1) + 1\times 2 \\ -4\times 1 + (-1)\times 0 + 2\times 2 & -4\times 0 + (-1)\times 2 + 2\times 1 & -4\times 1 + (-1)\times(-1) + 2\times 2 \end{bmatrix}$$

Simplifying these matrices, we find

$$AB = BA = \begin{bmatrix} 1 & 0 & 0 \\ 0 & 1 & 0 \\ 0 & 0 & 1 \end{bmatrix} = I$$

29. The matrix A is a non square matrix with m rows and n columns, and J is a square matrix. In order to be able to multiply these matrices in the order AJ, the matrix J must have n rows (as many rows as A has columns). Because J is a square matrix, it will have n columns. To be able to multiply in the order JA, the matrix A would have to have n rows (as many rows as J has columns). Since it is given that m is <u>not</u> equal to n, the matrices cannot be conformable for both orders of multiplication.

31. (a) "Row-column" multiplication gives

$$\begin{bmatrix} 1 & 2 \\ 0 & 0 \end{bmatrix} \times \begin{bmatrix} 1 & 2 \\ 0 & 0 \end{bmatrix} = \begin{bmatrix} 1\times 1 + 2\times 0 & 1\times 2 + 2\times 0 \\ 0\times 1 + 0\times 0 & 0\times 2 + 0\times 0 \end{bmatrix} = \begin{bmatrix} 1 & 2 \\ 0 & 0 \end{bmatrix}$$

which shows that the given matrix is idempotent.

31. (b) $\begin{bmatrix} \frac{1}{2} & \frac{1}{2} \\ \frac{1}{2} & \frac{1}{2} \end{bmatrix} \times \begin{bmatrix} \frac{1}{2} & \frac{1}{2} \\ \frac{1}{2} & \frac{1}{2} \end{bmatrix} = \begin{bmatrix} (\frac{1}{2}) \times (\frac{1}{2}) + (\frac{1}{2}) \times (\frac{1}{2}) & (\frac{1}{2}) \times (\frac{1}{2}) + (\frac{1}{2}) \times (\frac{1}{2}) \\ (\frac{1}{2}) \times (\frac{1}{2}) + (\frac{1}{2}) \times (\frac{1}{2}) & (\frac{1}{2}) \times (\frac{1}{2}) + (\frac{1}{2}) \times (\frac{1}{2}) \end{bmatrix}$

$= \begin{bmatrix} \frac{1}{2} & \frac{1}{2} \\ \frac{1}{2} & \frac{1}{2} \end{bmatrix}$

which shows that the given matrix is idempotent.

33. (a)
$\begin{array}{c} \text{Armchairs} \\ \text{Rockers} \end{array} \begin{bmatrix} \text{E} & \text{M} & \text{L} \\ 20 & 15 & 10 \\ 12 & 8 & 5 \end{bmatrix}$ This is the monthly production matrix

(b) To get the 6-months production figures, we multiply the matrix in part (a) by 6. This gives the matrix

$\begin{array}{c} \text{Armchairs} \\ \text{Rockers} \end{array} \begin{bmatrix} \text{E} & \text{M} & \text{L} \\ 120 & 90 & 60 \\ 72 & 48 & 30 \end{bmatrix}$

35. Let P be the production matrix in Problem 33(b), and let S be the matrix showing the sales for the 6 months. Then,

$P = \begin{bmatrix} 120 & 90 & 60 \\ 72 & 48 & 30 \end{bmatrix} \qquad S = \begin{bmatrix} 90 & 75 & 50 \\ 60 & 20 & 30 \end{bmatrix}$

The stock remaining at the end of 6 months is given by P - S. Thus, the answer is

$\begin{array}{c} \text{Armchairs} \\ \text{Rockers} \end{array} \begin{bmatrix} \text{E} & \text{M} & \text{L} \\ 30 & 15 & 10 \\ 12 & 28 & 0 \end{bmatrix}$

37. Let R and M be the following matrices:

$$R = \begin{bmatrix} 5 & 3 \\ 2 & 4 \\ 7 & 10 \end{bmatrix} \quad M = \begin{bmatrix} 100 & 200 & 300 & 400 & 300 \\ 50 & 100 & 200 & 200 & 300 \end{bmatrix}$$

The matrix RM will give the schedule of assembly requirements. Form the matrix RM by "row-column" multiplication.

$$\begin{bmatrix} 500 + 150 & 1000 + 300 & 1500 + 600 & 2000 + 600 & 1500 + 900 \\ 200 + 200 & 400 + 400 & 600 + 800 & 800 + 800 & 600 + 1200 \\ 700 + 500 & 1400 + 1000 & 2100 + 2000 & 2800 + 2000 & 2100 + 3000 \end{bmatrix}$$

Thus, we have the following schedule:

	July	Aug	Sept	Oct	Nov
Bolts	650	1300	2100	2600	2400
Clamps	400	800	1400	1600	1800
Screws	1200	2400	4100	4800	5100

39. (a) "Row-column" multiplication gives

$$C^2 = \begin{bmatrix} 0\times0 + 1\times1 + 1\times1 & 0\times1 + 1\times0 + 1\times0 & 0\times1 + 1\times0 + 1\times0 \\ 1\times0 + 0\times1 + 0\times1 & 1\times1 + 0\times0 + 0\times0 & 1\times1 + 0\times0 + 0\times0 \\ 1\times0 + 0\times1 + 0\times1 & 1\times1 + 0\times0 + 0\times0 & 1\times1 + 0\times0 + 0\times0 \end{bmatrix}$$

$$= \begin{bmatrix} 2 & 0 & 0 \\ 0 & 1 & 1 \\ 0 & 1 & 1 \end{bmatrix}$$

(b) It means that Tom can communicate with himself by two two-step communications.

(c) Two. The 1 in the second row, third column, means that Dick can communicate with Harry by a two-step communication. The 1 in the third row, second column means that Harry can communicate with Dick by a two-step communication.

41. If we do a "row-column" multiplication the result will be the matrix
$$\begin{bmatrix} -2 & -2 & 0 \\ 0 & 4 & 4 \end{bmatrix}$$

As the figure shows, the multiplication rotated the triangle about the y axis and doubled the length of each side.

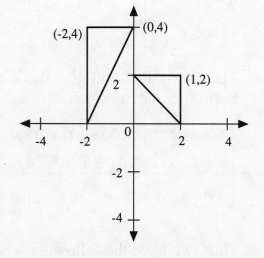

43. $r_2 \times P = \begin{bmatrix} -1 & 0 \\ 0 & 1 \end{bmatrix} \times \begin{bmatrix} a \\ b \end{bmatrix} = \begin{bmatrix} -1 \times a + 0 \times b \\ 0 \times a + 1 \times b \end{bmatrix} = \begin{bmatrix} -a \\ b \end{bmatrix}$

Thus, the point (a, b) is changed into the point (-a, b), the reflection of P across the y axis.

45. $R_2 \times P = \begin{bmatrix} -1 & 0 \\ 0 & -1 \end{bmatrix} \times \begin{bmatrix} a \\ b \end{bmatrix} = \begin{bmatrix} -1 \times a + 0 \times b \\ 0 \times a + (-1) \times b \end{bmatrix} = \begin{bmatrix} -a \\ -b \end{bmatrix}$

Thus, the point (a, b) is changed into the point (-a, -b), so P is rotated 180° around the origin.

47. $r_1 T = \begin{bmatrix} 2 & 3 & 3 \\ -1 & -1 & -2 \end{bmatrix}$ $r_2 T = \begin{bmatrix} -2 & -3 & -3 \\ 1 & 1 & 2 \end{bmatrix}$ $r_3 T = \begin{bmatrix} 1 & 1 & 2 \\ 2 & 3 & 3 \end{bmatrix}$

$R_1 T = \begin{bmatrix} -1 & -1 & -2 \\ 2 & 3 & 3 \end{bmatrix}$ $R_2 T = \begin{bmatrix} -2 & -3 & -3 \\ -1 & -1 & -2 \end{bmatrix}$ $R_3 T = \begin{bmatrix} 1 & 1 & 2 \\ -2 & -3 & -3 \end{bmatrix}$

The graph shows the effect of each multiplication.

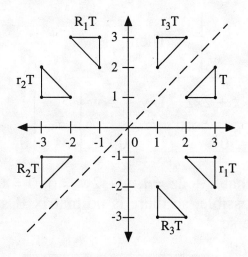

49. The sum of two 2×2 matrices is a 2×2 matrix, so S is closed with respect to addition.

51. Let the required matrix be B, where

$$B = \begin{bmatrix} a & b \\ c & d \end{bmatrix}$$

Then A + B is the matrix

$$\begin{bmatrix} 1+a & 3+b \\ 4+c & 6+d \end{bmatrix} = \begin{bmatrix} 1 & 3 \\ 4 & 6 \end{bmatrix}$$

This means that $1 + a = 1$, so $a = 0$; $3 + b = 3$, so $b = 0$; $4 + c = 4$, so $c = 0$; and $6 + d = 6$, so $d = 0$. Hence, B is the zero matrix:

$$B = \begin{bmatrix} 0 & 0 \\ 0 & 0 \end{bmatrix}$$

53. Since the sum of the two matrices is obtained by adding corresponding elements, A + B = B + A because the addition of the corresponding elements has the commutative property.

55. Suppose the matrix A does have an inverse, say B, where

$$B = \begin{bmatrix} x & y \\ z & w \end{bmatrix} \quad \text{Then}$$

$$AB = \begin{bmatrix} 1 \times x + 2 \times z & 1 \times y + 2 \times w \\ 0 & 0 \end{bmatrix} = \begin{bmatrix} 1 & 0 \\ 0 & 1 \end{bmatrix}$$

which requires that $x + 2z = 1$, $y + 2w = 0$, $0 = 0$, and $0 = 1$. The last equation is impossible, so there is no matrix B such that $AB = I$, that is, A has no inverse.

EXERCISE 8.2

STUDY TIPS Be sure you know how to represent a system of three linear equations in three unknowns by using the **augmented** matrix. The elementary operations that can be performed on a system of equations can be performed in simpler fashion on the augmented matrix to obtain the **solution** of the system. Look at the elementary row operations on page 443 and study the examples to see how these can be used to solve a system of equations.

1. Write the augmented matrix and reduce it to echelon form:

$$\begin{bmatrix} 1 & 1 & -1 & | & 3 \\ 1 & -2 & 1 & | & -3 \\ 2 & 1 & 1 & | & 4 \end{bmatrix} \sim \begin{bmatrix} 1 & 1 & -1 & | & 3 \\ 0 & 4 & -2 & | & 6 \\ 0 & 1 & -3 & | & 2 \end{bmatrix} \sim \begin{bmatrix} 1 & 1 & -1 & | & 3 \\ 0 & 3 & -2 & | & 6 \\ 0 & 0 & 7 & | & 0 \end{bmatrix}$$

$$R_1 - R_2 \to R_2 \qquad R_2 - 3R_3 \to R_3$$
$$2R_1 - R_3 \to R_3$$

The echelon form of the matrix shows that the system has been reduced to the equivalent system:
$$x + y - z = 3$$
$$3y - 2z = 6$$
$$7z = 0$$

The third equation gives $z = \mathbf{0}$, which is substituted into the second equation to give $y = \mathbf{2}$. Substituting these values into the first equation gives $x + 2 = 3$, so $x = \mathbf{1}$. The solution $x = \mathbf{1}$, $y = \mathbf{2}$, $z = \mathbf{0}$ checks in the given system.

SECTION 8.2 Solving Systems of Linear Equations

3. Write the augmented matrix and reduce it to echelon form:

$$\begin{bmatrix} 2 & -1 & 2 & | & 5 \\ 2 & 1 & -1 & | & -6 \\ 3 & 0 & 2 & | & 3 \end{bmatrix} \sim \begin{bmatrix} 2 & -1 & 2 & | & 5 \\ 0 & -2 & 3 & | & 11 \\ 0 & -3 & 2 & | & 9 \end{bmatrix} \sim \begin{bmatrix} 2 & -1 & 2 & | & 5 \\ 0 & -2 & 3 & | & 11 \\ 0 & 0 & 5 & | & 15 \end{bmatrix}$$

$R_1 - R_2 \to R_2$ $3R_2 - 2R_3 \to R_3$

$3R_1 - 2R_3 \to R_3$

The echelon form of the matrix corresponds to the system:
$$2x - y + 2z = 5$$
$$-2y + 3z = 11$$
$$5z = 15$$

The third equation gives z = **3**, which we substitute into the second equation to get -2y + 9 = 11. Thus, y = **-1**. Now, substitute back into the first equation to get 2x + 1 + 6 = 5, so that 2x = -2 and x = **-1**. The solution x = **-1**, y = **-1**, z = **3** checks in the given system.

5. Write the augmented matrix and reduce it to echelon form:

$$\begin{bmatrix} 3 & 2 & 1 & | & -5 \\ 2 & -1 & -1 & | & -6 \\ 2 & 1 & 3 & | & 4 \end{bmatrix} \sim \begin{bmatrix} 3 & 2 & 1 & | & -5 \\ 2 & -1 & -1 & | & -6 \\ 0 & 2 & 4 & | & 10 \end{bmatrix} \sim \begin{bmatrix} 3 & 2 & 1 & | & -5 \\ 0 & 7 & 5 & | & 8 \\ 0 & 1 & 2 & | & 5 \end{bmatrix}$$

$R_3 - R_2 \to R_3$ $2R_1 - 3R_2 \to R_2$

$R_3/2 \to R_3$

$$\sim \begin{bmatrix} 3 & 2 & 1 & | & -5 \\ 0 & 7 & 5 & | & 8 \\ 0 & 0 & 9 & | & 27 \end{bmatrix}$$

$7R_3 - R_2 \to R_3$

The echelon form of the matrix corresponds to the system:
$$3x + 2y + z = -5$$
$$7y + 5z = 8$$
$$9z = 27$$

The third equation gives z = **3**, which we substitute into the second equation to get 7y + 15 = 8, so that y = **-1**. Now, we substitute back into the first equation to get 3x - 2 + 3 = -5. Thus, 3x = -6 and x = **-2**. The solution x = **-2**, y = **-1**, z = **3** checks in the given system.

7. Write the augmented matrix and reduce it to echelon form:

$$\begin{bmatrix} 1 & 1 & 1 & | & 3 \\ 1 & -2 & 1 & | & -3 \\ 3 & 0 & 3 & | & 5 \end{bmatrix} \sim \begin{bmatrix} 1 & 1 & 1 & | & 3 \\ 0 & 3 & 0 & | & 6 \\ 0 & 3 & 0 & | & 4 \end{bmatrix} \sim \begin{bmatrix} 1 & 1 & 1 & | & 3 \\ 0 & 3 & 0 & | & 6 \\ 0 & 0 & 0 & | & 2 \end{bmatrix}$$

$R_1 - R_2 \to R_2$ $R_2 - R_3 \to R_3$
$R_3 - R_1 \to R_3$

The equation that corresponds to the 3rd row of the echelon form is $0 = 2$, which is impossible. Hence, the given system has no solution.

9. Write the augmented matrix and reduce it to echelon form:

$$\begin{bmatrix} 1 & 1 & 1 & | & 3 \\ 1 & -2 & 1 & | & -3 \\ 1 & 0 & 1 & | & 1 \end{bmatrix} \sim \begin{bmatrix} 1 & 1 & 1 & | & 3 \\ 0 & -3 & 0 & | & -6 \\ 0 & -1 & 0 & | & -2 \end{bmatrix} \sim \begin{bmatrix} 1 & 1 & 1 & | & 3 \\ 0 & -3 & 0 & | & 6 \\ 0 & 0 & 0 & | & 0 \end{bmatrix}$$

$R_2 - R_1 \to R_2$ $R_2 - 3R_3 \to R_3$
$R_3 - R_1 \to R_3$

The echelon form corresponds to the system:
$$x + y + z = 3$$
$$-3y = -6$$
$$0z = 0$$

The third equation is satisfied for $z = k$, where k is any real number. The second equation is satisfied by $y = 2$. Substituting these values into the first equation gives $x + 2 + k = 3$, so that $x = 1 - k$. The solution $x = 1 - k$, $y = 2$, $z = k$, where k is any real number, checks in the given system.

11. Let (m, n, p) satisfy the two equations

$$a_1 x + b_1 y + c_1 z = d_1 \quad (1)$$
$$a_2 x + b_2 y + c_2 z = d_2 \quad (2)$$

This means that

$$a_1 m + b_1 n + c_1 p = d_1 \quad (3)$$

and $a_2 m + b_2 n + c_2 p = d_2. \quad (4)$

By adding the last two equations, term by term, we get

$$(a_1 m + a_2 m) + (b_1 n + b_2 n) + (c_1 p + c_2 p) = d_1 + d_2 \quad (5)$$

or, by factoring each pair of terms in the parentheses,

$$(a_1 + a_2)m + (b_1 + b_2)n + (c_1 + c_2)p = d_1 + d_2 \quad (6)$$

This result says that (m, n, p) satisfies the equation

$$(a_1 + a_2)x + (b_1 + b_2)y + (c_1 + c_2)z = d_1 + d_2 \quad (7)$$

Thus, (m, n, p) satisfies the system consisting of equations (1) and (7).

SECTION 8.2 Solving Systems of Linear Equations

Now, suppose that (m, n, p) satisfies the system consisting of equations (1) and (7). Then, equations (3) and (6) are true. Equation (6) is equivalent to equation (5). By subtracting equation (3), term by term, from equation (5), we obtain equation (4). This means that (m, n, p) satisfies equation (2), and therefore satisfies the system consisting of equations (1) and (2).

This completes the proof that the system consisting of equations (1) and (2) is equivalent to the system consisting of equations (1) and (7).

13. Let x be the number of Type I machines, y be the number of Type II machines, and z be the number of Type III machines. Then, we use the given information and the table to obtain the following system:
$$20x + 24y + 30z = 760 \text{ (candy sales)}$$
$$10x + 18y + 10z = 380 \text{ (peanuts sales)}$$
$$30y + 30z = 660 \text{ (sandwich sales)}$$
The arithmetic can be simplified by dividing the first equation by 2, and the third equation by 30 to get the equivalent system:
$$10x + 12y + 15z = 380$$
$$10x + 18y + 10z = 380$$
$$y + z = 22$$
We write the augmented matrix and reduce it to echelon form:

$$\begin{bmatrix} 10 & 12 & 15 & 380 \\ 10 & 18 & 10 & 380 \\ 0 & 1 & 1 & 22 \end{bmatrix} \sim \begin{bmatrix} 10 & 12 & 15 & 380 \\ 0 & 6 & -5 & 0 \\ 0 & 1 & 1 & 22 \end{bmatrix} \sim \begin{bmatrix} 10 & 12 & 15 & 380 \\ 0 & 6 & -5 & 0 \\ 0 & 0 & 11 & 132 \end{bmatrix}$$
$$R_2 - R_1 \to R_2 \qquad\qquad 6R_3 - R_2 \to R_3$$

The echelon form corresponds to the system:
$$10x + 12y + 15z = 380$$
$$6y - 5z = 0$$
$$11z = 132$$
The third equation gives $z = \mathbf{12}$, which substituted into the second equation gives $6y - 60 = 0$, so $y = \mathbf{10}$. Then back-substitution in the first equation gives $10x + 120 + 180 = 380$. Thus, $10x = 80$ and $x = \mathbf{8}$. Consequently, Mechano has **8** Type I, **10** Type II, and **12** Type III machines. You can check that these numbers satisfy all the given data.

226 CHAPTER 8 MATHEMATICAL SYSTEMS AND MATRICES

15. For convenience, assume that 100 units of fertilizer are to be obtained. Suppose it takes x units of Type I, y units of Type II, and z units of Type III to get an 8-8-8 fertilizer. Then we have the system of equations:

$$0.06x + 0.08y + 0.12z = 8 \quad \text{(Chemical A)}$$
$$0.06x + 0.12y + 0.08z = 8 \quad \text{(Chemical B)}$$
$$0.08x + 0.04y + 0.12z = 8 \quad \text{(Chemical C)}$$

Now multiply each equation by 100, divide the resulting first two equations by 2 and divide the third equation by 4 to obtain

$$3x + 4y + 6z = 400$$
$$3x + 6y + 4z = 400$$
$$2x + y + 3z = 200$$

Write the augmented matrix and reduce it to echelon form:

$$\begin{bmatrix} 3 & 4 & 6 & | & 400 \\ 3 & 6 & 4 & | & 400 \\ 2 & 1 & 3 & | & 200 \end{bmatrix} \sim \begin{bmatrix} 3 & 4 & 6 & | & 400 \\ 0 & 2 & -2 & | & 0 \\ 0 & 5 & 3 & | & 200 \end{bmatrix} \sim \begin{bmatrix} 3 & 4 & 6 & | & 400 \\ 0 & 2 & -2 & | & 0 \\ 0 & 0 & 16 & | & 400 \end{bmatrix}$$

$R_2 - R_1 \to R_2$

$2R_1 - 3R_3 \to R_3$

$2R_3 - 5R_2 \to R_3$

The echelon form corresponds to the system:

$$3x + 4y + 6z = 400$$
$$2y - 2z = 0$$
$$16z = 400$$

The third equation gives z = **25**, which we substitute into the second equation to get y = **25**. Then, we substitute these values into the first equation to get $3x + 100 + 150 = 400$, which simplifies to $3x = 150$, giving x = **50**. Thus, for each hundred units, Gro-kwik must mix **50** units of Type I, **25** units of Type II, and **25** units of Type III. This means the final product will be **50%** Type 1, **25%** Type II, and **25%** Type III.

17. (a) $ad - bc = (1)(4) - (2)(2) = 4 - 4 = 0$. This matrix is *singular*.
    (b) $ad - bc = (2)(5) - (-3)(3) = 10 + 9 \neq 0$. This matrix is *nonsingular*.
    (c) $ad - bc = (0)(4) - (2)(2) = 0 - 4 \neq 0$. This matrix is *nonsingular*

19. x = **a**, y = **b**, z = **c**   21. x = **a**, y = **b**, z = **any real number**

23. $\begin{bmatrix} 1 & 0 & 0 & | & a \\ 0 & 1 & 0 & | & b \\ 0 & 0 & 0 & | & c \end{bmatrix}, c \neq 0;$ $\begin{bmatrix} 1 & 0 & 0 & | & a \\ 0 & 0 & 0 & | & b \\ 0 & 0 & 0 & | & c \end{bmatrix}, b \neq 0;$ $\begin{bmatrix} 0 & 0 & 0 & | & a \\ 0 & 1 & 0 & | & b \\ 0 & 0 & 1 & | & c \end{bmatrix}, a \neq 0$

SECTION 8.3   Clock and Modular Arithmetic   227

EXERCISE 8.3

STUDY TIPS Read the description of a mathematical system on page 453. This will help you in the rest of this chapter. If you are lucky enough to have a clock or a watch with the usual hour and minute hands, (rather than a digital watch) you will find it quite easy to understand addition and subtraction in clock arithmetic. If you are not that lucky, use the diagrams in Figures 8.1 and 8.2 to help you understand the basic operations. Table 8.5 gives you all the facts for addition in clock arithmetic. Definition 8.8 gives the meaning of subtraction in clock arithmetic and Example 2(b) shows how Table 8.5 can be used for subtraction problems. Multiplication in clock arithmetic is defined in Definition 8.9 and division in Definition 8.10. Be sure to note that division in clock arithmetic is not always possible just like in our number system division by 0 is not defined. (See Example 4c.)

1. $9 \oplus 7$ means 7 hours after 9 o'clock, so the answer is **4**.

3. $8 \oplus 3$ means 3 hours after 8 o'clock, so the answer is **11**.

5. $7 \oplus 8$ means 8 hours after 7 o'clock, so the answer is **3**.

7. $8 \oplus 11$ means 11 hours after 8 o'clock so the answer is **7**.

9. $8 \ominus 3$ means 3 hours before 8 o'clock, so the answer is **5**.

11. $9 \ominus 12$ means 12 hours before 9 o'clock, so the answer is **9**.

13. $8 \ominus 7$ means 7 hours before 8 o'clock, so the answer is **1**.

15. $n \oplus 7 = 9$ is true if $n = 9 \ominus 7 =$ **2**.

17. $2 \oplus n = 1$ is true if $n = 1 \ominus 2 =$ **11**.

19. $3 \ominus 5 = n$ is true if $3 = n \oplus 5$, so $n =$ **10**.

21. $1 \ominus n = 12$ is true if $1 = n \oplus 12$, so $n =$ **1**.

23. $4 \otimes 3 = 3 \oplus 3 \oplus 3 \oplus 3 =$ **12**.

25. $9 \otimes 2 = $ **6** because in ordinary arithmetic $9 \times 2 = 18$, which divided by 12, leaves a remainder of **6**.

27. $2 \otimes 8 = $ **4** because in ordinary arithmetic $2 \times 8 = 16$, which divided by 12, leaves a remainder of **4**.

29.

$\otimes$	1	2	3	4	5	6	7	8	9	10	11	12
1	1	2	3	4	5	6	7	8	9	10	11	12
2	2	4	6	8	10	12	2	4	6	8	10	12
3	3	6	9	12	3	6	9	12	3	6	9	12
4	4	8	12	4	8	12	4	8	12	4	8	12
5	5	10	3	8	1	6	11	4	9	2	7	12
6	6	12	6	12	6	12	6	12	6	12	6	12
7	7	2	9	4	11	6	1	8	3	10	5	12
8	8	4	12	8	4	12	8	4	12	8	4	12
9	9	6	3	12	9	6	3	12	9	6	3	12
10	10	8	6	4	2	12	10	8	6	4	2	12
11	11	10	9	8	7	6	5	4	3	2	1	12
12	12	12	12	12	12	12	12	12	12	12	12	12

31. $\frac{3}{5} = n$ only if $3 = 5 \otimes n$. The table above shows that $5 \otimes 3 = 3$, so $n = $ **3**.

33. $\frac{1}{11} = n$ only if $1 = 11 \otimes n$. The table above shows that $11 \otimes 11 = 1$, so $n = $ **11**.

35. $\frac{n}{5} = 8$ is true if $n = 5 \otimes 8 = $ **4**. (See the table above.)

37. $\frac{n}{2} \oplus 4 = 8$ is true if $\frac{n}{2} = 8 \ominus 4 = 4$, and $\frac{n}{2} = 4$ if $n = 2 \otimes 4 = $ **8** (See the table above.)

39. $\frac{2}{n} = 3$ is true only if $2 = 3 \otimes n$. The table above shows no entry of 2 in the row for 3. Therefore, there is **no solution**.

41. False; $4 - 2 = 2$ is not a multiple of 3.

SECTION 8.3 Clock and Modular Arithmetic

43. False; 7 - 6 = 1 is not a multiple of 5.

45. False; 9 - 8 = 1 is not a multiple of 10.

47. $3 + 4 \equiv 7 \equiv \mathbf{2}(\bmod\ 5)$ 49. $3 + 1 \equiv \mathbf{4}(\bmod\ 5)$

51. $4 \times 2 \equiv 8 \equiv \mathbf{3}(\bmod\ 5)$ 53. $2 \times 3 \equiv 6 \equiv \mathbf{1}(\bmod\ 5)$

55. $2 - 4 \equiv (-2) \equiv (5 - 2) \equiv \mathbf{3}(\bmod\ 5)$

57. $1 - 3 \equiv (-2) \equiv (5 - 2) \equiv \mathbf{3}(\bmod\ 5)$

59. $4 \times (3 + 0) \equiv [(4 \times 3) + (4 \times 0)](\bmod\ 5)$, so n = **0**.

61. $2 \times (0 + 3) \equiv [(2 \times 0) + (2 \times 3)](\bmod\ 5)$, so n = **0**.

63. $2 + 1 \equiv 3(\bmod\ 5)$, so n = **1**.   65. $2 \times 2 \equiv 4(\bmod\ 5)$, so n = **2**.

67. $7 - 3 \equiv 4(\bmod\ 5)$, and $7 \equiv 2(\bmod\ 5)$, so n = **2**.

69. $3 \equiv (7 - 4)(\bmod\ 5)$, and $7 \equiv 2(\bmod\ 5)$, so n = **2**.

71. $\frac{n}{2} \equiv 4(\bmod\ 5)$ is true if $n \equiv 2 \times 4 \equiv 8 \equiv 3(\bmod\ 5)$, so n = **3**.

73. $\frac{3}{4} \equiv n(\bmod\ 5)$ is true if $3 \equiv (4 \times n)(\bmod\ 5)$, which is true for n = **2** because $4 \times 2 \equiv 8 \equiv 3(\bmod\ 5)$.

75. Yes, all the entries in the table are elements of S.

77. Yes. The identity is 1 because, if a is an element of S, then $1 \times a \equiv a \times 1 \equiv a(\bmod\ 5)$.

79. A military clock has a 24 hour cycle.

83. Write the digits in the ISBN under the numbers 10, 9, 8, . . ., 2 as shown in the text. Multiply the two numbers in a vertical line and add the results as below.

$$\begin{array}{cccccccccc}10 & 9 & 8 & 7 & 6 & 5 & 4 & 3 & 2 \\ 0 & 0 & 6 & 0 & 4 & 0 & 6 & 1 & 3\end{array}$$

$$0 + 0 + 48 + 0 + 24 + 0 + 24 + 0 + 24 + 3 + 6 = 105$$

Then solve the congruence $105 + c \equiv 0 \pmod{11}$.
Since 110 is the multiple of 11 that is just greater than 105, and $105 + 5 = 110$, we have $c = \mathbf{5}$.

85. Follow the same procedure as in Problem 83:

$$\begin{array}{cccccccccc}10 & 9 & 8 & 7 & 6 & 5 & 4 & 3 & 2 \\ 0 & 3 & 1 & 2 & 8 & 7 & 8 & 6 & 7\end{array}$$

$$0 + 27 + 8 + 14 + 48 + 35 + 32 + 18 + 14 = 196$$

Now solve the congruence $196 + c \equiv 0 \pmod{11}$.
Since 198 is the multiple of 11 that is just greater then 196, and $196 + 2 = 198$, we have $c = \mathbf{2}$.

87. Let x be the blurred digit and follow the same procedure as in Problem 81:

$$\begin{array}{cccccccccc}10 & 9 & 8 & 7 & 6 & 5 & 4 & 3 & 2 \\ 0 & 0 & 3 & 0 & 5 & 8 & 9 & x & 4\end{array}$$

$0 + 0 + 24 + 0 + 30 + 40 + 36 + 3x + 8 = 138 + 3x$
Since the check number is $c = 2$, we must have
$$(138 + 3x) + 2 \equiv 0 \pmod{11}$$
or
$$140 + 3x \equiv 0 \pmod{11}$$
Now, 143 is the multiple of 11 that is just greater than 140, and $143 = 140 + 3$. Therefore, we must have $3x = 3$, so $x = \mathbf{1}$.

89.

$\oplus$	0	1	2	3	4
0	0	1	2	3	4
1	1	2	3	4	0
2	2	3	4	0	1
3	3	4	0	1	2
4	4	0	1	2	3

91. Yes, the table at the left is symmetric with respect to the diagonal from upper left to lower right.

SECTION 8.4 Groups and Fields

EXERCISE 8.4

STUDY TIPS This section is concerned with the rules and laws that are obeyed by an abstract mathematical system. Read the definitions carefully. They tell you what is meant by **closure** (all the elements in the table come from the original set of elements), **associativity**, **commutativity** (the table is symmetric with respect to a diagonal drawn from top left to bottom right), an **identity** (the row of elements by the identity in the table is the same as the top row and the column under the identity is the same as the left-most column) an **inverse**, and the **distributive property**. Be sure to study the examples that make use of these definitions.

After you familiarize yourself with these rules and properties we study two types of mathematical structures that obey these rules: **groups and fields.** Definitions 8.18 and 8.19 will tell you about groups and fields and the examples show how to determine whether or not a mathematical system is a indeed a group or a field. Make up your own examples of groups and fields. Look back at the material you have studied in the book. Which number systems form groups? Which form fields?

1. (a) a @ b = **a**   (b) b @ c = **c**   (c) c @ a = **b**

3. (a) b @ (a @ b) = b @ a = **a**   (b) (b @ a) @ b = a @ b = **a**   (c) Yes

5. (a) b @ c = **c**   (b) c @ b = **c**   (c) Yes

7. Yes; all the entries in the table are elements of S.

9. (a) Yes; if a and b are natural numbers, then a F b = a, which is a natural number.
   (b) Yes. a F (b F c) = a F b = a, and (a F b) F c = a F c = a. Thus, a F (b F c) = (a F b) F c.
   (c) No. a F b = a, and b F a = b, so if a ≠ b, then a F b ≠ b F a.

11. (a) No. For example, 1 + 3 = 4, which is not an odd number.
    (b) Yes; the product of two odd numbers is an odd number.
    (c) Yes; the sum of two even numbers is an even number.
    (d) Yes; the product of two even numbers is an even number.

13.

∩	∅	{a}	{b}	{a,b}
∅	∅	∅	∅	∅
{a}	∅	{a}	∅	{a}
{b}	∅	∅	{b}	{b}
{a, b}	∅	{a}	{b}	{a, b}

15. (a) ({b} ∩ {a, b}) ∩ {a} = {b} ∩ {a} = ∅
    (b) {b} ∩ ({a, b} ∩ {a}) = {b} ∩ {a} = ∅
    (c) Yes

17. Yes; all the entries in the above table (Problem 13) are elements of S.

19. (a)

L	1	2	3	4
1	1	2	3	4
2	2	2	3	4
3	3	3	3	4
4	4	4	4	4

   (b) The first row is the same as the L row, and the first column is the same as the L column, so the identity element is 1.

21. The table given for this problem shows that 4 is the identity element. This information is needed here.
    (a) There is no 4 in the first row of the table, so 1 has no inverse.
    (b) There is no 4 in the second row of the table, so 2 has no inverse.
    (c) There is no 4 in the third row of the table, so 3 has no inverse.
    (d) Since 4 S 4 = 4, the element 4 is its own inverse.

23. Yes; $1 \times a = a \times 1 = a$ for each element a of the set S, so **1** is the identity element.

25. (a) A is the identity element because if B is any subset of A, then
       A ∩ B = B ∩ A = B.
    (b) No; there is only the one identity element.

27. The 0 row of the table is the same as the @ row, and the 0 column is the same as the @ column, so 0 is the identity element.

29. (a) 3 F (4 L 5) = **3**, because 3 is the first of the numbers 3 and (4 L 5).
    (b) 4 F (5 L 6) = **4**, because 4 is the first of the numbers 4 and (5 L 6).

31. Suppose a, b, c are real numbers. Then a F (b L c) = a and (a F b) L (a F c) = a L a = a. This shows that the distributive property of F over L holds for all real numbers. (Keep in mind that F means to select the first of the numbers in the expression a F b.)

33. If a, b, c are real numbers, we may replace b - c by b + (-c). Then, since the distributive property of multiplication over addition holds, we have a(b - c) = a[b + (-c)] = ab + a(-c) = ab - ac . This shows that multiplication is distributive over subtraction.

35 Yes; fractions are real numbers and multiplication is distributive over addition for the set of real numbers.

37. **1.** The set S is closed under the operation $*$ because every entry in the table is an element of S.

   2. For all x, y, z that are elements of S: $x * (y * s) = (x * y) * z$. This can be checked for all possibilities. For instance,
   $$a * (b * c) = a * b = c \text{ and } (a * b) * c = c * c = c$$
   Thus, the set S has the associative property under the operation $*$.

   3. The bottom row is the same as the $*$ row, and the last column is the same as the $*$ column, so c is the identity element.

   4. The identity element c occurs once in each row and column of the table, so every element has an inverse. For instance, $a * b = b * a = c$, so a and b are inverses of each other.

   This shows that the set S under the operation $*$ meets the four requirements and is thus a group. (Note that it is a commutative group.)

39. The set of odd integers under multiplication is not a group because, except for the number 1, no number in this set has an inverse that is an element of the set. (The fractions are not in this set.)

41. The set of even integers under multiplication is not a group because it has no identity element. (The number 1 is not in this set.)

43. The set of positive integers under multiplication is not a group because, except for the number 1, no number in this set has an inverse that is an element of the set. (The fractions are not in this set.)

45. The set of integers under multiplication is not a group because, except for the numbers 1 and -1, no number in this set has an inverse that is an element of the set. (The fractions are not in this set.)

234 CHAPTER 8 MATHEMATICAL SYSTEMS AND MATRICES

47. We check the four requirements as follows:

    1. If a and b are real numbers, then a + b is a real number, so the set of real numbers is closed under addition.
    2. If a, b, c are real numbers then (a + b) + c = a + (b + c), so the set of real numbers has the associative property under addition.
    3. The number 0 is a real number and 0 + a = a + 0 = a for every real number a. Thus, 0 is the identity element for the set of real numbers under addition.
    4. If a is a real number, then -a is also a real number and a + (-a) = (-a) + a = 0. Thus, every real number has an inverse under addition.

    Since the set of real numbers under addition meets all four requirements, the system is a group. (Note that it is a commutative group because a + b = b + a.)

49. The set {-1, 0, 1} under multiplication is **not** a group because the number **0** has **no inverse** under multiplication.

51. The set {a, b, c, d, e} under the operation # defined by the table is **not** a group because there is **no identity element**. (There is no column of the table that reads a, b, c, d, e in that order.)

53. The set of positive even integers has no multiplicative inverses that are elements of the set, so this system is **not** a field.

55. The set of integral multiples of 2 is the set of even integers. This set has no multiplicative inverses that are elements of the set, so this system is **not** a field.

57. Check to see that a ♦ b is always an element of S.

59. Check the table to see if there is an element e in S such that the column under e is identical to the column at the far left and the row opposite e is identical to the top row. If there is, then **e** is the identity element. If there is no such element, then there is **no** identity element.

61. You have to check that x ∗ (y ∗ z) = (x ∗ y) ∗ z for all possible values of x, y, z from the set {a, b, c}. If you had to check all possible cases, there would be 27 of these because each of the 3 places has three possible values. However, since the operation has the commutative property, the number of cases to be checked is greatly reduced. Think about it.

63. $6 \times 9999 = 6(10,000 - 1)$
    $= 60,000 - 6 = \mathbf{59,994}$

65. $7 \times 59 = 7(60 - 1)$
    $= 420 - 7$
    $= \mathbf{413}$

67. $4 \times 9995 = 4(10,000 - 5) = 40,000 - 20 = \mathbf{39,980}$

69. Let x be the number you select, then follow the steps:

Think of a number:	x
Add 3 to it:	x + 3
Triple the result:	3x + 9
Subtract 9:	3x
Divide by the number with which you started:	$\frac{3x}{x} = 3$

EXERCISE 8.5

STUDY TIPS You should find game theory a very interesting application of matrices. This theory has important everyday applications in business, economics, and science. You will learn how to set up the payoff matrix for a given game and what strategies to use in playing the game. Remember, in a strictly determined game, the best strategy for the **row** player is to select the smallest number in the row (**Row: Low**). The best strategy for the column player is to select the highest number in the column (**Column: High**). Practice with the definitions that are given and study the examples carefully.

1. The 4 in the first row is the smallest element in its row and the largest in its column. The game is strictly determined. Row player should play row 1; column player should play column 1. Value of the game is 4.

3. There is no element that is the smallest in its row and the largest in its column. Hence, the game is not strictly determined.

5. The 4 in the first row is the smallest element in its row and the largest in its column. The game is strictly determined. Row player should play row 1; column player should play column 3. Value of the game is 4.

7. Either 4 in the third row is the smallest element in its row and the largest in its column. The game is strictly determined. Row player should play row 3; column player should play column 3. The value of the game is 4.

9. (a) There is no element that is the smallest in its row and the largest in its column, so there is no saddle point.
    (b) Row player's expected value is given by
    $$E_R = (p)(2) + (1 - p)(-1) = p(1) + (1-p)(4)$$
    Thus, $\qquad 3p - 1 = 4 - 3p$
    or $\qquad 6p = 5$ and $p = 5/6$.
    This means that the row player should play row 1 five-sixths of the time and row 2 one-sixth of the time. (The row player's expected value is
    $$E_R = (3)(5/6) - 1 = 5/2 - 1 = 3/2 = 1.5.)$$

11. (a) There is no element that is the smallest in its row and the largest in its column, so there is no saddle point.

    (b) Row player's expected value is given by
    $$E_R = (p)(3) + (1 - p)(-1) = (p)(-2) + (1 - p)(0)$$
    Thus, $\qquad 4p - 1 = -2p$
    or $\qquad 6p = 1$ and $p = 1/6$.
    This means that the row player should play row 1 one-sixth of the time and row 2 five-sixths of the time. (The row player's expected value is
    $$E_R = (4)(1/6) - 1 = -1/3.)$$

13. (a) There is no element that is the smallest in its row and the largest in its column, so there is no saddle point.
    (b) Since row 2 dominates row 3, row 3 may be discarded and we may use rows 1 and 2 for the payoff matrix. The row player's expected value is given by
    $$E_R = (p)(6) + (1 - p)(2) = (p)(0) + (1 - p)(4)$$
    Thus, $\qquad 4p + 2 = 4 - 4p$
    or $\qquad 8p = 2$ and $p = 1/4$
    This means that the row player should play row 1 one-fourth of the time and row 2 three-fourths of the time. The row player's expected value is
    $$E_R = (4)(1/4) + 2 = 3.$$

15. There is no element that is the smallest in its row and the largest in its column, so there is no saddle point. The expected row value is
    $$E_R = (p)(70) + (1 - p)(85) = (p)(80) + (1 - p)(75)$$
    Thus, $\qquad 85 - 15p = 5p + 75$
    or $\qquad 10 = 20p$ and $p = 1/2$.
    This means that you should study 2 hr half the time and 4 hr the other half the time. (Your expected score is $85 - (15)(1/2) = 77.5$.)

17. There is no element that is the smallest in its row and the largest in its column, so there is no saddle point. Since the first row is dominated by both the other rows, Ann should not use the first row. This leaves a 2-by-2 matrix for which the Ann's expected value is

$$E_R = (p)(10) + (1 - p)(15) = (p)(12) + (1 - p)(10)$$

Thus, $\qquad 15 - 5p = 2p + 10$

or $\qquad 5 = 7p \text{ and } p = 5/7.$

This means that Ann should buy no bonds, buy stocks with five-sevenths and money market funds with two-sevenths of her investment. Her expected return is found from

$$E_R = 15 - (5)(\tfrac{5}{7}) = 15 - \tfrac{25}{7} = 15 - 3\tfrac{4}{7} = 11\tfrac{3}{7}$$

that is, her expected return would be $11\tfrac{3}{7}\%$.

19. There is no element that is the smallest in its row and the largest in its column, so there is no saddle point. The expected row value is
$$E_R = (p)(60) + (1 - p)(30) = (p)(50) + (1 - p)(70)$$
Thus, $\qquad 30p + 30 = 70 - 20p$
or $\qquad 50p = 40 \text{ and } p = 4/5.$

This means that station R should price its gasoline at \$1.00 four-fifths of the time and at \$1.10 the other fifth of the time. (The expected row value would be $E_R = (30)(4/5) + 30 = 56$, that is, with this strategy, station R would expect to get 56% of the business.)

21. (a) The payoff matrix can be written as follows:

	Young	Old
Performance	70%	20%
Safety	40%	80%

(b) There is no saddle point. The expected row value is
$$E_R = (p)(70) + (1 - p)(40) = (p)(20) + (1 - p)(80)$$
Thus, $\qquad 30p + 40 = 80 - 60p$
or $\qquad 90p = 40 \text{ and } p = 4/9$

This means that 4/9 of the ads should be based on performance and 5/9 should be based on safety. (With this strategy, the expected row value would be ER = (30)(4/9) + 40 = 40/3 + 40 = 160/3, that is, the ads would be effective on $53\tfrac{1}{3}\%$ of the buyers.)

23. With the numbers in dollars, the required matrix can be written as

$$\text{Water?} \begin{array}{c} \\ \text{Yes} \\ \text{No} \end{array} \overset{\begin{array}{cc} \text{Freeze?} & \\ \text{Yes} & \text{No} \end{array}}{\begin{bmatrix} 6000 & -400 \\ -4000 & 4000 \end{bmatrix}}$$

There is no saddle point. The expected row value is

$$E_R = (p)(6000) + (1 - p)(-4000) = (p)(-400) + (1 - p)(4000)$$

Thus, $\quad\quad\quad\quad 10{,}000p - 4000 = 4000 - 4400p$

or $\quad\quad\quad\quad\quad\quad\quad\quad 14{,}400p = 8000$

and, after division by 1600, $\quad 9p = 5$ and $p = 5/9$.

This means that the farmer's optimal strategy would be to water five-ninths of the time and not to water four-ninths of the time. The corresponding expected payoff is

$$E_R = (10{,}000)(\tfrac{5}{9}) - 4000 = \frac{50000 - 36000}{9} = \frac{14000}{9}$$

So the farmer's expected payoff is $\dfrac{\$14000}{9}$ or about $1556.

25. If row i dominates row j, this means that, in the long run, playing row i is more profitable than playing row j. Thus, row j may be eliminated from the row player's options.

27. Both entries in row 1 are less than the corresponding entries in row 2. Therefore, row 2 dominates row 1.

29. With row 1 eliminated, the payoff matrix is

$$\begin{array}{c} \\ \text{Poems} \\ \text{Candies} \end{array} \overset{\begin{array}{cc} \text{Allergic} & \text{On a diet} \end{array}}{\begin{bmatrix} 2 & 5 \\ 3 & -2 \end{bmatrix}}$$

There is no saddle point. The expected row value is

$$E_R = (p)(2) + (1 - p)(3) = (p)(5) + (1 - p)(-2)$$

Thus, $\quad\quad\quad\quad\quad -p + 3 = 7p - 2$

or $\quad\quad\quad\quad\quad\quad\quad 5 = 8p$ and $p = 5/8$.

Hence, the student's optimal strategy is to send poems five-eighths of the time and candies three-eighths of the time. He should not send flowers.

PRACTICE TEST 8

STUDY TIPS Read the summary on pages 484-486 and be sure that you understand all the items before trying the practice test. If necessary, go back and review any section that you don't understand clearly. Do the test and check with the answers given in this manual. Correct your errors and try to get some help on problems that you do not understand.

1. We want to have

$$2\begin{bmatrix} 2 & x \\ 3 & y \end{bmatrix} + \begin{bmatrix} 2 & -1 \\ 3 & 2 \end{bmatrix} = \begin{bmatrix} 6 & 5 \\ 9 & 10 \end{bmatrix}$$

or

$$\begin{bmatrix} 4 & 2x \\ 6 & 2y \end{bmatrix} + \begin{bmatrix} 2 & -1 \\ 3 & 2 \end{bmatrix} = \begin{bmatrix} 6 & 5 \\ 9 & 10 \end{bmatrix}$$

or

$$\begin{bmatrix} 6 & 2x-1 \\ 9 & 2y+2 \end{bmatrix} = \begin{bmatrix} 6 & 5 \\ 9 & 10 \end{bmatrix}$$

For these two matrices to be equal, we must have

$$2x - 1 = 5 \quad \text{and} \quad 2y + 2 = 10$$

Therefore, $x = \mathbf{3}$ and $y = \mathbf{4}$

2. $$AB = \begin{bmatrix} -4+5 & 20-20 \\ -1+1 & 5-4 \end{bmatrix} = \begin{bmatrix} 1 & 0 \\ 0 & 1 \end{bmatrix} \qquad BA = \begin{bmatrix} -4+5 & 5-5 \\ -4+4 & 5-4 \end{bmatrix} = \begin{bmatrix} 1 & 0 \\ 0 & 1 \end{bmatrix}$$

3. The required answer can be found by multiplying the following matrices:

$$\begin{bmatrix} 20 & 25 & 10 \end{bmatrix} \begin{bmatrix} 1 & 2 & 1 & 1 \\ 1 & 3 & 2 & 1 \\ 4 & 4 & 2 & 2 \end{bmatrix}$$

$$= \begin{bmatrix} 20+25+40 & 40+75+40 & 20+50+20 & 20+25+20 \end{bmatrix}$$

	Frames	Wheels	Chains	Paints
= [	85	155	90	65]

240 CHAPTER 8 MATHEMATICAL SYSTEMS AND MATRICES

4. The required answer can be found by multiplying the following matrices:

$$\begin{matrix} \text{Materials} \\ \begin{bmatrix} 1 & 2 & 1 & 1 \\ 1 & 3 & 2 & 1 \\ 4 & 4 & 2 & 2 \end{bmatrix} \end{matrix} \begin{matrix} \text{Cost} \\ \begin{bmatrix} 2 \\ 1 \\ 0.75 \\ 0.50 \end{bmatrix} \end{matrix}$$

$$= [2 + 2 + 0.75 + 0.50 \quad 2 + 3 + 1.50 + 0.50 \quad 8 + 4 + 1.50 + 1.00]$$

$$= \begin{matrix} \text{Type I} & \text{Type II} & \text{Type III} \\ [5.25 & 7.00 & 14.50] \end{matrix}$$
(Costs in dollars)

5. $2A - 3B = \begin{bmatrix} 4 & 0 & 2 \\ 4 & -2 & 6 \\ 8 & 2 & 4 \end{bmatrix} - \begin{bmatrix} -6 & 3 & 0 \\ 12 & 9 & -6 \\ 3 & 6 & -3 \end{bmatrix} = \begin{bmatrix} 10 & -3 & 2 \\ -8 & -11 & 12 \\ 5 & -4 & 7 \end{bmatrix}$

6. $A + B = \begin{bmatrix} 0 & 1 & 1 \\ 6 & 2 & 1 \\ 5 & 3 & 1 \end{bmatrix}$

7. $AB = \begin{bmatrix} 2\times(-2) + 0\times4 + 1\times1 & 2\times1 + 0\times3 + 1\times2 & 2\times0 + 0\times(-2) + 1\times(-1) \\ 2\times(-2) + (-1)\times4 + 3\times1 & 2\times1 + (-1)\times3 + 3\times2 & 2\times0 + (-1)\times(-2) + 3\times(-1) \\ 4\times(-2) + 1\times4 + 2\times1 & 4\times1 + 1\times3 + 2\times2 & 4\times0 + 1\times(-2) + 2\times(-1) \end{bmatrix}$

$$= \begin{bmatrix} -3 & 4 & -1 \\ -5 & 5 & -1 \\ -2 & 11 & -4 \end{bmatrix}$$

$BA = \begin{bmatrix} (-2)\times2 + 1\times2 + 0\times4 & (-2)\times0 + 1\times(-1) + 0\times1 & (-2)\times1 + 1\times3 + 0\times2 \\ 4\times2 + 3\times2 + (-2)\times4 & 4\times0 + 3\times(-1) + (-2)\times1 & 4\times1 + 3\times3 + (-1)\times2 \\ 1\times2 + 2\times2 + (-1)\times4 & 1\times0 + 2\times(-1) + (-1)\times1 & 1\times1 + 2\times3 + (-1)\times2 \end{bmatrix}$

$$= \begin{bmatrix} -2 & -1 & 1 \\ -6 & -5 & 9 \\ 2 & -3 & 5 \end{bmatrix}$$

8. Use the matrix found for A + B in problem 6. To get $(A + B)^2$, multiply this matrix by itself. This gives

$$(A + B)^2 = \begin{bmatrix} 0\times0 + 1\times6 + 1\times5 & 0\times1 + 1\times2 + 1\times3 & 0\times4 + 1\times1 + 1\times1 \\ 6\times0 + 2\times6 + 1\times5 & 6\times1 + 2\times2 + 1\times3 & 6\times1 + 2\times1 + 1\times1 \\ 5\times0 + 3\times6 + 1\times5 & 5\times1 + 3\times2 + 1\times3 & 5\times1 + 3\times1 + 1\times1 \end{bmatrix}$$

$$= \begin{bmatrix} 11 & 5 & 2 \\ 17 & 13 & 9 \\ 23 & 14 & 9 \end{bmatrix}$$

9. We reduce the augmented matrix of the system to echelon form:

$$\begin{bmatrix} 1 & 3 & 0 & | & 19 \\ 0 & 1 & 3 & | & 10 \\ 3 & 0 & 1 & | & -5 \end{bmatrix} \sim \begin{bmatrix} 1 & 3 & 0 & | & 19 \\ 0 & 1 & 3 & | & 10 \\ 0 & -9 & 1 & | & -62 \end{bmatrix} \sim \begin{bmatrix} 1 & 0 & -9 & | & -11 \\ 0 & 1 & 3 & | & 10 \\ 0 & 0 & 28 & | & 28 \end{bmatrix}$$

The echelon form of the matrix corresponds to the system

$$\begin{aligned} x \quad\quad - 9z &= -11 \\ y + 3z &= 10 \\ 28z &= 28 \end{aligned}$$

The third equation gives $z = 1$, and, on back substitution, the other two equations give $y = 7$ and $x = -2$. The solution $(-2, 7, 1)$ checks in the given system.

10. Let n be the number of nickels, d the number of dimes, and q the number of quarters in the piggy bank. Then, we have the equations:

$$\begin{aligned} n + d + q &= 122 \quad \text{(There were 122 coins in all.)} \\ 5n + 10d + 25q &= 1500 \quad \text{(The total value was \$15.)} \\ 25q &= 20n \quad \text{(The total value of the quarters was 4} \\ &\quad\quad\quad\quad\quad \text{times the total value of the nickels.)} \end{aligned}$$

By dividing both sides of the second and third equations by 5, we get the system:

$$\begin{aligned} n + d + q &= 122 \\ n + 2d + 5q &= 300 \\ 4n \quad\quad - 5q &= 0 \end{aligned}$$

The augmented matrix of this system is

$$\begin{bmatrix} 1 & 1 & 1 & | & 122 \\ 1 & 2 & 5 & | & 300 \\ 4 & 0 & -5 & | & 0 \end{bmatrix} \sim \begin{bmatrix} 1 & 1 & 1 & | & 122 \\ 0 & 1 & 4 & | & 178 \\ 0 & -4 & -9 & | & -488 \end{bmatrix} \sim \begin{bmatrix} 1 & 1 & 1 & | & 122 \\ 0 & 1 & 4 & | & 178 \\ 0 & 0 & 7 & | & 224 \end{bmatrix}$$

$R_2 - R_1 \to R_2$ $\quad\quad 4R_2 + R_2 \to R_3$
$R_3 - 4R_1 \to R_3$

The system corresponding to the echelon form is

$$n + d + q = 122$$
$$d + 4q = 178$$
$$7q = 224$$

The third equation gives q = **32**, which we substitute into the second equation to get d + 128 = 178, so d = **50.** Then, back substitution into the first equation gives n = **40**. Thus, Sally had 40 nickels, 50 dimes, and 32 quarters.

11. The system corresponding to the given echelon form is

$$2x + y + z = 1$$
$$3y + 2z = 4$$
$$z = 2$$

Back-substitution in the second equation gives 3y + 4 = 4, so y = **0**. Then, back-substitution in the first equation gives 2x + 2 = 1 or 2x = -1, so x = **-1/2**. The solution (-1/2, 0, 2) checks in the system.

12. The third row of the matrix corresponds to the equation 0x + 0y + 0z = 2. There are **no** real numbers (x, y, z) to satisfy this equation, so the system has **no solution**.

13. The given matrix corresponds to the system

$$2x + y + z = 1$$
$$3y + 2z = 4$$
$$0z = 0$$

The third equation is satisfied by z = **k**, where k is any real number. Back-substitution into the second equation gives 3y + 2k = 4, so y = $\frac{4 - 2k}{3}$. Then back-substitution into the first equation gives

$$2x + \frac{4 - 2k}{3} + k = 1$$

or $6x + 4 - 2k + 3k = 3$,
which gives $6x = -k - 1$
and $x = -\frac{k + 1}{6}$

The system is satisfied by $(-\frac{k + 1}{6}, \frac{4 - 2k}{3}, k)$, where k is any real number.

14. (a) 3 ⊕ 11 = **2**   (11 hours after 3 o'clock is **2** o'clock.)
    (b) 8 ⊕ 9 = **5**   (9 hours after 8 o'clock is **5** o'clock.)
    (c) 3 ⊖ 9 = **6**   (9 hours before 3 o'clock is **6** o'clock.)
    (d) 5 - 12 = **5**   (12 hours before 5 o'clock is **5** o'clock.)

15. (a) 3 ⊗ 5 = **3**   (3 × 5 = 15 and 15 - 12 = **3**.)
    (b) 6 ⊗ 8 = **12**   (6 × 8 = 48, a multiple of **12**.)
    (c) Look at the clock arithmetic multiplication table and go along the row for 11 until you come to 3. The number in the top line is 9. Thus, in clock arithmetic, $\frac{3}{11}$ = **9**. You can check this by noting that 11 × 9 = 99, which leaves a remainder of 3 when divided by 12.
    (d) There is no solution. If you look at the multiplication table and go along the row for 6, you will see that there is no 5 in this row.

16. (a) True. 5 - 2 is a multiple of 3.
    (b) True. 9 - 5 is a multiple of 4.
    (c) False. 9 - 2 = 7 is not a multiple of 6.

17. (a) Since 3 + 2 = 5, the value of n in 3 + 2 ≡ n (mod 5) is **0**.
    (b) Since 4 × 3 = 12 = 2 × 5 + 2, n = **2**.
    (c) 2 - 4 ≡ n (mod 5) is true if 4 + n ≡ 2 (mod 5), which is true if 4 + n = 7, that is. if n = **3**.
    (d) $\frac{3}{2}$ ≡ n (mod 5) is true if 3 ≡ 2n (mod 5), which is true for 2n = 8 that is for n = **4**.

18. (a) 6 + n ≡ 1 (mod 7) is true if 6 + n = 8, that is, if n = 2. Since any multiple of 7 may be added to this solution, the general solution is n = **2 + 7k**, k any integer.
    (b) 3 - n ≡ 4 (mod 7) is true if 3 ≡ n + 4 (mod 7), which is true if n + 4 = 10, that is, if n = **6**. Since any multiple of 7 may be added to this solution, the general solution is n = **6 + 7k**, k any integer.

19. (a) 2n ≡ 1 (mod 3) is true if 2n = 4 + 3k, k any integer. Thus, the general solution is n = **2 + 3k/2**, k any integer.
    (b) $\frac{n}{2}$ ≡ 2 (mod 3) is true if n ≡ 4 (mod 3) = 1 (mod 3), so the general solution is n = **1 + 3k**, k any integer.

20. Yes. The set S is closed with respect to the operation * because all the entries in the table are elements of S.

21. (a) ($ * ¢) * # = # * # = #   (b) $ * (¢ * #) = $ * ¢ = #
    (c) ($ * #) * (% * ¢) = $ * $ = %

22. Yes. The operation is commutative because the table is symmetric to the main diagonal from upper left to lower right. This means that if a and b are any two elements of S, then a * b = b * a.

23. The identity element is **#** because the # column is the same as the column under * and the # row is the same as the top row of the table. This means that if a is any element of s, then a * # = # * a = a.

24. (a) The inverse of # is **itself** because # is the identity element.
    (b) Because $ * ¢ = #, the inverse of $ is **¢**.
    (c) Because % * % = #, the inverse of % is **%** itself.
    (d) Because ¢ * $ = #, the inverse of ¢ is **$**.

25. (a) There is no identity element because no column reads exactly the same as the column under @.
    (b) Since there is no identity element, no element has an inverse.
    (c) No. The table is not symmetric to the diagonal from upper left to lower right.

26. (a) Since the operation S selects the second of the two numbers in a S b, a S (b L C) = b L c. Also, a S b = b and a S c = c, so that
    (a S b) L (a S c) = b L c. Thus, it is true that
    a S (b L c) = (a S b) L (a S c)
    Therefore, S is distributive over L.

    (b) Since b S c = c, it follows that a L (b S c) = a L c. Also, it is true that (a L b) S (a L c) = a L c, because a L c is the second number. Thus, a L (b S c) = (a L b) S (a L c), so L is distributive over S.

27. With the operation of addition mod 3, the set S = {0, 1, 2} has the five properties required to make it a commutative group.
    1. S has the closure property because the sum (mod 3) of any two elements of S is an element of S.
    2. S has the associative property because ordinary addition does.
    3. 0 is the identity element.
    4. Every element of S has an inverse. The inverse of 0 is 0 and the elements 1 and 2 are inverses of each other. [1 + 2 = 2 + 1 = 3 and 3 ≡ 0 (mod 3).]
    5. S has the commutative property because ordinary addition does.

28. With the operation of multiplication mod 3, the set T = {1, 2} has the five properties required to make it a commutative group.
    1. T has the closure property because the product of any two elements (mod 3) is an element of T.
    2. T has the associative property because ordinary multiplication does.
    3. The identity element is 1.
    4. Since $1 \times 1 \equiv 1$ (mod 3) and $2 \times 2 \equiv 1$ (mod 3), the elements 1 and 2 are their own inverses.
    5. T has the commutative property because ordinary multiplication does.

29. Yes. The system has the six required properties (closure, associative, commutative, identity, distributive of multiplication over addition, and inverses, except there is no inverse for zero with respect to multiplication), so the system is a field.

246 CHAPTER 8 MATHEMATICAL SYSTEMS AND MATRICES

30. Let $S = \{0, 1\}$. We check the requirements for S to be a group under the operation $\oplus$.
    1. S has the closure property because all the entries in the table are elements of S.
    2. You can check that the operation $\oplus$ corresponds to addition (mod 2), so S has the associative property under this operation.
    3. The identity element is 0 because $0 \oplus 0 = 0$, and $0 \oplus 1 = 1 \oplus 0 = 1$.
    4. Each row and column of the table has an identity element, so each element of S has an inverse. The table shows that 0 and 1 are both their own inverses.
    Therefore, the set S with the operation $\oplus$ is a **group**.

31. You can check as in problem 30 that the set S under the operation $\otimes$ has the closure and associative properties. The identity element is 1. However, the element 0 has no inverse, so the system is **not** a group.

32. In problems 30 and 31, we have checked all the requirements for a field except the distributive property of $\otimes$ over $\oplus$. Since the operation $\oplus$ corresponds to addition (mod 2) and the operation $\otimes$ corresponds to multiplication (mod 2), the system does have this distributive property. Thus, the system is a field.

33. (a) There is no element that is both the least in its row and the greatest in its column, so there is no saddle point and the game is not strictly determined.

    (b) The 1 in the third row, third column is the least in its row and the greatest in its column, so this is a saddle point. The game is strictly determined and its value is 1.

    (c) The 2's in the bottom row are the least in the row and the greatest in their respective columns, so these are saddle points. The game is strictly determined and its value is 2.

34. The -6 in the first row is the saddle point. The row player should play row 1 and the column player should play column 2. The payoff for the row player is -6.

35. (a) With the given percents written as decimals, the payoff matrix is

$$\begin{array}{c} \text{This Year} \\ \text{Next Year} \end{array} \begin{array}{c} \\ H \\ C \end{array} \begin{bmatrix} \overset{H}{0.6} & \overset{C}{0.4} \\ 0.3 & 0.7 \end{bmatrix}$$

(b) There is no saddle point, so this is not a strictly determined game. If E_R is the expected value for HART and p is the probability that the first row is what will happen, then
$$E_R = (p)(0.6) + (1 - p)(0.3) = (p)(0.4) + (1 - p)(0.7)$$
Thus, $\qquad 0.3p + 0.3 = 0.7 - 0.3p$
or $\qquad\qquad\qquad 0.6p = 0.4$ and $p = 2/3$
Hence, $E_R = (0.3)(2/3) + 0.3 = 0.2 + 0.3 = 0.5$, so that the expected percentage of persons using HART next year is 50%.

CHAPTER 9
COUNTING TECHNIQUES

EXERCISE 9.1

STUDY TIPS To study the ideas of probability in the next chapter, we need to learn how to count the number of ways in which an event can occur. Two counting techniques are presented in this section: tree diagrams and the sequential counting principle (SCP), pages 493-495. Tree diagrams are used when there are not very many outcomes. Example 1 gives you a good illustration. Study the Problem Solving on page 496 and the illustrative examples to understand how the SCP is used.

1. 3.

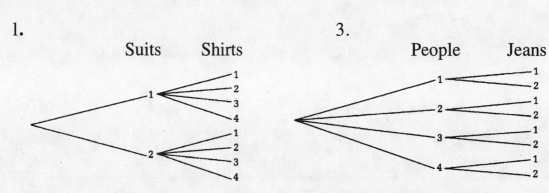

8 different outfits   8 different outcomes

5. (a) $1 \times 3 \times 3 = $ **9**   (b) $3 \times 6 \times 7 = $ **126**   (c) $5 \times 1 \times 5 = $ **25**

7. (a) $4 \times 3 = $ **12**   (b) $4 \times 3 \times 4 \times 3 = $ **144**

9. $5 \times 6 = $ **30**   11. $26^3 = $ **17,576**   13. $7 \times 6 \times 9 = $ **378**

15. $8 \times 2 = $ **16**   17. $9 \times 10 = $ **90**   19. $9 \times 10^8 = $ **900,000,000**

21. $40 \times 39 \times 38 = $ **59,280**

23. (a) **10**   (b) **2**   (c) **4**   (d) $10 \times 2 \times 4 = $ **80**

25. (a) **19**   (b) $19 \times 8 = 152$

27. $10 \times 9 \times 8 = $ **720**

29. $10 \times 6 \times 8 = $ **480**

SECTION 9.1   The Sequential Counting Principle   249

31. (a, @), (a, &), (a, %), (b, @), (b, &), (b, %), (c, @), (c, &), (c, %),
    (@, a), (&, a), (%, a), (@, b), (&, b), (%, b), (@, c), (&, c), (%, c)

33. If a single event can occur in m ways or in n ways, then the total number of ways in which the event can occur is m + n. (This assumes that the m ways and the n ways are all distinct.)

35. $4 \times 3 \times 2 \times 1 = $ **24**

37. Yes. Only 17,576 different sets of initials are possible. Thus, if there are 27,000 people, at least two people must have the same set of initials.

39. $12 \times 12 = $ **144**   41.   It will point to 3 if no slippage occurs.

EXERCISE 9.2

STUDY TIPS   An ordered arrangement of r objects selected from a set of n distinct objects is called a **permutation** and is denoted by P(n, r). The tree diagram on page 504 shows the permutations of the letters a, b, c taken three at a time. Be sure to memorize Definitions 9.1 and 9.2. Two important facts are shown on page 505: **P(n, n) = n!** and **P(n, r) = n(n-1)(n-2)···(n-r+1)**. By defining 0! = 1, we get the useful formula

$$P(n, r) = \frac{n!}{(n-r)!}, r \leq n$$

Note that this formula is used when the order of the objects to be counted is important. (For example, if you are selecting two persons and the first one is to be the boss and the second the worker, the order is important.) Note the complementary counting principle on page 507. Study the examples carefully to see how all these ideas are used.

1. $4 \times 3 \times 2 \times 1 = $ **24**   3. $6 \times 5 \times 4 \times 3 \times 2 \times 1 = $ **720**

5. $5 \times 4 \times 3 \times 2 \times 1 = $ **120**   7. $8! = $ **40,320**

9. $9! = $ **362,880**   11. $\frac{11!}{8!} = \frac{11 \times 10 \times 9 \times (8!)}{8!} = $ **990**

13. $\frac{9!}{5!4!} = \frac{9 \times 8 \times 7 \times 6 \times (5!)}{5! \times 4 \times 3 \times 2 \times 1} = 9 \times 7 \times 2 = $ **126**

15. $P(10,2) = 10 \times 9 = $ **90**.   17. $\frac{P(5,2)}{2!} = \frac{5 \times 4}{2 \times 1} = $ **10**

19. $3 \cdot P(8, 5) = 3 \times 8 \times 7 \times 6 \times 5 \times 4 = $ **20,160**

21. $5! = \mathbf{120}$

23. $P(10, 4) = 10 \times 9 \times 8 \times 7 = \mathbf{5040}$

25. $P(13, 3) = 13 \times 12 \times 11 = \mathbf{1716}$

27. $P(26, 2) = 26 \times 25 = \mathbf{650}$

29. $P(5, 3) = 5 \times 4 \times 3 = \mathbf{60}$

31. $3 \times 3 = \mathbf{9}$

33. There is just one way to have no boys. Therefore, there are $2^5 - 1 = \mathbf{31}$ ways to have at least one boy.

35. There is just one way to have no tails. All the coins must come up heads.

37. The number of ways in which the six coins can fall is $2^6 = 64$. The number of ways of getting no heads is 1 and the number of ways of getting exactly one head is 6. Thus, the number of ways of getting at least 2 heads is $64 - 1 - 6 = \mathbf{57}$.

39. There are **50** multiples of 2; there are **20** multiples of 5; and there are **10** multiples of 2×5. Use the equation
$$n(A \cup B) = n(A) + n(B) - n(A \cap B)$$
with $n(A) = \mathbf{50}$, $n(B) = \mathbf{20}$, and $n(A \cap B) = \mathbf{10}$. Then,
$$n(A \cup B) = 50 + 20 - 10 = \mathbf{60}$$

41. (1) $n! = n \times (n-1)!$ for $n > 1$. This formula holds for $n = 1$ only if $0!$ is defined to be 1.

 (2) The formula $P(n, r) = \dfrac{n!}{(n-r)!}$ holds for $n = r$ only if $0!$ is defined to be 1.

43. The number of elements in the union of two sets is the sum of the number of elements in each of the sets diminished by the number of elements common to the two sets.

45. $5! = \mathbf{120}$

47. **Three**, including a possible tie for fourth place.

49. If one of the persons is seated, then the others can be seated in $(n-1)!$ ways. Excepts for rotation around the table, this is the total number of ways in which the n persons could be seated.

51. After A is seated, there are two ways to seat B next to A (so that B does not face A). Then there are two ways of seating the remaining two persons. Thus, there are $2 \times 2 = \mathbf{4}$ ways of seating the four persons.

EXERCISE 9.3

STUDY TIPS A set of r objects selected from a set of n objects, with the order disregarded is called a **combination** and is denoted by C(n, r). Note the important formulas $C(n, r) = \frac{P(n,r)}{r!} = \frac{n!}{r!(n-r)!}$. When you work with these formulas, use "cancellation" wisely. Thus, to simplify $\frac{7!}{5!2!}$, write the 7! as 7•6•5! and "cancel" the 5!. Do not multiply it out. Study the examples carefully before you try the exercises and remember that combinations are used when the order is **not** important.

1. $C(5, 2) = \frac{5 \times 4}{2 \times 1} = \mathbf{10}$, $P(5, 2) = 5 \times 4 = \mathbf{20}$

3. $C(7, 3) = \frac{7 \times 6 \times 5}{3 \times 2 \times 1} = \mathbf{35}$, $P(7, 3) = 7 \times 6 \times 5 = \mathbf{210}$

5. $C(9, 6) = \frac{9 \times 8 \times 7 \times 6 \times 5 \times 4}{6 \times 5 \times 4 \times 3 \times 2 \times 1} = \frac{9 \times 8 \times 7}{3 \times 2 \times 1} = \mathbf{84}$,

   $P(9, 6) = 9 \times 8 \times 7 \times 6 \times 5 \times 4 = \mathbf{60,480}$

7. $C(5, 4) = \frac{5 \times 4 \times 3 \times 2}{4 \times 3 \times 2 \times 1} = \mathbf{5}$ 9. $C(10, 2) = \frac{10 \times 9}{2 \times 1} = \mathbf{45}$

11. $C(8, 2) = \frac{8 \times 7}{2 \times 1} = \mathbf{28}$ 13. $C(12, 8) = C(12, 4) = \frac{12 \times 11 \times 10 \times 9}{4 \times 3 \times 2 \times 1} = \mathbf{495}$

15. (a) $C(10, 3) = \frac{10 \times 9 \times 8}{3 \times 2 \times 1} = \mathbf{120}$

    (b) $C(10, 0) + C(10, 1) + C(10, 2) = 1 + 10 + \frac{10 \times 9}{2 \times 1} = \mathbf{56}$

    (c) $C(10, 0) = \mathbf{1}$ (d) $C(10, 10) = \mathbf{1}$

17. $C(5, 4) = C(5, 1) = \mathbf{5}$

19. (a) $C(5, 3) = C(5, 2) = \frac{5 \times 4}{2 \times 1} = \mathbf{10}$

    (b) $C(5, 0) + C(5, 1) + C(5, 2) + C(5, 3) = 1 + 5 + 10 + 10 = \mathbf{26}$

252   CHAPTER 9   COUNTING TECHNIQUES

21. $C(24,3) = \dfrac{24 \times 23 \times 22}{3 \times 2 \times 1} = 4 \times 23 \times 22 = \mathbf{2024}$

23. $C(10,3) = \dfrac{10 \times 9 \times 8}{3 \times 2 \times 1} = 5 \times 3 \times 8 = \mathbf{120}$

25. $C(52, 5) = \dfrac{52 \times 51 \times 50 \times 49 \times 48}{5 \times 4 \times 3 \times 2 \times 1} = \mathbf{2{,}598{,}960}$

27. $C(100,5) = \dfrac{100 \times 99 \times 98 \times 97 \times 96}{5 \times 4 \times 3 \times 2 \times 1} = \mathbf{75{,}287{,}520}$

29. $C(4, 1) + C(4, 2) + C(4, 3) + C(4, 4) = 2^4 - 1 = \mathbf{15}$

31. $C(6, 1) + C(6, 2) + 6(6, 3) + C(6, 4) + C(6, 5) + C(6, 6) = 2^6 - 1 = \mathbf{63}$

33. This is the number of ways that 4 of the 8 people can be selected without regard to order:
$$C(8,4) = \dfrac{8 \times 7 \times 6 \times 5}{4 \times 3 \times 2 \times 1} = \mathbf{70}$$

35. The important difference is that permutations take account of order and combinations do not.

37. The order of the numbers is important, so a permutation is being used.

39. $n = 6$:

| | 1 | 1+5 ↓ 6 | 5+10 ↓ 15 | 10+10 ↓ 20 | 10+5 ↓ 15 | 5+1 ↓ 6 | 1 |

$n = 7$:

| | 1 | 1+6 ↓ 7 | 6+15 ↓ 21 | 15+20 ↓ 35 | 20+15 ↓ 35 | 15+6 ↓ 21 | 6+1 ↓ 7 | 1 |

41. (a) $(a + b)^4 = a^4 + 4a^3b + 6a^2b^2 + 4ab^3 + b^4$
    (b) $(a + b)^5 = a^5 + 5a^4b + 10a^3b^2 + 10a^2b^3 + 5ab^4 + b^5$

43. $C(5, 0) = \mathbf{1}$ 45. $C(5, 3) = \dfrac{5 \times 4 \times 3}{3 \times 2 \times 1} = \mathbf{10}$ 47. $C(5, 5) = \mathbf{1}$

49. The left side is the sum of the number of ways in which there could be 0 heads and n tails, 1 head and n - 1 tails, 2 heads and n - 2 tails, and so on, to n heads and 0 tails. The right side is exactly the number of ways in which n coins can fall either heads or tails. Thus,
$$C(n, 0) + C(n, 1) + C(n, 2) + \cdots + C(n, n) = \mathbf{2^n}$$

SECTION 9.4 Miscellaneous Counting Problems

EXERCISE 9.4

STUDY TIPS — In this section we give you a diagram (Page 519) to help you decide when to use permutations and when to use combinations. Examples 1-4 will tell you how to apply these counting methods for sets of distinct objects. The remainder of this section deals with sets of nondistinct objects and the examples illustrate how to do the counting for such sets.

1. (a) $P(52, 3) = 52 \times 51 \times 50 = \mathbf{132{,}600}$
   (b) $C(52, 3) = \dfrac{P(52, 3)}{3!} = \mathbf{22{,}100}$

3. (a) $50 \times 50 \times 50 = \mathbf{125{,}000}$ (The plays do not have to be all different.)
   (b) $C(50, 3) = \dfrac{P(50, 3)}{3!} = \mathbf{19{,}600}$

5. $C(5, 3) = \dfrac{5 \times 4 \times 3}{3 \times 2 \times 1} = \mathbf{10}$

7. (a) $C(7, 3) = \dfrac{7 \times 6 \times 5}{3 \times 2 \times 1} = \mathbf{35}$

   (b) Peter can select the 3 hours in $C(4, 3) = \mathbf{4}$ ways and the courses in $P(7, 3) = 7 \times 6 \times 5 = \mathbf{210}$ ways, so he can arrange a schedule in $C(4, 3)P(7, 3) = 4 \times 210 = \mathbf{840}$ ways.

9. (a) $P(14, 2)P(10, 2) = 14 \times 13 \times 10 \times 9 = \mathbf{16{,}380}$
   (b) $P(12, 2)P(10, 2) = 12 \times 11 \times 10 \times 9 = \mathbf{11{,}880}$

11. (a) $C(7, 3)C(8, 1) = \dfrac{7 \times 6 \times 5}{3 \times 2 \times 1} \times \dfrac{8}{1} = \mathbf{280}$
    (b) $C(5, 3)C(2, 1) = \dfrac{5 \times 4 \times 3}{3 \times 2 \times 1} \times \dfrac{2}{1} = \mathbf{20}$

13. There are 11 letters in all: 1 T, 3 A's, 2 L's, 1 H, 2 S's, and 2 E's, so the answer is $\dfrac{11!}{1!\,3!\,2!\,1!\,2!\,2!} = \mathbf{831{,}600}$

15. There are 9 letters in all: 2 R's, 2 E's, 2 D's, 2 I's, and 1 V, so the answer is $\dfrac{9!}{2!\,2!\,2!\,2!\,1!} = \mathbf{22{,}680}$

17. The contractor can select 2 components from the first subcontractor in C(7, 2) ways, 3 components from the second subcontractor in C(5, 3) ways, and 2 components from the third subcontractor in C(2, 2) ways. Thus, he can select his 7 components in
$$C(7,2)C(5,3)C(2,2) = \frac{7\times 6}{2\times 1} \times \frac{5\times 4\times 3}{3\times 2\times 1} \times \frac{2\times 1}{2\times 1} = 21 \times 10 \times 1 = \mathbf{210} \text{ ways.}$$

19. We have to choose the least value of n for which C(n, 5) ≥ 21, that is, for which $\frac{n(n-1)(n-2)(n-3)(n-4)}{5!} \geq 21$. To simplify the calculation, multiply both sides by 5! = 120:
$$n(n-1)(n-2)(n-3)(n-4) \geq 2520.$$
Try n = 6. The left side becomes $6 \times 5 \times 4 \times 3 \times 2 = 720$. Too small! Try n = 7. The left side becomes $7 \times 6 \times 5 \times 4 \times 3 = 2520$. Thus, to carry out the program, the network must have at least **7** different movies.

21. She can select 2 of the 3 courses and the order in which they come in the morning. Then she can select 2 of the 4 morning hours for these 2 courses. That leaves her the choice of 1 of the 2 afternoon hours for the third course. Thus, she has
$$P(3, 2)C(4, 2)C(2, 1) = 3\times 2 \times \frac{4\times 3}{2\times 1} \times \frac{2}{1} = 6 \times 6 \times 2 = \mathbf{72} \text{ choices.}$$

23. Answers will vary. Read the text to see what each of the SCP, P(n, k) and C(n, k) formulas calculates.

25. $144 = 2^4 3^2$. The factor 2 can be used 0, 1, 2, 3, or 4 times, and the factor 3 can be used 0, 1, or 2 times. So the number of exact divisors is $(4 + 1)(2 + 1) = 5 \times 3 = \mathbf{15.}$

27. The 2 can be a factor 0, 1, 2, . . ., or a times; (a+ 1) choices.
    The 3 can be a factor 0, 1, 2, . . ., or b times; (b + 1) choices.
    The 5 can be a factor 0, 1, 2, . . ., or c times; (c+ 1) choices.
    The 7 can be a factor 0, 1, 2, . . ., or d times; (d +1) choices.
    Thus, the number of exact divisors is
    $$\mathbf{(a + 1)(b + 1)(c + 1)(d + 1).}$$

PRACTICE TEST 9

STUDY TIPS Your main task is to know when to use the formulas given in the text. Once again, remember: permutations are used when the order is important and combinations when the order is **not** important. Example: If two people are sent to do a job and the first one that arrives makes twice as much money as the second one, the order is important! If two people are sent to do a job and they are both assigned to a cleaning crew, the order is not important. For situations in which the number of outcomes is small, make a tree diagram.

1.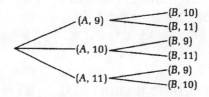

2. $2 \times 3 \times 5 = \mathbf{30}$ different meals are possible.

3. (a) Since each die can fall in 6 different ways, $6 \times 6 = \mathbf{36}$ different results are possible.
   (b) To get a sum of 5, one die must come up 1 and the other 4, or one must come up 2 and the other 3. Since each of these events can occur in 2 ways, the total number of ways is $2 \times 2 = \mathbf{4}$.

4. There are 2 black jacks and 26 red cards, so the total number of ways to get a black jack followed by a red card is $2 \times 26 = \mathbf{52}$

5. You have a choice of 3 flights followed by a choice of 5 flights, so the answer is $3 \times 5 = \mathbf{15}$.

6. (a) $7! = 7 \times 6 \times 5 \times 4 \times 3 \times 2 \times 1 = \mathbf{5040}$
   (b) $\dfrac{7!}{4!} = \dfrac{7 \times 6 \times 5 \times 4 \times 3 \times 2 \times 1}{4 \times 3 \times 2 \times 1} = 7 \times 6 \times 5 = \mathbf{210}$

7. (a) $3! \times 4! = 3 \times 2 \times 1 \times 4 \times 3 \times 2 \times 1 = \mathbf{144}$
   (b) $3! + 4! = 3 \times 2 \times 1 + 4 \times 3 \times 2 \times 1 = 6 + 24 = \mathbf{30}$

8. (a) $P(5, 5) = 5! = 5 \times 4 \times 3 \times 2 \times 1 = \mathbf{120}$
   (b) $P(6, 6) = 6! = 6 \times 5 \times 4 \times 3 \times 2 \times 1 = \mathbf{720}$

9. (a) $P(8, 2) = 8 \times 7 = \mathbf{56}$ (b) $P(7, 3) = 7 \times 6 \times 5 = \mathbf{210}$

10. They can be arranged in $P(4, 4) = 4! = \mathbf{24}$ ways.

11. Each couple can be seated in 2 ways, and the 3 couples can be arranged in $P(3, 3)$ ways, so the answer is $2^3 \times 3! = \mathbf{48}$.

12. There are 2 white birds to choose from and 3 ways to select them (first, second or third) and there are $P(4,2)$ ways of selecting two non-white birds. Thus, the number of ways to select exactly one white bird is $2 \times 3 \times P(4,2) = 6 \times 12$ or $\mathbf{72}$.

13. 24 of the numbers are divisible by 2, 9 of the numbers are divisible by 5, and 4 of the numbers are divisible by both 2 and 5. Thus, the number divisible by 2 or by 5 is
$$24 + 9 - 4 = \mathbf{29}.$$

14. The number of different sums is $C(4, 2) = \dfrac{4 \times 3}{2 \times 1} = \mathbf{6}$.

15. (a) $C(5, 2) = \dfrac{5 \times 4}{2 \times 1} = \mathbf{10}$ (b) $C(6, 4) = C(6, 2) = \dfrac{6 \times 5}{2 \times 1} = \mathbf{15}$

16. (a) $C(6, 0) = \mathbf{1}$
    (b) $C(5, 4) = C(5, 1) = \mathbf{5}$ and $C(5, 3) = C(5, 2) = \dfrac{5 \times 4}{2 \times 1} = \mathbf{10}$. Thus,
    $$\dfrac{C(5, 4)}{C(5, 3)} = \dfrac{5}{10} = \mathbf{\dfrac{1}{2}}.$$

17. The number of subsets is $C(6, 3) = \dfrac{6 \times 5 \times 4}{3 \times 2 \times 1} = \mathbf{20}$.

18. The number of different schedules is $P(4, 4) = 4! = \mathbf{24}$.

19. The number of different sets is $C(52, 2) = \dfrac{52 \times 51}{2 \times 1} = \mathbf{1326}$.

20. The number of different sums is
$$C(5,1) + C(5,2) + C(5,3) = \dfrac{5}{1} + \dfrac{5 \times 4}{2 \times 1} + \dfrac{5 \times 4 \times 3}{3 \times 2 \times 1} = 5 + 10 + 10 = \mathbf{25}.$$

21. The number of choices is $C(8, 4) = \dfrac{8 \times 7 \times 6 \times 5}{4 \times 3 \times 2 \times 1} = \mathbf{70}$.

SECTION 9.4   PRACTICE TEST 9   257

22. There are 8 letters: 2 B's, 4 O's, 1 G and 1 A, so the number of distinct arrangements is
$$\frac{8!}{2!4!1!1!} = \frac{8\times7\times6\times5\times4\times3\times2\times1}{2\times1\times4\times3\times2\times1\times1\times1} = \mathbf{840}.$$

23. There are 11 letters: 4 M's, 4 A's, 2 D's and 1 I, so the number of distinct arrangements is
$$\frac{11!}{4!4!2!1!} = \frac{11\times10\times9\times8\times7\times6\times5\times4\times3\times2\times1}{4\times3\times2\times1\times4\times3\times2\times1\times2\times1\times1} = \mathbf{34{,}650}.$$

24. There are 5 choices for the foreman, followed by C(4, 2) choices for the helpers. Thus, the total number of choices is
$$5C(4, 2) = 5 \times \frac{4\times3}{2\times1} = \mathbf{30}.$$

25. There are 1977 marks in all: 1189 a's, 460 d's, and 328 n's, so the number of distinct arrangements is
$$\frac{\mathbf{1977!}}{\mathbf{1189!460!328!}}.$$

CHAPTER 10
PROBABILITY

EXERCISE 10.1

STUDY TIPS In the study of probability, tossing a coin and drawing a card from a deck are activities called **experiments** and the set of all possible outcomes is the **sample space**. In this section, we define the probability of an event. Be sure that you learn Definition 10.1 and understand that the probability of an event is always a number between **0** and **1**. If the event cannot occur its probability is **0** and if the event is certain to occur its probability is **1**. To learn how to compute the probability of an event, study the Problem Solving, pages 533-534. Examples 2-7 illustrate how to find the probability of an event when the events being considered are all equally likely. Make sure you remember the sample space in Example 6, it will be used again!

1. There are 6 possible outcomes of which exactly one is favorable, so
    $$P(E) = \frac{1}{6}.$$

3. There are 2 favorable outcomes and 6 possible outcomes, so
    $$P(E) = \frac{2}{6} = \frac{1}{3}.$$

5. There are 10 possible outcomes of which exactly one is favorable, so
    $$P(E) = \frac{1}{10}.$$

7. There are 9 favorable outcomes and 10 possible outcomes, so $P(E) = \frac{9}{10}$.

9. There are no favorable outcomes possible, so $P(E) = 0$.

11. There are 4 aces and 52 cards in all, so $P(\text{ace}) = \frac{4}{52} = \frac{1}{13}$.

13. There are 13 spades out of 52 cards, so $P(\text{spade}) = \frac{13}{52} = \frac{1}{4}$.

SECTION 10.1   Sample Spaces and Probability

15. There are 13 spades and 9 picture cards that are not spades, so there are $13 + 9 = 22$ favorable outcomes out of 52 possible outcomes. Therefore, $P(E) = \dfrac{22}{52} = \dfrac{11}{26}$.

17. (a) There is 1 favorable outcome out of 5 possible outcomes, so
$$P(E) = \dfrac{1}{5}.$$
(b) There are 3 favorable outcomes out of 5 possible outcomes so
$$P(E) = \dfrac{3}{5}.$$
(c) There are 4 favorable outcomes out of 5 possible outcomes, so
$$P(E) = \dfrac{4}{5}.$$

19. (a) There is 1 favorable outcome out of 3 possible outcomes, so
$$P(E) = \dfrac{1}{3}.$$
(b) There is 1 favorable outcome out of 3 possible outcomes, so
$$P(E) = \dfrac{1}{3}.$$

21. There is 1 favorable outcome and 8 possible outcomes, so
$$P(E) = \dfrac{1}{8}.$$

23. There are 3 favorable outcomes and 8 possible outcomes, so
$$P(E) = \dfrac{3}{8}.$$

25. (a) There are 650 favorable outcomes and 850 possible outcomes, so
$$P(E) = \dfrac{650}{850} = \dfrac{13}{17}.$$
(b) There are 200 favorable outcomes and 850 possible outcomes, so
$$P(E') = \dfrac{200}{850} = \dfrac{4}{17}.$$

This problem can also be done by using the result of part (a) and the equation $P(E') = 1 - P(E)$ to get the same answer:
$$P(E') = 1 - \dfrac{13}{17} = \dfrac{4}{17}.$$

260 CHAPTER 10 PROBABILITY

27. (a) There are 660 favorable outcomes and 1500 possible outcomes, so
$$P(F) = \frac{660}{1500} = \frac{11}{25}.$$
(b) There are 840 favorable outcomes and 1500 possible outcomes, so
$$P(M) = \frac{840}{1500} = \frac{14}{25}.$$

29. (a) There are 160 favorable outcomes and 1500 possible outcomes, so
$$P(E) = \frac{160}{1500} = \frac{8}{75}.$$
(b) There are 300 favorable outcomes and 1500 possible outcomes, so
$$P(E) = \frac{300}{1500} = \frac{1}{5}.$$

31. There are 6 favorable outcomes and 36 possible outcomes, so
$$P(E) = \frac{6}{36} = \frac{1}{6}.$$

33. There are 6 favorable outcomes and 36 possible outcomes, so
$$P(E) = \frac{6}{36} = \frac{1}{6}.$$

35. There are 18 favorable outcomes and 36 possible outcomes, so
$$P(E) = \frac{18}{36} = \frac{1}{2}.$$

37. The probability formula does not apply if the events are not all equally likely to occur. For example, if a die is weighted so that a 6 is twice as likely to come up as any other number, then to calculate the probability that an even number comes up, it would be wrong to use the fact that three of the six faces are even so that the probability is 1/2. Instead, the 6 face must be given a weight of 2 and the other faces weights of 1. Then the weight of the even faces is 1 + 1 + 2 = 4 and the weight of all the faces is 1 + 1 + 1 + 1 + 1 + 2 = 7. This means that the probability that an even number comes up is 4/7.

39. There were 465,300 successes and 472,000 operations, so the probability of a success is $P(S) = \frac{465300}{472000} = \frac{4653}{4720}$ or about **0.986**.

41. There were 506,000 successes and 508,000 operations, so the probability of a success is $P(S) = \frac{506}{508} = \frac{253}{254}$ or about **0.996**.

SECTION 10.2 Counting Techniques and Probability

EXERCISE 10.2

STUDY TIPS This section discusses the computation of probabilities by using a tree diagram and by using permutations and combinations. Keep in mind that permutations are used when the **order** is important and combinations when the order is **not** important. Note that some problems can be worked using permutations, combinations or the SCP. Work them the way you understand best! (See Examples 4 and 5.) Study all the examples very carefully so that you learn how to work the problems.

1. The total number of possibilities is $3 \times 2 \times 3 = 18$ and the number favorable is 1. Therefore, the probability is $\frac{1}{18}$.

3. In 1 throw, P(even number) = 1/2, so in 3 throws,
$$P(\text{even number}) = (\frac{1}{2})^3 = \frac{1}{8}.$$

5. The total number of possible choices for the committee of four is $C(25,4)$ and the total number of choices of the remaining committee members if both Helen and Patty are selected is $C(23,2)$. So the probability that both Helen and Patty will be selected is

$$P = \frac{C(23,2)}{C(25,4)} = \frac{23 \times 22}{2 \times 1} \times \frac{4 \times 3 \times 2 \times 1}{25 \times 24 \times 23 \times 22} = \frac{1}{50}.$$

7. (a) Suppose that Mr. C. Nile chooses a restaurant. The probability that Mr. D. Mented goes to that restaurant is **1/5**.
   (b) The probability that they miss each other is 1 - 1/5 = **4/5**.

9. (a) To get 35¢, a person must get first a dime and then a quarter or first a quarter and then a dime. Since there are 2 dimes and 2 quarters, there are $2 \times (2 \times 2) = 8$ ways of getting 35¢. There are 7 coins, so there are $7 \times 6 = 42$ ways of getting 2 coins. Thus,
$$P(35¢) = \frac{8}{42} = \frac{4}{21}.$$
   (b) To get 50¢, a person must get the 2 quarters in either order, and there are 2 ways of doing this. Hence,
$$P(50¢) = \frac{2}{42} = \frac{1}{21}.$$

11. $P(\text{2 kings}) = \dfrac{P(4, 2)}{P(52, 2)} = \dfrac{4 \times 3}{52 \times 51} = \dfrac{1}{13 \times 17} = \dfrac{1}{\mathbf{221}}$

13. There are 13 spades and 3 kings (other than the king of spades), so there are 13×3 ways of getting a spade and one of the 3 kings in that order. The number of ways of drawing 2 cards is 52×51. Thus, the probability of drawing a spade and one of the other 3 kings, in that order, is
$$\dfrac{13 \times 3}{52 \times 51} = \dfrac{1}{4 \times 17} = \dfrac{1}{\mathbf{68}}.$$

15. Since there are 26 red cards and 52 cards in all, the probability of drawing 2 red cards in 2 draws is
$$P(\text{2 red cards}) = \dfrac{P(26,2)}{P(52,2)} = \dfrac{26 \times 25}{52 \times 51} = \dfrac{\mathbf{25}}{\mathbf{102}}.$$

17. (a) There are 5 $50 bills and 3 $10 bills, so there are $5 \times 3 = \mathbf{15}$ ways of getting a $50 bill and a $10 bill in that order.

    (b) There are $5 + 4 + 3 + 2 + 1 = 15$ bills in all, so there are 15×14 ways of selecting 2 bills. Thus, the probability of getting a $5 bill and a $10 bill, in that order, is
    $$\dfrac{5 \times 3}{15 \times 14} = \dfrac{1}{\mathbf{14}}.$$

    (c) Since there are 2 ways to get a $50 bill and a $10 bill (without regard to order), the probability is twice that found in part (b):
    $$2 \times \dfrac{1}{14} = \dfrac{1}{\mathbf{7}}.$$

19. The probability that at least one of the two tires is defective is
    $1 - P(\text{neither is defective}) = 1 - (0.98)^2 = 1 - 0.9604 = \mathbf{0.0396}.$

21. The probability is $0.10 \times 0.05 = \mathbf{0.005}$

SECTION 10.2 Counting Techniques and Probability

23. (a) Since the order does not matter, we can use combinations rather than permutations. There are 4 kings and 4 aces and 44 other cards in the deck of 52 cards. Therefore, the probability of getting 2 kings, 2 aces, and one other card is

$$\frac{C(4,2)C(4,2)C(44,1)}{C(52,5)}.$$

(b) The probability of getting 3 kings and 2 aces is

$$\frac{C(4,3)C(4,2)}{C(52,5)}.$$

25. The number of different combinations of 2 of the 5 applicants is $C(5, 2) = 10$, and there are 4 ways of selecting 1 of the 4 men. Thus, the probability that 1 man and the 1 woman are hired is

$$\frac{4}{10} = \frac{2}{5}.$$

27. There are 4 possible royal flushes, so the probability of getting a royal flush is

$$P = \frac{4}{C(52,5)} = \frac{1}{649740} \approx 0.0000015.$$

29. There are 13 sets of four of a kind and if one of these occurs, the fifth card can be any one of the remaining 48 cards. Thus, the probability of getting four of a kind is

$$P = \frac{13 \times 48}{C(52,5)} = \frac{1}{4165} \approx 0.00024.$$

31. The number of combinations of 5 cards in a given suit is $C(13, 5)$ and 10 of these are straights (including a royal flush). Thus, the number of flushes that are not straights is $C(13, 5) - 10$. The probability of getting such a flush is

$$P = \frac{4 \times [C(13,5) - 10]}{C(52,5)} = \frac{5148 - 40}{C(52,5)} = \frac{5108}{C(52,5)} = \frac{1277}{649740} \approx 0.0020.$$

33. No. The coin is probably weighted to come up heads. **Bet on heads.**

35. No. The coin is probably weighted to come up heads. **Bet on heads.**

37. The number of combinations of 150 women that could be selected is C(40,000, 150) and the number of combinations of 150 men is C(36,000, 150). The number of combinations of 300 persons that could be selected is C(76,000, 300). Hence, the probability that 150 men and 150 women will be selected is

$$\frac{C(40,000, 150)C(35,000, 150)}{C(76,000, 300)}.$$

39. The number of possible combinations of six women is C(16, 6) and of six men is C(14, 6). The number of combinations of 12 people selected from the 30 is C(30,12). Thus, the probability that the final jury will consist of six men and six women is

$$\frac{C(16,6)C(14,6)}{C(30,12)}.$$

SECTION 10.3 Computation of Probabilities

EXERCISE 10.3

STUDY TIPS This section offers additional work with the computation of probabilities. Properties 1 and 2, page 549, are restatements of ideas mentioned earlier. If you have two events that may overlap, then you have to use Property 3, which says that the probability of A **or** B is the probability of A plus the probability of B minus the probability of A **and** B. The examples show how this property is used. To learn how to use diagrams to solve probability problems, look at the Discovery part of Exercise 10.3.

1. $E = \emptyset$, so $P(E) = \mathbf{0}$ (Property **1**)

3. Since the sum is always between 0 and 13, $P(E) = \mathbf{1}$. (Property **2**)

5. Let A be the event that the number is even and B the event that the number is greater than 7. Then
$$P(A) = \frac{5}{10},\ P(B) = \frac{3}{10},\ P(A \cap B) = \frac{2}{10}.$$
Now, we use Equation (3), $P(A \cup B) = P(A) + P(B) - P(A \cap B)$, to get $P(A \cup B) = \frac{5}{10} + \frac{3}{10} - \frac{2}{10} = \frac{6}{10} = \frac{\mathbf{3}}{\mathbf{5}}.$

7. Since E is certain to occur, $P(E) = \mathbf{1}$.

9. There are 14 favorable outcomes and 52 possible outcomes, so
$$P(E) = \frac{14}{52} = \frac{\mathbf{7}}{\mathbf{26}}.$$

11. There are 13 diamonds and 52 cards in all, so
$$P(E) = \frac{13}{52} = \frac{\mathbf{1}}{\mathbf{4}}.$$

13. There are 12 picture cards and 52 cards in all, so
$$P(E) = \frac{12}{52} = \frac{\mathbf{3}}{\mathbf{13}}.$$

15. Let N be the event that the stock remained unchanged. Since 9 stocks remained unchanged, $P(N) = \frac{9}{50}$. The event that the stock did not change is N', so the probability that a randomly selected stock did change is

$$P(N') = 1 - P(N) = 1 - \frac{9}{50} = \mathbf{\frac{41}{50}}.$$

17. Let E_4 be the event that more than 3 persons are in line. From the table, we read that $P(E_4) = \mathbf{0.20}.$

19. The table gives 0.10 as the probability that 0 persons are in line, and 0.15 that 1 person is in line, so the probability that less than 2 persons are in line is 0.10 + 0.15 = 0.25. Since the event E_4 that more than 3 persons are in line and the event, say E_2, that less than 2 persons are in line are mutually exclusive, $E_2 \cap E_4 = \emptyset$. Therefore, the probability that more than 3 or less than 2 persons are in line is

$$P(E_2 \cup E_4) = P(E_2) + P(E_4) = 0.25 + 0.20 = \mathbf{0.45}.$$

21. Using Table 10.6, we find the probability that a person who is alive at 20 is still alive at age 65 is

$$P(E) = \frac{493}{926}.$$

Hence, the probability that a person who is alive at age 20 is not alive at age 65 is

$$P(E') = 1 - P(E) = 1 - \frac{493}{926} = \mathbf{\frac{433}{926}}.$$

23. From Table 10.6, we find the probability that a person who is alive at age 25 will be alive at age 70 to be

$$P(E) = \frac{386}{890}.$$

Hence, the probability that a person who is alive at age 25 is not alive at age 70 is

$$P(E') = 1 - P(E) = 1 - \frac{386}{890} = \frac{504}{890} = \mathbf{\frac{252}{445}}.$$

25. Problem 24 assumes that none of the persons in the table lived to age 100. Hence, the probability that a person who is alive at age 55 will live less than 80 more years, is $P(E) = \mathbf{1}.$

SECTION 10.3 Computation of Probabilities

27. The table gives the number of correct forms with no itemized deductions as $15 + 40 = 55$. Since the total number of forms is 100, the probability that a form had no itemized deductions and was correctly filled out is
$$P(E) = \frac{55}{100} = \frac{11}{20}.$$

29. The table gives the number of short forms as 20, so the number of long forms was 80. Hence, the probability that a form was not a short form is
$$P(E) = \frac{80}{100} = \frac{4}{5}.$$

31. The green and yellow lights are on for a total of 65 seconds and the entire cycle takes 85 seconds. Thus, the probability of finding the light green or yellow is
$$P(E) = \frac{65}{85} = \frac{13}{17}.$$

33. The probability of an event being 0, means that the event **cannot occur**.

35. If A and B have no elements in common, so that $A \cap B = 0$, then $P(A \cup B) = P(A) + P(B)$.

37. We can tabulate John Dough's points as follows:

Item	Points
Age 27	5
Time at Address, 3 yr	5
Age of Auto, 2 yr	16
Monthly Auto Payment $200	0
Housing Cost $130/month	10
Checking Account Only	2
No Finance Co. Reference	15
Major Credit Cards 1	5
Debt to Income Ratio 12%	20
Total Score	78

    From the table showing the probability of repayment, we estimate that John's probability is between 0.81 and 0.84, say **about 0.83**.

39. Answer will depend on the number of points you score.

41. The Venn diagram shows the data given for this problem.

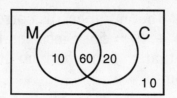

M: Married
C: College Graduate

The number of persons in M ∪ C is 10 + 60 + 20 = 90, so the probability that a person selected at random is married or a college graduate is $P(E) = \frac{90}{100} = \frac{9}{10}$.

43. The Venn diagram shows how the voters voted. Note that the numbers add up to 950, so 50 persons did not vote at all.

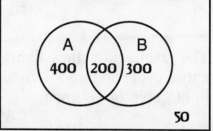

The diagram shows that 400 voted for A and not B. Thus, the probability that an eligible voter selected at random voted for A and not for B is $P(E) = \frac{400}{1000} = \frac{2}{5}$.

45. Since 200 voted for both A and B, the probability that an eligible voter selected at random voted for both A and B is $P(E) = \frac{200}{1000} = \frac{1}{5}$.

47. Since 50 did not vote at all, the probability that an eligible voter selected at random did not vote at all is $P(E) = \frac{50}{1000} = \frac{1}{20}$.

SECTION 10.4   Conditional Probability   269

EXERCISE 10.4

STUDY TIPS This section deals with the computation of probabilities using a subset of the sample space. This type of probability is called **conditional** probability. The problems can be worked using the formula in Definition 10.2, page 557, or using the given information to reduce the sample space. For your convenience, we work the examples both ways, labeling the techniques used Method 1 and Method 2. Note that you get the same answer, so pick the technique you understand best and try to stick to it.

1. Let F be the event that a 5 came up and let E be the event that an even number came up. Then $F \cap E = \emptyset$, so $P(F \cap E) = 0$ and $P(F \mid E) = \mathbf{0}$.

3. (a) Let A be the event that the numbers coming up were identical, and let B be the event that the sum was 8. There is only one way for the sum to be 8 and the numbers to be identical: that is for both numbers to be 4. Since the probability that both numbers were 4's is 1/36, and the probability that the numbers were identical is 1/6, we have $P(B \mid A) = \dfrac{P(A \cap B)}{P(A)} = \dfrac{1/36}{1/6} = \dfrac{\mathbf{1}}{\mathbf{6}}$.

   (b) It is impossible to get a sum of 9 if the two numbers coming up are identical, so   P(9 | numbers identical) = **0**.

   (c) If the two numbers coming up are identical, then the sum is even, so   P(even sum | numbers identical) = **1**.

   (d) It is impossible to get an odd sum if the two numbers coming up are identical, so   P(odd sum | numbers identical) = **0**.

5. $P[BB \mid (BB \text{ or } BG)] = \dfrac{P[BB \text{ and } (BB \text{ or } BG)]}{P(BB \text{ or } BG)} = \dfrac{1/4}{1/2} = \dfrac{\mathbf{1}}{\mathbf{2}}$.

7. If the first child is a girl and there are exactly two girls, then we must have GGB or GBG. Since there are $2^3 = 8$ possibilities,
$$P(\text{GGB or GBG}) = 1/4.$$
The probability that the first child is a girl is 1/2. Therefore, the probability that there are exactly 2 girls given that the first child is a girl is
$$P(\text{exactly 2 G} \mid \text{1st child is G}) = \frac{P(\text{GGB or GBG})}{P(\text{1st child is G})} = \frac{1/4}{1/2} = \frac{1}{2}.$$

9. (a) From the table, we have the probability that an employee missed 6-10 days, given that the employee is a woman,
$$P = \frac{20\%}{100\%} = \frac{1}{5}.$$
(b) Suppose there were 100 men and 100 women employees. The table shows that 40 men and 20 women missed 6-10 days. Thus, the probability that the employee is a woman, given that the employee missed 6-10 days is
$$P = \frac{20}{60} = \frac{1}{3}.$$

11. (a) The table shows that 55% of the students received neither an A nor a B. Thus, the probability that a randomly selected student received neither an A nor a B is $P = \dfrac{55\%}{100\%} = \dfrac{11}{20}$.

(b) The table shows that 40% of the students received either a B or a D. If 200 students were randomly selected, 40% of 200 = **80** would be expected to have received a B or a D.

(c) The table shows that 50% of the students received a grade of A, B or F, and 30% received a grade of B. Thus, if it is known that one of the students did not receive a grade of C or D, the probability that this student received a B is $P = \dfrac{30\%}{50\%} = \dfrac{3}{5}$.

13. If the first card was a king, there are 3 kings left in the remaining 51 cards, so the probability that the second card is a king is
$$P = \frac{3}{51} = \frac{1}{17}.$$

15. There are 25 low-risk stocks, of which 5 are computer stocks, so the probability that the person selected a computer stock, given that the person selected a low-risk stock is
$$P = \frac{5}{25} = \frac{1}{5}.$$

SECTION 10.4 Conditional Probability

17. $P(B \mid A) = \dfrac{P(A \cap B)}{P(A)} = \dfrac{0.4}{0.6} = \dfrac{2}{3}$

19. (a) $P = \dfrac{72}{72 + 168} = \dfrac{72}{240} = \dfrac{3}{10}$

    (b) $P = \dfrac{72 + 84}{72 + 84 + 168 + 76} = \dfrac{156}{400} = \dfrac{39}{100}$

21. 1294 is the only one of the four dates whose first three digits form a number divisible by 3, so Billy had to choose from 1492, 1249, and 1429. The probability that he guessed right is **1/3**.

23. (a) The table shows that 135 of the 200 patients have improved. Thus, the probability that a patient chosen at random has improved is $P(I) = \dfrac{135}{200} = \dfrac{\mathbf{27}}{\mathbf{40}}$.

    (b) The table shows that 70 of the 100 patients taking the experimental drug have improved, so the probability that one of these patients chosen at random has improved is $P(I \mid E) = \dfrac{70}{100} = \dfrac{\mathbf{7}}{\mathbf{10}}$.

25. Answers will vary.

27. The total number of crimes listed in the table is 128.

    (a) The probability that the victim of one of these crimes was a male is
    $P(\text{male}) = \dfrac{5 + 18 + 52}{128} = \dfrac{\mathbf{75}}{\mathbf{128}}.$

    (b) The probability that the victim was a female is
    $P(\text{female}) = \dfrac{2 + 9 + 42}{128} = \dfrac{\mathbf{53}}{\mathbf{128}}.$

    (c) Male; [P(male) > P(female)]

29. $P(\text{assault} \mid \text{female}) = \dfrac{9}{2 + 9 + 42} = \dfrac{\mathbf{9}}{\mathbf{53}}$

CHAPTER 10 PROBABILITY

EXERCISE 10.5

STUDY TIPS To determine whether two events are **independent**, that is, whether the occurrence of one of them affects the probability of the other, we have to study Definitions 10.3 and 10.4. When specified events are independent, the probability of the event that all of them occur is just the product of their separate probabilities as illustrated in Examples 1-4. Be sure to learn what is meant by a **stochastic process**, and study Examples 5-7. For experiments involving repeated trials of the same experiment with two outcomes we need the **binomial** probability formula, studied in the Using Your Knowledge part of Exercise 10.5.

1. $P(E_1 \mid E_2) = P(E_1)$, so E_1 and E_2 are **independent**.

3. (a) $\frac{1}{4} \times \frac{1}{8} \times \frac{1}{3} = \frac{1}{96}$ (b) $\frac{1}{4} \times \frac{1}{8} \times \frac{2}{3} = \frac{1}{48}$ (c) $\frac{3}{4} \times \frac{7}{8} \times \frac{2}{3} = \frac{7}{16}$

5. (a) $P(HTH) = \frac{1}{2} \times \frac{1}{2} \times \frac{1}{2} = \frac{1}{8}$

   (b) $P(HHH) = P(HHT) = P(HTH) = P(THH) = \frac{1}{8}$. Therefore, the probability of at least 2 heads is $\frac{4}{8} = \frac{1}{2}$.

   (c) $P(HHH) = \frac{1}{8}$. Therefore the probability of at most two heads is $1 - P(HHH) = 1 - \frac{1}{8} = \frac{7}{8}$.

7. (a) There are 13 spades and 52 cards in all, so the probability that the first card is a spade is $\frac{13}{52} = \frac{1}{4}$.

   (b.) Since the first card is returned to the deck, the probability that the second card is a spade is also $\frac{1}{4}$.

   (c) The probability that both cards are spades is $\frac{1}{4} \times \frac{1}{4} = \frac{1}{16}$.

   (d) The probability that the first card is not a spade is $1 - \frac{1}{4} = \frac{3}{4}$. This is also the probability that the second card is not a spade. Thus, the probability that neither card is a spade is $\frac{3}{4} \times \frac{3}{4} = \frac{9}{16}$.

SECTION 10.5 Independent Events

9. (a) P(no girls) = $\frac{1}{8}$ and P(1 girl) = $\frac{3}{8}$, so P(M) = $\frac{1}{8} + \frac{3}{8} = \frac{1}{2}$.

    (Note that the probability that the first child is a girl and the other two are boys is $\frac{1}{2} \times \frac{1}{2} \times \frac{1}{2} = \frac{1}{8}$. This is also the probability that the second child is a girl and the other two are boys and also is the probability that the third child is a girl and the other two are boys. Hence, the probability that there is exactly one girl is $\frac{3}{8}$.)

    (You can verify this by making a tree diagram.)

    (b) P(B) = 1 - P(no boys) - P(no girls) = $1 - \frac{1}{8} - \frac{1}{8} = \frac{3}{4}$

    (c) M is the event that the family has at most 1 girl, and B is the event that the family has children of both sexes, so M ∩ B is the event that the family has exactly 1 girl. From part (a), P(1 girl) = 3/8, so P(M ∩ B) = **3/8**.

    (d) In part (a), we found P(M) = 1/2 and from part (b), P(B) = 3/4. Since
    $$\frac{3}{8} = \frac{1}{2} \times \frac{3}{4}, \quad P(M \cap B) = P(M)P(B)$$
    which shows that M and B are **independent**.

11. (a) P(3 successful) = $\frac{1}{4} \times \frac{2}{3} \times \frac{1}{2} = \frac{1}{12}$

    (b) P(0 successful) = $(1 - \frac{1}{4})(1 - \frac{2}{3})(1 - \frac{1}{2}) = \frac{3}{4} \times \frac{1}{3} \times \frac{1}{2} = \frac{1}{8}$

13. (a) P(2 tails) = $\frac{1}{2} \times \frac{1}{2} = \frac{1}{4}$ (b) P(h and 6) = $\frac{1}{2} \times \frac{1}{6} = \frac{1}{12}$

    (c) P(h and even number) = $\frac{1}{2} \times \frac{1}{2} = \frac{1}{4}$

15. $\frac{579}{854} \times \frac{579}{781} \approx$ **0.503**   17. $(0.90)^5 \approx$ **0.59**

19. (a) P(all 3 fail) = $(\frac{1}{20})^3 = \frac{1}{8000}$

    (b) P(exactly 2 fail) = $C(3, 1) \times \frac{1}{20} \times \frac{1}{20} \times \frac{19}{20} = \frac{57}{8000}$

21. $C(3, 1)(0.7)(0.3)(0.3) = 3 \times 0.7 \times 0.9 =$ **0.189**

23.

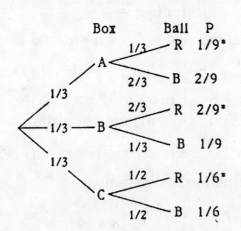

Add the starred items to get

$P = 1/9 + 2/9 + 1/6$

$= 2/18 + 4/18 + 3/18$

$= 9/18$

$= 1/2.$

25. Draw a tree diagram, letting F be the fair coin and U the unbalanced coin:

```
Coin      Face up    P
        1/2   H     1/4*
   F
1/2     1/2   T     1/4

        3/4   H     3/8*
1/2  U
        1/4   T     1/8
```

Use the starred numbers as weights to get

$$P(U \mid H) = \frac{P(U \cap H)}{P(H)}$$

$$= \frac{3/8}{3/8 + 1/4} = \frac{3/8}{5/8} = \frac{3}{5}.$$

27. (a) The probability of either event does not depend on the probability of the other event. The occurrence of either event does not depend on the occurrence or non-occurrence of the other event.

(b) If A and B are independent events, then $P(A \cap B) = P(A) \cdot P(B)$.

29. $C(50, 25)$ is the number of ways to get exactly 25 heads and 2^{50} is the total number of ways in which the coin can fall. Thus,

$$P(\text{exactly 25 heads}) = \frac{C(50,25)}{2^{50}}.$$

31. The numbers of ways of getting 0, 1, or 2 heads (without regard to order) out of the 6 tosses are C(6, 0), C(6, 1), and C(6, 2), respectively. Thus, for the biased coin, the probability of getting less than 3 heads is

$$C(6, 0)(1/3)^6 + C(6, 1)(1/3)^5(2/3) + C(6, 2)(1/3)^4(2/3)^2$$
$$= \frac{1}{729} + \frac{6}{1} \times \frac{2}{729} + \frac{6 \times 5}{2 \times 1} \times \frac{4}{729}$$
$$= \frac{1 + 12 + 60}{729} = \frac{73}{729}.$$

Thus, the probability of getting at least 3 heads is
$$1 - \frac{73}{729} = \mathbf{\frac{656}{729}}.$$

33. The number of ways (without regard to order) of getting exactly two 3's out of 5 tosses is C(5, 2), so the probability is

$$C(5, 2) \times (\frac{1}{6})^2(\frac{5}{6})^3 = \frac{5 \times 4}{2 \times 1} \times \frac{5^3}{6^5} = \frac{5^4}{3 \times 6^4} = \mathbf{\frac{625}{3888}}.$$

35. $P = \frac{1}{3} \times \frac{1}{2} \times \frac{3}{4} = \mathbf{\frac{1}{8}}$

37. $P = 1 - 0.98 = \mathbf{0.02}$

39. $P = (0.02)^3 = \mathbf{0.000008}$

EXERCISE 10.6

STUDY TIPS Most students are extremely interested in this section because they want to know the odds for many events. Learn Definition 10.5, page 575, so that you get a correct idea of what is meant by the odds in favor of an event. In terms of probabilities, an equivalent definition is as follows: The odds in favor of an event E are given by P(E) to P(Not E). A simpler idea is to think of the odds as the ratio of favorable to unfavorable occurrences for an event. Examples 1-4 illustrate the idea of odds. Definitions 10.6 (Fair Bet) and 10.7 (Expected Value) are important in any type of betting problem. Study the Problem Solving on page 579 and the remaining examples before you try to do any of the exercises.

1. There is 1 favorable outcome and there are 5 unfavorable outcomes, so the odds are **1 to 5** in favor of getting a 2.

3. There are 4 aces and 48 other cards, so the odds are 4 to 48 or **1 to 12** in favor of getting an ace.

5. You can get TT, TH, HT, HH, so there is 1 favorable outcome and 3 unfavorable outcomes. Thus, the odds in favor of getting 2 tails are **1 to 3**.

7. There are 5 vowels and 21 other letters, so the odds in favor of getting a vowel are **5 to 21**.

9. There are 3 odd numbers and 3 even numbers, so the odds against getting an odd number are **1 to 1**.

11. There are 12 picture cards and 40 other cards in the deck, so the odds against getting a picture card are 40 to 12 or **10 to 3**.

13. The probability of winning $5 is 1/50, so the odds in favor of winning $5 are **1 to 49**.

15. $P = \dfrac{3}{3+2} = \dfrac{3}{5}$

17. $\dfrac{630}{100000} = 0.0063$; the odds in favor of being a victim are **63 to 9937**.

19. The odds in favor of having complications are **1 to 1**.

SECTION 10.6 Odds and Mathematical Expectation

21. The probability is $\dfrac{10}{4867 + 10} = \dfrac{\mathbf{10}}{\mathbf{4877}}$.

23. The probability is $\dfrac{3}{21 + 3} = \dfrac{\mathbf{1}}{\mathbf{8}}$.

25. The odds in favor of growing up incompetent are 0.33 to (1 - 0.33) or **33 to 67**.

27. The probability that the sum is even is 1/2, and this is also the probability that the sum is not even, so the expected value is
$$E = \$10 \times \tfrac{1}{2} - \$20 \times \tfrac{1}{2} = \$5 - \$10 = \mathbf{-\$5}.$$

29. The probability of winning is 1/10,000, so a fair price to pay is $\$21,500 \times 0.0001 = \mathbf{\$2.15}$.

31. If the dice are fair, the expected value is 0. The man's expected value is $\$5 \times \tfrac{1}{6} - \$3 \times \tfrac{5}{6} = -\dfrac{\$10}{6}$. This means that the man is expected to lose $\$1\tfrac{2}{3}$ per roll on the average, so the dice are **not fair**.

33. For the first location, $E_1 = (\$100,000)(2/3) - (\$50,000)(1/3)$
    $= \$50,000$
    For the second location, $E_2 = (\$150,000)(2/5) - (\$80,000)(3/5)$
    $= \$12,000$
    Since $E_1 > E_2$, the **first location should be chosen**.

35. (a) $E_1 = (\$20,000)(1/5) - (\$10,000)(4/5) = \mathbf{-\$4000}$
    (b) $E_2 = (\$30,000)(1/5) + (\$5,000)(4/5) = \mathbf{\$10,000}$
    (c) Since continuing the campaign would result in an expected loss of $4000, and discontinuing the campaign would result in an expected gain of $ 10,000, the campaign should be **discontinued**.

37. The mathematical expectation is that you lose $0.58 per bet on the average, so it is **not a fair bet**. (For a fair bet, the mathematical expectation is 0.)

278 CHAPTER 10   PROBABILITY

39. (a) The curve shows that the probability of winning is **0.80**.
    (b) The probability of losing is 1 - 0.80 = **0.20.**

41. The curve shows that the probability of winning is 0.9, so the probability of losing is 1 - 0.9 = 0.1. If $x is put up against the man's $10, the expected value is E = ($x)(0.9) - ($10)(0.1). For the bet to be fair, E must be 0, so we have
$$0.9x - 1 = 0 \text{ or } 9x - 10 = 0.$$
Thus, x = 10/9 = 1.111 ..., and (to the nearest cent) **$1.11** should be put up against the man's $10.

43. The probability of a black or a red number is 36/38 and if one of these colors comes up, player wins 50c and loses 50c, so he breaks even. The probability of 0 or 00 is 2/38 and if 0 or 00 turns up, then the player gets his money back (but gains nothing) on one color and loses 50c on the other color Thus, the expected value is
$$E = (0)(\frac{36}{38}) - (\$0.50)(\frac{2}{38})(\frac{36}{38}) = -\frac{\$9}{361} \approx \mathbf{-\$0.025.}$$

PRACTICE TEST 10

STUDY TIPS   How can you "ACE" the test you are about to take? Here is a study tip you can use when taking the practice tests in the book or those in this manual.
1. **A**nswer all the problems in the practice test without notes or references.
2. **C**orrect any wrong answers and work that particular problem again.
3. **E**xamine your understanding of each problem by analyzing the objective of each question and avoiding the common errors associated with that type of question.

1. (a) This event is certain, so P(E) = **1**.

   (b) 4 of the 6 numbers are greater than 2, so the probability is $\frac{4}{6} = \mathbf{\frac{2}{3}}$.

2. (a) There are 5 possible outcomes on the first draw and 4 possible outcomes on the second draw, so the sample space has 5 × 4 or **20** elements.
   (b) There are 5 possible outcomes on each of the two draws, so the sample space has 5 × 5 or **25** elements.

3. (a) Of the 5 balls 2 are even numbered, so the probability is $\frac{2}{5}$.

   (b) There is just 1 ball numbered 2, so the probability is $\frac{1}{5}$.

   (c) The probability is $1 - \frac{1}{5} = \frac{4}{5}$.

4. (a) The probability that the first card is red is 26/52 = 1/2. If the first card is red, there are 25 red cards out of the 51 cards left, so the probability that the second card is red is 25/51. Thus, the probability that both cards are red is
$$\left(\frac{1}{2}\right)\left(\frac{25}{51}\right) = \frac{25}{102}.$$

   (b) The probability that the first card is not an ace is 48/52 = 12/13. The probability that the second card is not an ace is then 47/51. Thus, the probability that neither card is an ace is
$$\left(\frac{12}{13}\right)\left(\frac{47}{51}\right) = \frac{188}{221}.$$

5. (a) 26 of the 52 cards are red, so the probability that both cards will be red is $\frac{1}{2} \times \frac{1}{2} = \frac{1}{4}$.

   (b) If neither card is red then both cards must be black. The probability that both cards are black is the same as the probability that both cards are red, that is, $\frac{1}{4}$.

6. P(no heads) = $(\frac{1}{2})^5 = \frac{1}{32}$. Therefore, P(at least one head) = $1 - \frac{1}{32} = \frac{31}{32}$.

7. (a) There are 5 white and 3 black balls out of a total of 10 balls, so P(a white or a black ball) = $\frac{8}{10} = \frac{4}{5}$.

   (b) This is the same as part (a), so the probability is $\frac{4}{5}$.

8. Let E stand for passing English and M stand for passing math. Then, P(E ∪ M) = 0.9, P(E) = 0.8, and P(E ∩ M) = 0.6 are all given. Use the equation P(E ∪ M) = P(E) + P(M) - P(E ∩ M) to get 0.9 = 0.8 + P(M) - 0.6, which gives P(M) = **0.7**.

9. There are 12 face cards and a total of 52 cards in the deck, so
$$P(3 \text{ face cards}) = \frac{12}{52} \times \frac{11}{51} \times \frac{10}{50} = \frac{3}{13} \times \frac{11}{51} \times \frac{1}{5} = \frac{33}{3315} = \frac{11}{1105} \approx \mathbf{0.01}.$$

10. This is just the probability that the second die comes up 6, that is, $\dfrac{1}{6}$.

11. In order to get a sum of 11 and have the second die come up an even number, this die must come up 6. Since there are 3 even numbers, the probability that the die comes up 6, given that it is an even number is 1/3. To get a sum of 11, the first die must come up 5. The probability of this is 1/6. Thus, the probability that the sum is eleven given that the second die comes up an even number is
$$P = \frac{1}{6} \times \frac{1}{3} = \frac{1}{18}.$$

12. (a) There are 3 ways to get a sum of 10: 1st die 4, 2nd die 6; 1st die 5, 2nd die 5; 1st die 6, 2nd die 4. Since the dice can come up in 36 ways, the probability of getting a sum of 10 is $\dfrac{3}{36} = \dfrac{1}{12}$.

    (b) 3 of the 6 numbers are odd, so the probability of getting an odd number is $\dfrac{3}{6} = \dfrac{1}{2}$.

    (c) Let A be the event that the sum is 10 and B be the event that the first die comes up an odd number. If the first number is odd, then the sum of 10 can be obtained only if both numbers are 5's. From part (a), we see that the probability of this is 1/36. Thus, we have
$$P(A \cap B) = \frac{1}{36} \text{ and } P(A)P(B) = \frac{1}{12} \times \frac{1}{2} = \frac{1}{24}.$$
    Since these two results are not equal, A and B are **not independent**.

13. (a) $P = (0.04)^3 = \mathbf{0.000064}$ (b) $P = (1 - 0.04)^3 = (0.96)^3 = \mathbf{0.884736}$

14. There are 2 courses and 3 possible hours. This makes $2 \times 3 = 6$ possible arrangements. Thus, the probability that Roland will have English at 8 A.M. and history at 3 P.M. is **1/6**.

15. $P = (1 - 0.97)^2 = (0.03)^2 = \mathbf{0.0009}$

16. (a) There are 4 kings and 48 other cards, so the odds in favor of selecting a king are 4 to 48 or (dividing by 4) **1 to 12**.

    (b) The odds against selecting a king are just reversed from the answer in part (a). The answer is **12 to 1.**

17. (a) Since $P(E) = 3/7$, $P(E') = 1 - 3/7 = 4/7$. Thus, the odds in favor of E are **3 to 4**.
    (b) The odds against E are **4 to 3** [just the reverse of the answer in part (a)].

18. (a) If the odds in favor are 3 to 7, the odds against the event are **7 to 3**.
    (b) The probability that the event will not occur is $\dfrac{7}{7+3} = \dfrac{7}{10}$.

19. The expected value of this game is
$$E = (\$5)(\tfrac{1}{2}) + (\$5)(\tfrac{1}{4}) = \$2.50 + \$1.25 = \$3.75$$
so we should be willing to pay **$3.75** to play this game.

20. The mathematical expectation of being an instant winner is
$$E = (\$2)(\tfrac{1}{10}) + (\$5)(\tfrac{1}{50}) + (\$25)(\tfrac{1}{600}) + (\$50)(\tfrac{1}{1200})$$
$$= \$0.20 + \$0.10 + \tfrac{\$1}{24} + \tfrac{\$1}{24} = \$0.30 + \tfrac{\$1}{12} = 38\tfrac{1}{3} \text{ cents}.$$

CHAPTER 11
STATISTICS

EXERCISE 11.1

STUDY TIPS Statistics studies different ways of organizing and reporting data. The simplest way to organize a list of numbers is to make a frequency distribution. If the number of items is too large, we can group the data as is done in the Getting Started. Keep in mind that if a data value falls on the upper limit it should be included in the next higher class. For better understanding we can make a picture of the data using a histogram and a frequency polygon. Example 2 on page 592, the explanation that follows it and the Problem Solving on page 594 show you how to construct histograms and frequency polygons.

1. Descriptive statistics is the science of collecting, organizing and summarizing data.

3. A sample usually consists of just a part of the population.

5. (a) The population of the U. S.
   (b) The 1006 households surveyed

7. (a) No
   (b) The sample includes only those viewers that are willing to pay for the call.

9. (a) Make a card for each student, number the cards and mix them up. Then draw ten cards at random and select the corresponding students.
   (b) Make a card for each student, number the cards and:
    1. Pick only even numbered cards.
    2. Pick only odd numbered cards.
    3. Don't mix the cards and pick the first 10.

SECTION 11.1   Sampling and Frequency Distributions

11.(a)
Hours	Tally Marks	Frequency
0				2
1					3
2				2
3					3
4					3
5				2
6				2
7				2
8						4
9			1
10				2
11		0
12				2
13		0
14			1
15			1

The answers to parts (b), (c), (d), and (e) are obtained from the frequency distribution.

(b) **8**   (c) **4**   (d) **15**

(e) $11/30 =$ **36.7%**

13 (a)

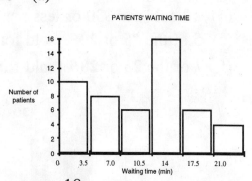

(b) There were 18 out of 50 patients, that is, $\frac{18}{50}$ or **36%**, who waited 7.0 min. or less.

(c) There were 26 out of 50 patients, that is, $\frac{26}{50}$ or **52%** who had to wait more than 10.5 min.

15.(a)
Age	Tally Marks	Frequency
6			1
7			1
8					3
9					3
10				2

(b)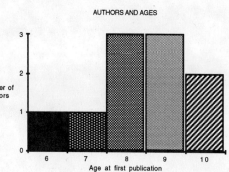

(c) 2 out of 10, that is, **20%** were less than 8 when they published their first books.

17. (a)

Price	Tally Marks	Frequency
$0 < P \le 10$	||	2
$10 < P \le 20$	||||	5
$20 < P \le 30$	||||	5
$30 < P \le 40$	||||	4
$40 < P \le 50$			1
$50 < P \le 60$	||||	4
$60 < P \le 70$	||||	4

(b) There is **no** most frequent price interval. *Five* sold between $10 and $20 and *five* sold between $20 and $30.

(c) **9** sold for more than $40 per share.

(d) **12** sold for $30 or less per share

(e) 5 of the 25 or **20%** sold for prices between $20\frac{1}{8}$ and $30 per share.

(f) 7 of the 25 or **28%** sold for $20 or less per share.

19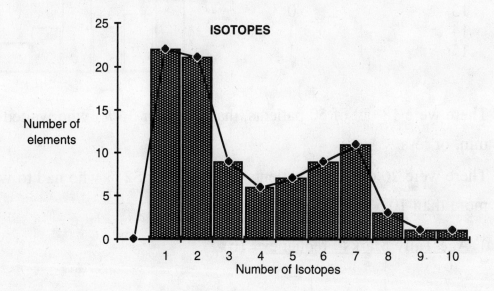

SECTION 11.1 Sampling and Frequency Distributions

21. (a)

Letter	Freq.	Letter	Freq.
a	10	n	13
b	4	o	14
c	3	p	2
d	4	q	0
e	18	r	7
f	2	s	10
g	3	t	17
h	9	u	5
i	12	v	2
j	0	w	5
k	1	x	0
l	4	y	2
m	4	z	0

(b) **e** is the most frequently occurring letter.

(c) Vowels

a	10
e	18
i	12
o	14
u	5
	59

Thus, 59 of the 151 letters, or about **39.1%** are vowels.

23.

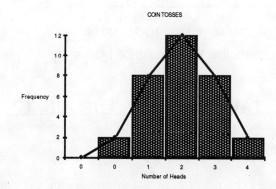

25. (a)

Concentration	Tally Marks	Freq.
0.00-0.04	IIII I	6
0.05-0.09	IIII IIII III	13
0.10-0.14	IIII	5
0.15-0.19	IIII	4
0.20-0.24	II	2

(b) The concentration is greater than 0.14 on 6 of the 30 days, that is **20%** of the days.

27. (a)

WEEKLY SALARY	TALLY MARKS	FREQUENCY
$600 to $2400	IIII IIII	9
$2400 to $4200		0
$4200 to $6000	II	2
$6000 to $7800	I	1

(b)

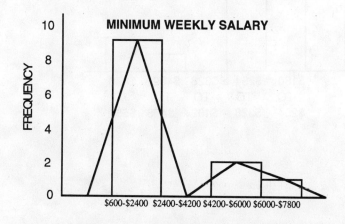

286 CHAPTER 11 STATISTICS

29. (a)

SALARY(THOUSANDS)	TALLY MARKS	FREQUENCY
20 to 34	‖‖‖ ‖‖‖ ‖‖‖ ‖‖‖	20
34 to 48			1
48 to 62			1
62 to 76	||	2
76 to 90			1

(b)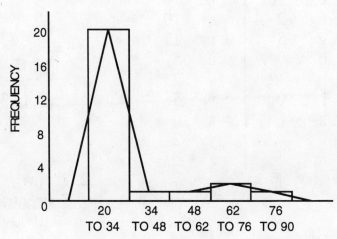

31. (a) **$74** (b)

PRICE	TALLY MARKS	FREQUENCY
$180 TO $254	‖‖‖	5
$254 TO $328	‖‖‖ ||	7
$328 TO $402		0
$402 TO $476			1
$476 TO $550	||	2

(c)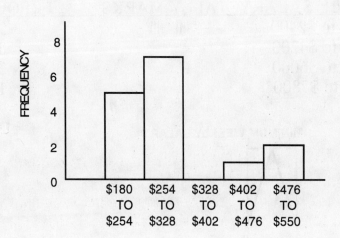

SECTION 11.1 Sampling and Frequency Distributions

33. (a) **3** years
    (b)
AGE	TALLY MARKS	FREQUENCY
109 TO 112	₩₩ ₩₩ ΙΙ	12
112 TO 115	₩₩ ₩₩	10
115 TO 118	ΙΙ	2
118 TO 121		0
121 TO 124	Ι	1

(c)

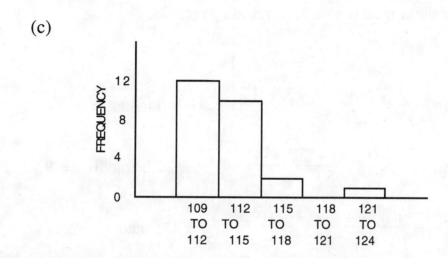

35. The upper and lower class limits, respectively, are the least and the greatest values in that class. Each class boundary is the midpoint between the upper limit of the respective class and the lower limit of the next class.

37.
Digit	Frequency	Digit	Frequency
0	1	5	4
1	4	6	3
2	5	7	3
3	6	8	5
4	4	9	5

39. 3,205,000 - 3,104,729 =**100,271** is the approximate jump in sales.

41. Most of the bar (from 0 to 3,000,000) is omitted in each case. Thus, it is not possible to get a good visual picture of the relative sales.

EXERCISE 11.2

STUDY TIPS The preceding section emphasized how to organize and present data. If we want to describe an entire sample or a population by a number or an average, we can use three measures of central tendency: the **mean**, the **mode**, and the **median**. To do this, you must know Definitions 11.2, 11.3 and 11.4. Study the examples carefully and make sure you learn the differences between the mean, median and mode by reading the comparison of these measurements given on page 608.

1. (a) Mean $= \dfrac{1 + 5 + 9 + 13 + 17}{5} = \dfrac{45}{5} = 9$; median = **9**

   (b) Mean $= \dfrac{1 + 3 + 9 + 27 + 81}{5} = \dfrac{121}{5} = \mathbf{24.2}$; median = **9**

   (c) Mean $= \dfrac{1 + 4 + 9 + 16 + 25}{5} = \dfrac{55}{5} = \mathbf{11}$; median = **9**

   (d) Mean = median for part (a) only. Median = 9 for all three. None has a mode.

3. The sum of the 20 scores is 121, so the mean $= \dfrac{121}{20} = \mathbf{6.05}$. The median is the average of the 10th and the 11th scores: median $= \dfrac{6 + 7}{2} = \mathbf{6.5}$. The mode is the most frequent score, **8**. The mode is the least representative of the three measures.

5. Betty is **correct**. She took account of the number of students making each score, and Agnes did not.

SECTION 11.2 Measures of Central Tendency

7.
Number of Letters	Frequency
2	1
3	5
4	4
5	5
6	0
7	3
8	2
9	0
10	0
11	1

(a) There are two modes: **3** and **5**.

(b) The median is **5**. (The sum of the frequencies is 21, so we add the numbers in the frequency column until we first get a sum of 11 or more. This occurs in the row where the number of letters is 5.)

(c) Mean = $\dfrac{1\times 2 + 5\times 3 + 4\times 4 + 5\times 5 + 3\times 7 + 2\times 8 + 1\times 11}{21}$

$= \dfrac{106}{21} \approx \mathbf{5.05}$

(d) **No**. There is too much repetition for this to be representative of ordinary English writing.

9. (b) and (c) Answers will vary, but should be approximately **8**.

11. $81 \times 20 = 1620$ and $1620 - 1560 = \mathbf{60}$

13. The new mean is $\dfrac{20\times(\$90) + 30\times(\$80)}{50} = \dfrac{\$4200}{50} = \mathbf{\$84}$ per week.

15. (a) The sum of the last five items is **0.34**
    (b) **$25,000 - $34,999**
    (c) The mean of the lower income levels in thousands of dollars is
    $\dfrac{0 + 10 + 15 + 20 + 25 + 35 + 50 + 80 + 120}{9} = \dfrac{355}{9}$
    so the mean is about **$39,444**.
    (d) The sum of the first three proportions is 0.36 which means that 36% have incomes less than $20,000.

17. Statements **(b)** and **(c)** are both true.

19. The answer is **(c)**. 70, which is about half way between 95 and 50, is the most reasonable estimate of the average score. Without additional information, we would assume a more or less uniform distribution of the scores.

21. The median of a set of scores is the middle number (if there is one) when the scores are arranged in order of magnitude. If there is no middle number, the median is the average of the two middle numbers. The median is not a good measure as it gives no indication of how the scores are spread.

23. (a) $\frac{12}{5} = $ **2.4** (b) The mode is **2** because there are 2 people in line three of the five minutes.

25. The mean in thousands of dollars is
$$\frac{100 + 50 + 25 + 2 \times 10 + 6 \times 6}{11} = \frac{231}{11} = 21$$ so Scrooge is using the **mean** of **all** the salaries.

27. In thousands of dollars, we have $8 = \frac{10 + 6}{2}$, so B. Crooked is using the **mean** of **one** secretary's and **one** worker's salaries.

SECTION 11.3   Measures of Dispersion   291

EXERCISE 11.3

STUDY TIPS   After learning about individual measurements, and averages, we are interested in finding out how far from these averages you may be. Two measures of dispersion, the **range** and the **standard deviation** will supply that answer. The range is easy to find, since it is simply the difference between the least and the greatest of the numbers in a given set. For the standard deviation you must use the formula in Definition 11.6. (If you are allowed to use a calculator, with a s_{n-1} button, the calculator will do it for you.) Make sure you understand the calculations in Example 1 and 2, since the standard deviation is used in many everyday applications such as polls, test scores and so on.

1. (a) Range = 21 - 3 = **18**   (b)

x	$x - \bar{x}$	$(x - \bar{x})^2$
3	-7	49
5	-5	25
8	-2	4
13	3	9
21	11	121
5)50		4)208
10 = $\bar{x}$		52

$\sqrt{52}$ = **7.21** = s

3. (a) Range = 25 - 5 = **20**   (b)

x	$x - \bar{x}$	$(x - \bar{x})^2$
5	-10	100
10	-5	25
15	0	0
20	5	25
25	10	100
5)75		4)250
15 = $\bar{x}$		62.5

$\sqrt{62.5}$ = **7.91** = s

5. (a) Range = 9 - 5 = **4** (b)

x	$x - \bar{x}$	$(x - \bar{x})^2$
5	-2	4
6	-1	1
7	0	0
8	1	1
9	2	4

5)35   4)10
7 = $\bar{x}$ 2.5

$\sqrt{2.5}$ = **1.58** = s

7. (a) Range = 9 - 1 = **8** (b)

x	$x - \bar{x}$	$(x - \bar{x})^2$
1	-4	16
2	-3	9
3	-2	4
5	0	0
7	2	4
8	3	9
9	4	16

7)35   6)58
5 = $\bar{x}$ 9.6667

$\sqrt{9.6667}$ = **3.11** = s

9. (a) Range = 3 - (-3) = **6** (b)

x	$x - \bar{x}$	$(x - \bar{x})^2$
-3	-3	9
-2	-2	4
-1	-1	1
0	0	0
1	1	1
2	2	4
3	3	9

7)0   6)28
0 = $\bar{x}$ 4.6667

$\sqrt{4.6667}$ = **2.16** = s

SECTION 11.3 Measures of Dispersion

11. (a) Mode = **8** (b) Median = (6 + 7)/2 = **6.5** (c) Mean = **6**

(e) The table shows that 14 of the 20 scores are within 1 standard deviation from the mean. This is 14/20 or **70%**.

(f) **100%** of the scores lie within 2 standard deviations from the mean.

(d)

x	x - $\bar{x}$	(x - $\bar{x}$)^2
0	-6	36
0	-6	36
1	-5	25
2	-4	16
4	-2	4
4	-2	4
5	-1	1
6	0	0
6	0	0
6	0	0
7	1	1
8	2	4
8	2	4
8	2	4
8	2	4
9	3	9
9	3	9
9	3	9
10	4	16
10	4	16
20)120		19)198
6 = $\bar{x}$		10.421

$\sqrt{10.421} = 3.23 = s$

13. (a) Mode = **110** (b) Median = **110** (c) Mean = **108**

(e) 102, 103, and 113 are more than 1 standard deviation from the mean. This is 3/10 or **30%**.

(d)

x	x - $\bar{x}$	(x - $\bar{x}$)^2
102	-6	36
103	-5	25
105	-3	9
106	-2	4
110	2	4
110	2	4
110	2	4
110	2	4
111	3	9
113	5	25
10)1080		9)124
108 = $\bar{x}$		13.778

$\sqrt{13.778} = 3.71 = s$

15.
x	x - $\bar{x}$	(x - $\bar{x}$)²
6	-1	1
2	-5	25
17	10	100
3	-4	16
5	-2	4
9	2	4

6)42   5)150
7 = $\bar{x}$ 30

Range = 17 - 2 = **15**
Mean = **7**
Standard Deviation = $\sqrt{30}$ = **5.48**

17.
x	x - $\bar{x}$	(x - $\bar{x}$)²
6	-1	1
7	0	0
9	2	4
12	5	25
3	-4	16
5	-2	4

6)42   5)50
7 = $\bar{x}$ 10

Range = 12 - 3 = **9**
Mean = **7**
Standard Deviation = $\sqrt{10}$ = **3.16**

19. If s = 0, then (x - $\bar{x}$)² = 0 for all the numbers in the set. Thus, x = $\bar{x}$ for all these numbers, so the numbers are **all the same**.

21. $s = \sqrt{np(1-p)} = \sqrt{180(\frac{1}{6})(1-\frac{1}{6})} = \sqrt{180(\frac{1}{6})(\frac{5}{6})} = \sqrt{25} = \mathbf{5}$.

23. Use the same formula as in Problem 21 above:

$s = \sqrt{400(\frac{1}{4})(1-\frac{1}{4})} = \sqrt{400(\frac{1}{4})(\frac{3}{4})} = \sqrt{25 \times 3} = 5\sqrt{3} = \mathbf{8.66}$.

SECTION 11.4 The Normal Distribution   295

EXERCISE 11.4

STUDY TIPS The Getting Started shows how the data in a normal distribution are spread out. Memorize the percents given in Figure 11.8. They are discussed in the Problem Solving on page 621, illustrated in Example 1-3 and used in the following section. The **z-scores** in part B are used to compare scores in different subjects. Their distribution and the procedure to find the probability that a specified score lies between two given scores, is covered in part C. If you have heard about **percentiles** and want to learn more about them, the Using Your Knowledge on page 627 discusses the subject.

1. (a) From the figure for this problem, we see that $\bar{x}$ = **100 in.**

   (b) Since $\bar{x}$ = 100 and $\bar{x}$ + s = 110, we get s = **10 in.**
   (c) Since the curve is a normal curve, **68%** are within 1 standard deviation from the mean.
   (d) 13.5% of 20,000 = 2700, so about **2700** stalks are between 80 and 90 in. tall.

3. In this problem, we are given that $\bar{x}$ = 120 sec and s = 15 sec.
   (a) 2.5% of 1000 = **25**   (b) 2.5 % of 1000 = **25**
   (c) 68% of 1000 = **680**

5. We are given $\bar{x}$ = 50 and s = 10.
   A: 2.5% get this grade. 2.5% of 500 = 12.5, so **12 or 13** get A.
   B: 13.5% get this grade. 13.5% of 500 = 67.5, so **68 or 67** get B.
   C: 68% get this grade. 68% of 500 = 340, so **340** get C.
   D: 13.5% get this grade. 13.5% of 500 = 67.5, so **68 or 67** get D.
   F: 2.5% get this grade. 2.5% of 500 = 12.5, so **12 or 13** get F.
   (Of course, the teacher's choices must add up to 500.)

7. 21,000 is 2 standard deviations below the mean, so about 2.5% of the tires gave out before running 21,000 mi.   2.5% of 200 = **5**.

9. Since s = 0.5 in., and $\bar{x}$ = 20 ft, almost all the measurements fall between 20 ft - 3 × 0.5 in. and 20 ft + 3 × 0.5 in., that is, **between 19 ft, 10.5 in and 20 ft, 1.5 in.**

11. If the lifetimes are normally distributed, 2.5% will last for 40 or more days. Thus, of the 8000 ball bearings, about **200** would last for 40 or more days, so the purchasing director would decide to buy.

13. For the SAT, the z-score is $\dfrac{570 - 455}{112} \approx 1.027$.

    For the ACT, the z-score is $\dfrac{25 - 17.3}{7.9} \approx 0.975$

    Thus, the **SAT** score is the **higher**.

15. We are given that $\bar{x} = 5$ and $s = 1.25$.

    (a) $z = \dfrac{6 - 5}{1.25} = 0.8$ (b) $z = \dfrac{7 - 5}{1.25} = 1.6$ (c) $z = \dfrac{7.5 - 5}{1.25} = 2$

17. For the German test, $\bar{x} = 75$ and $s = 20$, so $z = \dfrac{85 - 75}{20} = \dfrac{1}{2}$.

    For the English test, $\bar{x} = 80$ and $s = 15$, so $z = \dfrac{85 - 80}{15} = \dfrac{1}{3}$.

    Thus, the score in **German** was the **better** score.

19. Here $\bar{x} = 100$ and $s = 15$.

    (a) $z = \dfrac{110 - 100}{15} = \dfrac{2}{3} = 0.667$. From Table II, we find the probability that a randomly selected score is between the mean and 0.667 standard deviation above the mean to be **0.249**.

    (b) $z = \dfrac{130 - 100}{15} = 2$. Again, from Table 11, we find the probability that a randomly selected score is between the mean and 2 standard deviations above the mean to be **0.477**.

21. (a) 55 and 145 are each 3 standard deviations from the mean. From Table II, we find the probability that a randomly selected score falls between the mean and 3 standard deviations above the mean to be 0.499. This means that the probability that the score is between 100 and 145 is 0.499. Because of the symmetry of the normal curve, the probability that the score is between 55 and 100 is also 0.499. Therefore, the probability that the score is between 55 and 145 is $0.499 + 0.499 = \mathbf{0.998}$.

    (b) $z = \dfrac{100 - 60}{15} = 2.67$. So 60 is 2.67 standard deviations below the mean. Table II gives 0.496 as the probability that a randomly selected score is between the mean and 2.67 standard deviations from the mean. Since the probability that such a score is below the mean is 0.5, the probability that the score is below 60 is $0.5 - 0.496 = \mathbf{0.004}$.

SECTION 11.4 The Normal Distribution

23. In Problem 19, we found 0.249 to be the probability that a randomly selected score is between 100 and 110, and 0.477 to be the corresponding probability for the score to be between 100 and 130. Consequently, the probability that the score is between 110 and 130 is
0.477 - 0.249 = **0.228**.

25. In order for a student to be among the tallest 20%, the z-score corresponding to his height must be greater than 0.80. Since x = 5 ft 7 in. or 67 in., and s = 3 in., the height x in. of the shortest student in this group must be such that
$$\frac{x - 67}{3} > 0.80.$$
This gives x - 67 > 2.4 and x > 69.4. Thus, the shortest student in the tallest group is **about 5 ft $9\frac{1}{2}$ in. tall**.

27. **Yes**. See Figure 11.10. Here both curves have the mean 0. The standard deviation for A is 1/3 unit while that for B is 1 unit.

29. **No**. If curve A in Figure 11.10 is moved two units to the right, this would be an example. The standard deviation for A would be 1/3 unit and that for B would be 1 unit, while the mean for A would be 2 and that for B would be 0.

31. Her percentile = $\frac{40}{50} \times 100 =$ **80**.

33. (a) There were 2 scores less than 90, so the percentile is
$\frac{2}{10} \times 100 =$ **20**.
   (b) There were 6 scores less than 97, so the percentile is
$\frac{6}{10} \times 100 =$ **60**.
   (c) There were 9 scores less than 100, so the percentile is
$\frac{9}{10} \times 100 =$ **90**.
   (d) There were no scores less than 83, so the percentile is **0**.

298 CHAPTER 11 STATISTICS

35. (a) 40 and 60 are each 2 standard deviations from the mean, so we use Chebyshev's theorem with h = 2:

$$(1 - \frac{1}{h^2})N = (1 - \frac{1}{4})(100) = (\frac{3}{4})(100) = 75.$$

Thus, **at least 75** of the measurements must be between 40 and 60.

(b) 35 and 65 are each 3 standard deviations from the mean, so we use the theorem with h = 3:

$$(1 - \frac{1}{h^2})N = (1 - \frac{1}{9})(100) = \frac{800}{9} = 88\frac{8}{9}.$$

Thus, **at least 89** of the measurements must be between 35 and 65.

(c) 43 and 57 are each $\frac{7}{5}$ths of a standard deviation from the mean, so we use the theorem with h = $\frac{7}{5}$.

$$(1 - \frac{25}{49})(100) = \frac{2400}{49} = 48\frac{48}{49}$$

Thus, **at least 49** of the measurements must be between 43 and 57.

37.

x	x - x̄	(x - x̄)²
1	-5	25
1	-5	25
1	-5	25
2	-4	16
6	0	0
10	4	16
11	5	25
11	5	25
11	5	25
9)54		8)182
6 = x̄		22.75

$$s = \sqrt{22.75} = 4.77$$

Three of the numbers (2, 6, and 10) lie within 1 standard deviation from the mean, and all 9 of the numbers lie within 2 standard deviations from the mean. The theorem makes no prediction for 1 standard deviation; it predicts at least 75% for 2 standard deviations.

39. **No**; the theorem predicts the least number that must lie within 2 standard deviations from the mean, and it predicts the least number that must lie within 3 standard deviations from the mean, but it does not predict how many must lie between 2 and 3 standard deviations from the mean. However, since at least 75% of the numbers must lie within 2 standard deviations from the mean, the number of numbers between 2 and 3 standard deviations from the mean cannot be more than 25% of all the numbers.

SECTION 11.5 Statistical Graphs

EXERCISE 11.5

STUDY TIPS Have you seen all those graphs in newspapers and magazines? In this section we show you how to construct your own **line graphs, bar graphs** and **circle graphs** to give accurate pictures of data. But these graphs can also be distorted to present a wrong impression of the data. Learn how that is done in the Discovery, page 640.

1.

3.

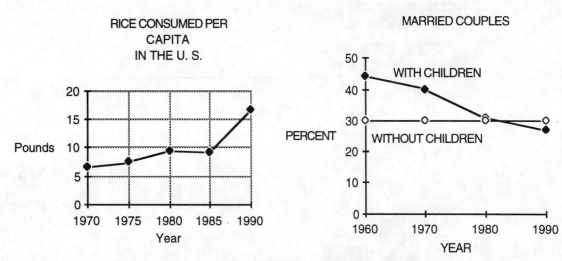

5.

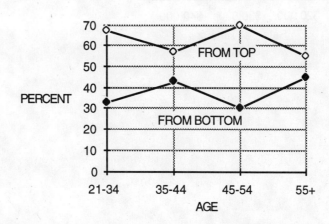

7.

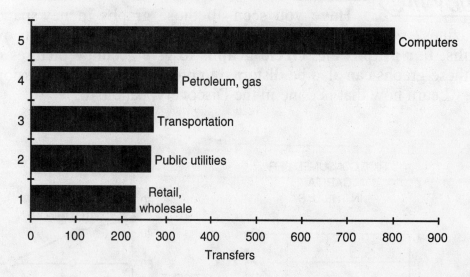

9. (a) 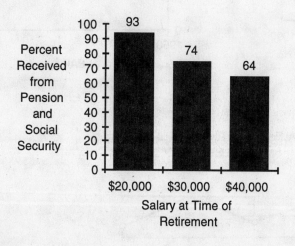 (b) 93% of $20,000 = **$18,600**

11. (a)

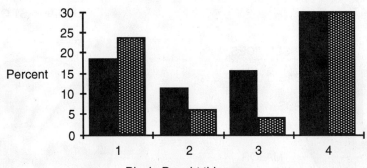

Where 35 mm cameras were bought

Black: Bought this year.
Gray: Bought 2 years ago

1: Discount stores 2: Sears
3: Department stores 4: Specialty stores

(b) At specialty stores (c) Department stores
(d) The specialty stores

13 MEETING HOURS PER WEEK 15.

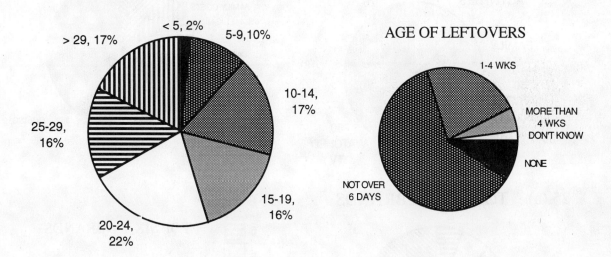

17.

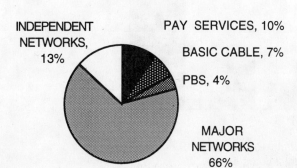

19(a)

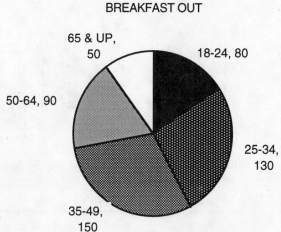

(b) Cater to the **35-49** age group.

21

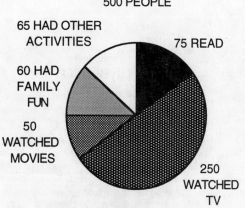

23 Retired Couples' Budget

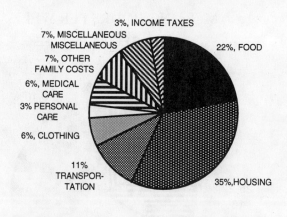

25.(a) TOP SIX TV BRANDS

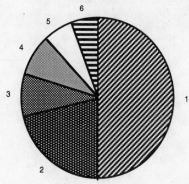

(b)

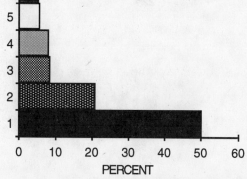

SECTION 11.5 Statistical Graphs 303

25. (c) The **circle graph** makes the stronger impression. The area corresponding to Brand 1 overshadows the remainder of the chart.

27. (a) **Yes.** To give a correct visual impression, only the height should be doubled. (If both the height and the radius are doubled, the volume is multiplied by 8)

(b) MIGHTY MIDGET BUSINESS

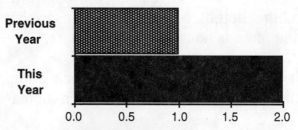

29.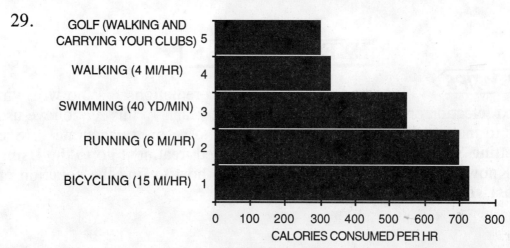

31.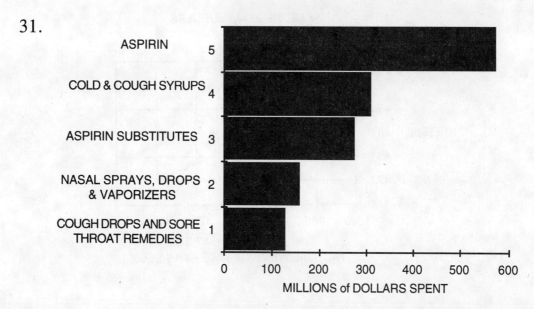

304 CHAPTER 11 STATISTICS

33. (a) The chart shows about **$16,000,000** for the 1984 sales.
    (b) The sales seemed to start leveling off in **1986**.

35. (a) Use a **large scale** on the **vertical** axis.
    (b) Use a **very small** scale on the **vertical** axis.

37. The area of the bar in a bar graph indicates the amount of the item that is graphed. For this reason, the bars are usually shaded. In a histogram, the height of the bar corresponds to the frequency in question and there is no space between bars.

39. The graph on the **right** has a very much compressed vertical scale, which diminishes the visual effect of each increase or decrease.

EXERCISE 11.6

STUDY TIPS Everybody wants to make predictions. Who will win the next election or go to the Super Bowl next year? In this section we use graphs to make predictions, and they turn out to be pretty accurate! (See the Getting Started.) For a more sophisticated treatment go to the Using Your Knowledge on pages 645-646 and learn how to find the equation of the **least-squares** line for a set of data.

1.

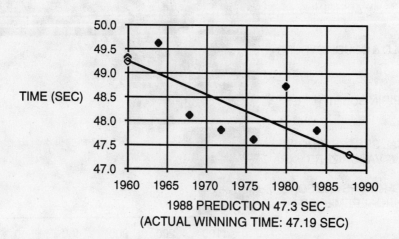

SECTION 11.6 Making Predictions 305

3.

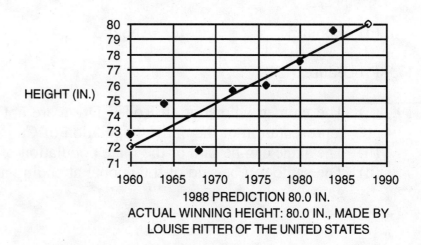

5. (a) GASOLINE MILEAGE

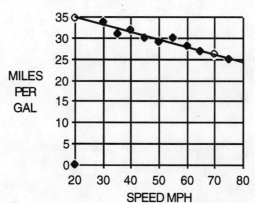

(b) The small circle on the 70 vertical line indicates about **26** mi/gal.

7. (a)

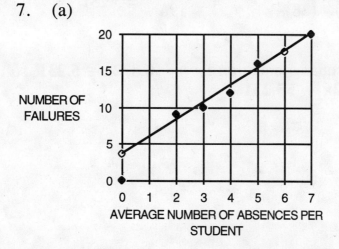

(b) The open circle on the vertical 6 line says that about **18** would fail.

9. $\dfrac{12}{50} \times 3000 = \mathbf{720}$ 11. $\dfrac{3}{150} \times 10{,}000 = \mathbf{200}$ 13. $0.727 \times 586 \approx \mathbf{426}$

15. (a) About **13** (b) About **9** (c) About **9**

(d) About **3** (e) About **3** (f) About **3**

17. Procedure **(d)**

19. (a) The women who shop in Rodeo Drive are not necessarily a good representation of the entire population of California.
(b) The same can be said of the male population of Berkeley.
(c) The same can be said of the people attending an Oakland baseball game.

21.

x	y	x^2	xy
1	5.2	1	5.2
2	3.4	4	6.8
3	2.2	9	6.6
4	1.22	16	4.88
5	-0.01	25	-0.05
6	0.40	36	2.40
7	-0.20	49	-1.40
28	12.21	140	24.43

In the table, x is the number of the Olympic with 1960 as number 1, and y is the number of seconds over 50. We use the formulas given to find the slope and the y intercept of the least-squares line.

$$m = \frac{7(24.43) - (28)(12.21)}{7(140) - (28)^2} = \frac{-170}{196} = -0.872$$

$$b = \frac{(140)(12.21) - (28)(24.43)}{7(140) - (28)^2} = \frac{1025.36}{196} = 5.231$$

Thus, the least-squares equation is **y = -0.872x + 5.231**, so that the time **t = -0.872x + 55.231.**

SECTION 11.6 Making Predictions

23. We use x and y just as in the solution of problem 21.

x	y	x^2	xy
1	11.2	1	11.2
2	9.5	4	19.0
3	10.0	9	30.0
4	8.6	16	34.4
5	5.7	25	28.5
6	4.8	36	28.8
7	5.9	49	41.3
28	55.7	140	193.2

$$m = \frac{7(193.2) - (28)(55.7)}{7(140) - (28)^2} = \frac{-207.2}{196} = -1.057$$

$$b = \frac{(140)(55.7) - (28)(193.2)}{7(140) - (28)^2} = \frac{2388.4}{196} = 12.186$$

Thus, the least-squares equation is **y = -1.057x + 12.186**, and the time **t = -1.057x + 62.186.**

25.

x	y	x^2	xy
0	8.6	0	0
1	6.3	1	6.3
2	6.1	4	12.2
10	4.4	100	44.0
15	3.8	225	57.0
16	2.3	256	36.8
20	1.3	400	26.0
64	32.8	986	182.3

In the table, x = the year - 1965, and y = number of seconds - 45. We use the same formulas as in problem 21 to get the least-squares line.

$$m = \frac{7(182.3) - (64)(32.8)}{7(986) - (64)^2} = \frac{-823.1}{2806} = -0.293$$

$$b = \frac{(986)(32.8) - (64)(182.3)}{7(986) - (64)^2} = \frac{20673.6}{2806} = 7.368$$

Thus, the least-squares equation is **y = -0.293x + 7.368** and the time **t = -0.293x + 52.368.**

EXERCISE 11.7

STUDY TIPS Sometimes the graph of a set of data will look like a scattered set of points, which explains why we call such a graph a scattergram. If most of the points lie close to the least-squares line and this line slopes up, we have a **positive correlation** but if the line slopes down we have a **negative correlation.** (See Figure 11.30).

1. Expect a **positive** correlation. 3. Expect **no** correlation.

5. Expect a **negative** correlation. 7. Expect a **positive** correlation.

9. 11.

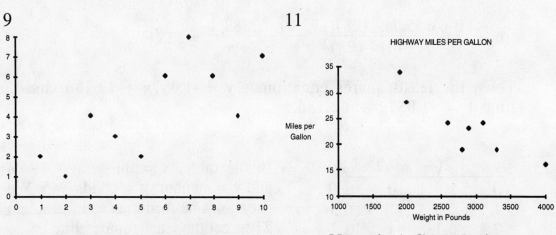

Positive Correlation   Negative Correlation

13. 15.

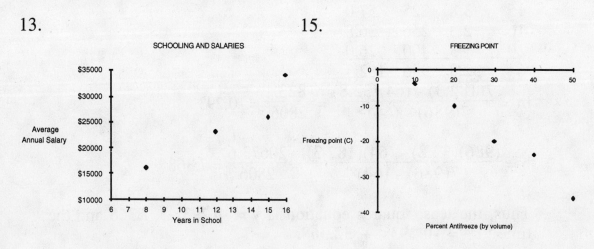

Positive Correlation   Negative Correlation

SECTION 11.7 Scattergrams and Correlation

17. (a) There was no outage during the **7th** week.
    (b) The greatest number of outages occurred during the **11th** week.
    (c) A decline in the number of outages seemed to start after the **2nd** week.
    (d) **No correlation** is shown by the chart.

19. (a) **No**. It is very unlikely that airbags cause traffic accidents.
    (b) Answers may vary. It is possible that the number of cars with airbags being sold is increasing at a rate that makes them representative of the entire population of cars.

21. To find the sums needed to calculate the correlation coefficient, we arrange the data in order of increasing x (weight of auto in hundreds of pounds) and y (number of miles per gallon) and make the following table:

x	y	x^2	y^2	xy
19	34	361	1156	646
20	28	400	784	560
26	24	676	576	624
28	19	784	361	532
29	23	841	529	667
31	24	961	576	744
33	19	1089	361	627
40	16	1600	256	640
226	187	6712	4599	5040

Since there are 8 number pairs given, we use the formula for the correlation coefficient with n = 8 as in the following computation:

Numerator = 8(5040) - (226)(187) = -1942

Denominator = $\sqrt{8(6712) - (226)^2} \sqrt{8(4599) - (187)^2}$
$= \sqrt{2620} \sqrt{1823} \approx 2185$

Therefore, $r = \dfrac{-1942}{2185} \approx \mathbf{-0.89}$.

310 CHAPTER 11 STATISTICS

23. We let x be the number of years of schooling and y be the number of hundreds in the average salary, and use the given data to make the following table:

x	y	x^2	y^2	xy
8	168	64	28,224	1344
12	233	144	54,289	2796
15	258	225	66,564	3870
16	339	256	114,921	5424
51	998	689	263,998	13,434

Since 4 number pairs are given, we use n = 4 in the formula for the correlation coefficient.

Numerator = 4(13,434) - (51)(998) = 2838

Denominator = $\sqrt{4(689) - (51)^2} \sqrt{4(263,998) - (998)^2}$

$= \sqrt{155}\sqrt{59988} \approx 3049$

Therefore, $r = \frac{2838}{3049} \approx \mathbf{0.93}$.

25. To simplify the computation, we let x = height - 60 and y = test score. Then we make the following table:

x	y	x^2	y^2	xy
1	80	1	6400	80
2	85	4	7225	170
4	70	16	4900	280
5	60	25	3600	300
7	60	49	3600	420
7	100	49	10000	700
8	35	64	1225	280
10	75	100	5625	750
12	95	144	9025	1140
13	45	169	2025	585
69	705	621	53625	4705

There are 10 number pairs given, so we use n = 10 in the formula for the correlation coefficient.

Numerator = 10(4705) - (69)(705) = -1595

Denominator = $\sqrt{10(621) - (69)^2} \sqrt{10(53625) - (705)^2}$

$= \sqrt{1449}\sqrt{39225} \approx 7539$

Therefore, $r = \frac{-1595}{7539} \approx \mathbf{-0.21}$.

PRACTICE TEST 11

STUDY TIPS There are so many items in this chapter that you should compare your notes with the summary on pages 651-652 to make sure you understand the terminology and the procedures and formulas that are used to solve the corresponding types of problems. Since this manual gives step-by-step procedures in solving the Exercise Problems, compare your answers with the ones here and be sure you find out how to correct any errors you make. Take advantage of the Practice Test given in the book since detailed solutions to this test appear next.

1. **No.** Not all members of the student body had the same chance of being chosen (only those in the English Department).

2.

Score	Tally Marks	Frequency
$60 < s \le 65$	II	2
$65 < s \le 70$	III	3
$70 < s \le 75$	III	3
$75 < s \le 80$	II	2
$80 < s \le 85$	III	3
$85 < s \le 90$	LHI II	7
$90 < s \le 95$	III	3
$95 < s \le 100$	II	2

3.(a) and (b). Histogram and frequency polygon for the data in problem 2.

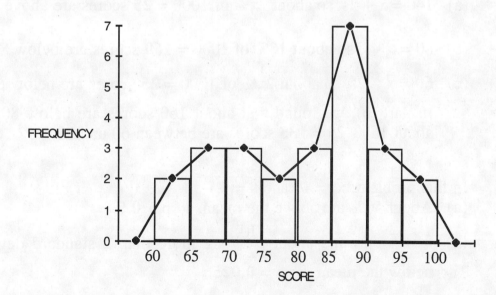

4. (a) $\bar{x} = \dfrac{69 + 70 + 71 + 73 + 78 + 82 + 82}{7} = \dfrac{525}{7} = \mathbf{75}$

   (b) The mode is **82**.   (c) The median is **73**.

5. (a) Range = 82 - 69 = **13**

   (b)
x	x - $\bar{x}$	(x - $\bar{x}$)²
69	-6	36
70	-5	25
71	-4	16
73	-2	4
78	3	9
82	7	49
82	7	49
7)525		6)188
75 = $\bar{x}$		s = $\sqrt{188/6} \approx$ **5.60**

6. We are given $\bar{x}$ = 128, s = 8.

   (a) $\bar{x}$ - 3s = 128 - 24 = 104, and $\bar{x}$ + 3s = 128 + 24 = 152. Thus, the total number of heads should be **between 104 and 152.**

   (b) 112 = 128 - 16 = $\bar{x}$ - 2s, so p = **0.025**.

7. We are given n = 1000, $\bar{x}$ = 100, s = 20.

   (a) 140 = $\bar{x}$ + 2s, so about $2\tfrac{1}{2}$% of 1000 = **25** scores are above 140.

   (b) 80 = $\bar{x}$ - s, so about 16% of 1000 = **160** scores are below 80.

   (c) 60 = $\bar{x}$ - 2s, so about $2\tfrac{1}{2}$% of 1000 = **25** scores are below 60.

   In part (b), we found that about 160 scores are below 80. Thus, about 160 - 25 = **135** scores are between 60 and 80.

8. In this problem, $\bar{x}$ = 10 and s = 1.
   (a) About 50% are above the mean, so p = **0.5**.

   (b) 8 = $\bar{x}$ - 2s and about $2\tfrac{1}{2}$% are more than 2 standard deviations below the mean, so p = **0.025**.

9. Here, n = 1000, $\bar{x}$ = 50, and s = 5.

   (a) $z = \dfrac{58 - 50}{5} = \dfrac{8}{5} = \mathbf{1.6}$ (b) $z = \dfrac{62 - 50}{5} = \dfrac{12}{5} = \mathbf{2.4}$

10. French: $z = \dfrac{88 - 76}{18} = \dfrac{2}{3}$ Psychology: $z = \dfrac{90 - 80}{16} = \dfrac{5}{8}$

    To compare these scores, change them to a common denominator: $\dfrac{2}{3} = \dfrac{16}{24}$ and $\dfrac{5}{8} = \dfrac{15}{24}$, so that $\dfrac{2}{3} > \dfrac{5}{8}$. Thus, Agnes' score in **French** was the **better** score.

11. The z-score corresponding to a score of 62 is $\dfrac{62 - 50}{5} = \dfrac{12}{5} = 2.4$. Table II gives the probability p = **0.492**.

12.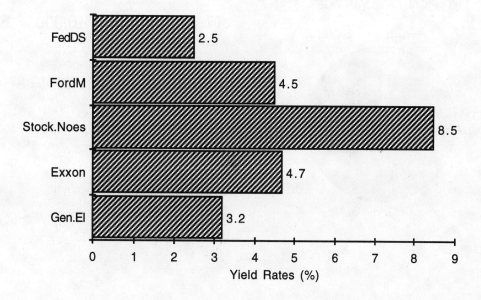

13. (a)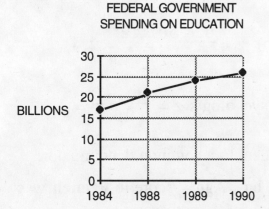

14.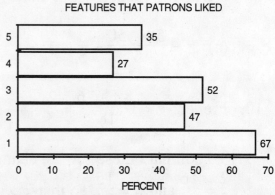

1. Low calorie entrees
2. Varied portion sizes
3. Cholesterol-free entrees
4. All-you-can-eat specials
5. Self-service soup bar

(b)　About **$19 billion**.

15.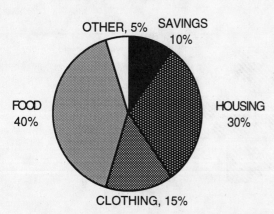

16. About **50%**

17. About $\dfrac{3}{150} \times 10{,}000 = \mathbf{200}$

18. About $\dfrac{4000}{50000} \times 100 = \mathbf{8}$

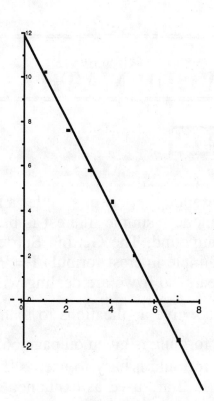

19. The graph shows that for x = 7, y = - **1.8**.

20. (a) Would expect a **positive** correlation.
    (b) Would expect a **positive** correlation.
    (c) Would expect **no** correlation.
    (d) Would expect a **negative** correlation.

CHAPTER 12
YOUR MONEY AND YOUR MATH

EXERCISE 12.1

STUDY TIPS Do you know the difference between simple and compound interest? If you are borrowing money, simple interest is best but if you are collecting interest ask for compound. The Getting Started shows you why. Make sure you learn the simple interest formula I = Prt and keep in mind that the time must be in years, so if we are dealing with quarterly interest, then $t = \frac{4}{12}$. The examples give applications to loans, taxes and discounts. The compound interest formula is given on page 662. If you have a calculator with a y^x key, this formula is easy to use. If a calculator is not available, tables like Table 12.2 are used as explained in Example 4.

1. I = Prt = ($3000)(0.08)(1) = **$240**

3. I = Prt = ($2000)(0.09)(3) = **$540**

5. I = Prt = ($4000)(0.10)(1/2) = **$200**

7. I = Prt = ($2500)(0.10)(1/4) = **$62.50**

9. I = Prt = ($16,000)(0.07)(5/12) = **$466.67**

11. (a) Sales tax = (0.06)($40.20) = **$2.41**   (b) $40.20 + $2.41 = **$42.61**

13. FICA tax = (0.0751)($24,000) = **$1802.40**

15. Maybelle's estimated tax was
    $3052.50 + (0.28)($25,850 - 20,350) = $3052.50 + (0.28)(5500)
    = $3052.50 + $1540
    = **$4592.50**

17. (a) Discount = (0.20)($200) = **$40**   (b) $200 - $40 = **$160**
    (c) $160 + (0.05)($160) = $160 + $8 = **$168**

19. (a) Discount = (0.25)($500) = **$125**   (b) $500 - $125 = **$375**

SECTION 12.1 Interest, Taxes and Discounts

21. (a) Discount = (0.05)($14.20) = **$0.71**   (b) $14.20 - $0.71 = **$13.49**

23. Amount = ($100)(1.5938) = **$159.38**; interest = **$59.38**

25. Amount = ($2580)(2.7731) = **$7154.60**; interest = **$4574.60**

27. Amount = ($12,000)(2.1829) = **$26,195**;   interest = **$14,195**

29. Amount = ($20,000)(1.2682) = **$25,364**;   interest = **$5364**

31. Amount = ($40,000)(97.0172) = **$3,880,690**; interest = **$3,840,690**

33. (a) Amount = ($1000)(2.8543) = **$2854.30**
    (b) Amount = ($1000)(2.8983) = **$2898.30**

35. (a) Amount = ($3000)(1.3310) = **$3993**   (b) Interest = **$993**

37. ($1000)(1.8061 - 1.7908) = $1000(0.0153) = **$15.30**

39. Simple interest means that the interest does not earn additional interest, while compound interest means that the interest earns interest at the same rate as the original principal. For instance, $100 at 10% simple interest would earn $20 in two years, while $100 at compound interest would earn $21 in two years.

41. Table 12.2 gives an accumulated amount of 1.0609 for n = 2 and 3% per period, so the effective rate is 1.0609 - 1 = 0.0609 = **6.09%**.

43. Table 12.2 gives an accumulated amount of 1.0824 for n = 4 and 2% per period, so the effective rate is 1.0824 - 1 = 0.0824 = **8.24%**.

45. Table 12.2 gives an accumulated amount of 1.0931 for n = 4 and $2\frac{1}{4}$%

    per period, so the effective rate is 1.0931 - 1 = 0.0931 = **9.31%**.

47. Table 12.4 gives an accumulated amount of 1.1956 for n = 12 and $1\frac{1}{2}$%

    per period, so the effective rate is 1.1956 - 1 = 0.1956 = **19.56%**.

318 CHAPTER 12 YOUR MONEY AND YOUR MATH

EXERCISE 12.2

STUDY TIPS If you have a credit card or are planning to get one, this is a "must" section. The Getting Started gives hints, tips and terminology associated with the different options regarding credit cards. The examples show applications to finance charges, interest and other fees that you have to pay on credit card accounts. Make sure you understand and practice with these examples before you go on to the problems.

1. Since the couple's combined gross income must be at least $40,000, Susie must earn **at least $15,000** per year for them to qualify.

3. $1\frac{1}{2}\%$ of ($100 - $10) = (0.015)($90) = $1.35
   New balance = $90 + $1.35 + $50 = **$141.35**

5. $1\frac{1}{2}\%$ of ($134.39 - $25) = (0.015)($109.39) = $1.64
   New balance = $109.39 + $1.34 + $73.98 = **$185.01**

7. $1\frac{1}{2}\%$ of ($378.93 - $75) = (0.015)($303.93) = $4.56
   New balance = $303.93 + $4.56 + $248.99 = **$557.48**

9. (a) $1\frac{1}{2}\%$ of $85 = (0.015)($85) = **$1.28**
   (b) $85 + $1.28 + $150 = **$236.28**
   (c) 5% of $236.38 = (0.05)($236.28) = **$11.81**

11. (a) $1\frac{1}{2}\%$ of $344 = (0.015)($344) = **$5.16**
    (b) $344 + $5.16 + $60 = **$409.16**
    (c) 5% of $409.16 = (0.05)($409.16) = **$20.46**

13. (a) $1\frac{1}{2}\%$ of $80.45 = (0.015)($80.45) = **$1.21**
    (b) $80.45 + $1.21 + $98.73 = **$180.39**
    (c) The new balance is under $200, so the minimum payment is **$10.**

SECTION 12.2 Credit Cards and Consumer Credit

15. (a) $1\frac{1}{2}\%$ of $55.90 = (0.015)(\$55.90) = \$0.84$

    (b) $\$55.90 + \$0.84 + \$35.99 = \$92.73$

    (c) The new balance is under $200, so the minimum payment is **$10.**

17. (a) $1\frac{1}{2}\%$ of $34.76 = (0.015)(\$34.76) = \$0.52$

    (b) $\$34.76 + \$0.52 + \$87.53 = \$122.81$

    (c) The new balance is under $200, so the minimum payment is **$10.**

19. Finance charge $= 1\frac{1}{2}\%$ of $190 = (0.015)(\$190) = \$2.85$

21. Finance charge $= 1\frac{1}{2}\%$ of $90 = (0.015)(\$90) = \$1.35$

23. (a) Balance = $9000 - $1600 = $7400. Add-on interest is
    $(\$7400)(0.09)(3) = \$1998.$

    (b) Total to be paid in 36 months is $7400 + $1998 = **$9398.**
    $\frac{9398}{36} = 261.06$, so the monthly payment (to the nearest dollar) is **$261**

25. (a) Balance = $2400 - $400 = $2000. Add-on interest is
    $(\$2000)(0.15)(1\frac{1}{2}) = \$450.$

    (b) Total to be paid in 18 months is $2000 + $450 = $2450.
    $\frac{2450}{18} = 136.11$, so the monthly payment (to the nearest dollar) is **$136.**

27. (a) Balance = $500 - $100 = $400. Add-on interest is
    $(\$400)(0.10)(1\frac{1}{2}) = \$60.$

    (b) Total to be paid in 18 months is $400 + $60 = $460. Since
    $\frac{460}{18} = 25.56$, the monthly payment (to the nearest dollar) is **$26.**

29. This procedure gives the longest possible time between the purchase date and the date when payment must be made to avoid a finance charge.

31. ($47)(24) = $1128. Interest $1128 - $1000 = $128. Since this is for 2 years, the interest rate is $\frac{128}{(1000)(2)} = 0.064 =$ **6.4%**.

33. ($50)(24) = $1200. Interest = $1200 - $1000 = $200. Since this is for 2 years, the interest rate is $\frac{200}{(1000)(2)} = 0.10 =$ **10%**.

EXERCISE 12.3

STUDY TIPS How can you compare rates on different types of credit accounts? By using the true annual percentage rate called the APR. Example 1 tells you how to use an APR table, a portion of which appears as Table 12.3, page 678. If you pay off an account early you will want to learn how to calculate the payoff amount. The rule used to calculate this payoff is called the **rule of 78** and Example 2 illustrates how this rule works.

1. $\frac{194}{25} = 7.76 =$ finance charge per $100. The nearest table entry for 12 payments is 7.74 corresponding to an APR of **14%**.

3. $\frac{264}{15} = 17.60 =$ finance charge per $100. The nearest table entry for 24 payments is 17.51 corresponding to **16%**.

5. $\frac{210}{15} = 14 =$ finance charge per $100. The nearest table entry for 18 payments is 13.99 corresponding to **17%**.

7. $\frac{1570}{45} = 34.89 =$ finance charge per $100. The nearest table entry for 48 payments is 34.81 corresponding to $\mathbf{15\frac{1}{2}\%}$.

9. $\frac{1800}{50} = 36 =$ finance charge per $100. The nearest table entry for 48 payments is 36.03 corresponding to **16%**.

SECTION 12.3 Annual Percentage Rate and Rule of 78   321

11. (a) Here, r = 4 and n = 12, so the unearned finance charge is
$$\frac{4\times5}{12\times13} \times \$15.60 = \frac{5}{39} \times \$15.60 = \mathbf{\$2}.$$
(b) There are 4 payments remaining, so it will take (4 × $25) - $2, that is, **$98** to pay off the loan.

13. (a) Here, r = 6 and n = 12, so the unearned finance charge is
$$\frac{6\times7}{12\times13} \times \$31.20 = \mathbf{\$8.40}.$$
(b) There are 6 payments remaining, so it will take (6 × $45) - $8.40 = **$261.60** to pay off the loan.

15. (a) Here, r = 5 and n = 20, so the unearned finance charge is
$$\frac{5\times6}{20\times21} \times \$58.50 = \frac{1}{14} \times \$58.50 = \mathbf{\$4.18}.$$
(b) There are 5 payments remaining, so it will take (5 × $10) - $4.18 = **$45.82** to pay off the loan.

17. (a) Finance charge = (12 × $27) - $300 = **$24.**
(b) Finance charge per $100 is $\frac{\$24}{3}$ = $8. The nearest table entry for 12 payments is $8.03 corresponding to $\mathbf{14\tfrac{1}{2}\%}$.

19. (a) Finance charge = (18 × $63) - $1000 = **$134.**
(b) Finance charge per $100 is $\frac{\$134}{10}$ = $13.40. The nearest table entry for 18 payments is $13.57 corresponding to $\mathbf{16\tfrac{1}{2}\%}$.

21. (a) Here, r = 4 and n = 12, so the unearned finance charge is
$$\frac{4\times5}{12\times13} \times \$31.20 = \frac{5}{39} \times \$31.20 = \mathbf{\$4}.$$
(b) There are 4 payments remaining, so it will take (4 × $50) - $4 = **$196** to pay off the loan.

23. (a) Finance charge = (0.10)($720) = **$72.**
    (b) Monthly payment = $\dfrac{\$720 + \$72}{12}$ = **$66.**
    (c) Here r = 4 and n = 12, so the interest refund is
    $\dfrac{4 \times 5}{12 \times 13} \times \$72 = \dfrac{5}{39} \times \$72 =$ **$9.23.**
    (d) There are 4 payments remaining, so it will take
    (4 × $66) − $9.23 = **$254.77** to pay off the loan.

25. (a) Finance charge = (0.10)($800)($1\tfrac{1}{2}$) = **$120.**
    (b) Monthly payment = $\dfrac{\$800 + \$120}{18}$ = **$51.11.**
    (c) Here, r = 3 and n = 18, so the interest refund is
    $\dfrac{3 \times 4}{18 \times 19} \times \$120 = \dfrac{2}{57} \times \$120 =$ **$4.21**.
    (d) It will take (3 × $51.11) − $4.21 = **$149.12** to pay off the loan.

27. APR = $\dfrac{2mI}{P(n+1)} = \dfrac{2 \times 12 \times 194}{2500 \times 13} = 0.143 =$ **14.3%** (**0.3%** more than the answer to problem 1)

29. APR = $\dfrac{2mI}{P(n+1)} = \dfrac{2 \times 12 \times 264}{1500 \times 25} = 0.169 =$ **16.9%** (**0.9%** more than the answer to problem 3)

31. APR = $\dfrac{2mI}{P(n+1)} = \dfrac{2 \times 12 \times 210}{1500 \times 19} = 0.177 =$ **17.7%** (**0.7%** more than the answer to problem 5)

33. APR = $\dfrac{2mI}{P(n+1)} = \dfrac{2 \times 12 \times 1570}{4500 \times 49} = 0.171 =$ **17.1%** (**1.6%** more than the answer to problem 7)

35. APR = $\dfrac{2mI}{P(n+1)} = \dfrac{2 \times 12 \times 1800}{5000 \times 49} = 0.176 =$ **17.6%** (**1.6%** more than the answer to problem 9)

SECTION 12.4 Buying a House 323

EXERCISE 12.4

STUDY TIPS There are many costs that people overlook or are not aware of when buying a house. The Getting Started discusses some of these costs. Keep in mind that an innocent sounding term like "a point" can be worth $800 on an $80,000 loan. Read the examples, redo the calculations and understand the terminology before you attempt the problems.

1. (a) The maximum they can spend according to the first criterion is 2.5 x $40,000 = **$100,000**, so they **can** afford this house.
   (b) $\frac{\$40000}{52}$ = $769.23 > $750, so, according to the second criterion, they **can** afford this house.
   (c) 0.28 x $\frac{40000}{12}$ = 933.33 > $750, so, according to the third criterion, they **can** afford this house.

3. (a) Amount of loan = (0.80)($77,000) = **$61,600**.
   (b) Down payment = $77,000 - $61,600 = **$15,400**.
   (c) For an FHA loan, down payment = $77,000 - $67,500 = **$9500**.

5. Table 12.4 gives $8.35 as the monthly payment per $1000 loan at 8% for 20 years. Thus, the mortgage payment is 30 × $8.35 = $250.50. Insurance and taxes amount to $560/12 = $46.67 per month. Thus, the total monthly payment is $250.50 + $46.67 = **$297.17.**

7. Table 12.4 gives $8.39 as the monthly payment per $1000 loan at 9% for 25 years. Thus, the mortgage payment is 45 x $8.39 = $377.55 Insurance and taxes amount to $750/12 = 62.50 per month. Thus, the total monthly payment is $377.55 + $62.50 = **$440.05.**

9. Table 12.4 gives $8.78 as the monthly payment per $1000 loan at 10% for 30 years. Thus, the mortgage payment is 73 x $8.78 = $640.94. Insurance and taxes amount to $1220/12 =$101.67 per month. Thus, the total monthly payment is $640.94 + $101.67 = **$742.61.**

11. (a) 95% of $60,000 = (0.95)($60,000) = $57,000, so the down payment is $60,000 - $57,000 = **$3000.**
    (b) Table 12.4 gives $9.52 as the monthly payment per $1000 loan at 11% for 30 years. Hence, the mortgage payment is 57 × $9.52 = $542.64. Taxes and insurance amount to $1500/12 = $125 per month. The total monthly payment is $542.64 + $125 = **$667.64.**

13. (a) Minimum down payment is 3% of $25,000 + 5% of $50,000 = $750 + $2500 = **$3250.**
    (b) Table 12.4 gives $10.29 as the monthly payment per $1000 loan at 12% for 30 years. The maximum loan is $75,000 - $3250 = $71,750. Hence, the mortgage payment is 71.75 × $10.29 = $738.31. Taxes and insurance amount to $360/12 = $30 per month. The total monthly payment is $738.31 + $30 = **$768.31.**

15. (a) Table 12.4 gives $8.78 as the monthly payment per $1000 loan at 10% for 30 years. So, the mortgage payment is 50 × $8.78 = **$439**.
    (b) 30 × 12 = **360**
    (c) 360 × $439 = **$158,040**
    (d) $158,040 - $50,000 = **$108,040**
    (e) If 50,000 = 80% of x, then x = $\frac{50000}{0.80}$ = 62,500. Thus, the price of the house was $62,500, which is **less** than the total interest.

17. (a) The down payment is 20% of $50,000 = $10,000. The loan fee is 1% of ($50,000 - $10,000) = $400 which makes the closing costs total $1185. Thus, the cash payment is $10,000 + $1185 = **$11,185.**
    (b) Insurance and taxes amount to $25 + $50 = $75 per month. Table 12.4 gives $11.85 as the monthly payment per $1000 loan at 14% for 30 years, so the mortgage payment is 40 × $11.85 = $474. The total monthly payment is thus $474 + $75 = **$549.**
    (c) Since the closing costs are $1185, there would be 1.185 × $11.85 = $14.04 added to the amount found in part (b) to give a total of **$563.04.**

19. (a) The down payment is 20% of $120,000 = $24,000. The loan amount is $120,000 - $24,000 = $96,000, so the loan fee is 1% of $96,000 = $960. This makes the total closing costs $1630. The down payment plus closing costs is thus $24,000 + $1630 = **$25,630.**
    (b) Table 12.4 gives $8.05 as the monthly payment per $1000 loan at 9% for 30 years. The mortgage payment is 96 × $8.05 = $772.80. Since taxes and insurance come to $100 per month, the total monthly payment is $100 + $772.80 = **$872.80**
    (c) To include the closing costs, add 1.63 × $8.05 = $13.12 to the amount in part (b) to get a total of **$885.92** per month.

21. (a) The down payment is 10% of $84,000 = $8400. The loan amount is $84,000 - $8400 = $75,600, so the loan fee is 2% of $75,600 = $1512. This makes the total closing costs come to $2067. The down payment plus closing costs is thus $8400 + $2067 = **$10,467.**
    (b) Table 12.4 gives $10.75 as the monthly payment per $1000 loan at 10% for 15 years. The mortgage payment is 75.6 × 10.75 = **$812.70.**

23. Answers will vary.

25. Table 12.4 gives $8.05 as the monthly payment per $1000 loan at 9% for 30 years. For a $100,000 loan, the monthly payment is $805, and in 30 years the total payments would be 360 × $805 = $289,800. Thus, the interest paid would be $289,800 - $100,000 = **$189,800**, which is **greater** than the price of the house.
    For a $50,000 loan, the monthly payment is $402.50 and in 30 years, the total payments would be 360 × $402.50 = $144,900. Thus, the interest paid would be $144,900 - $50,000 = **$94,900**, which is **more** than the price of the house.

PRACTICE TEST 12

STUDY TIPS The Practice Test on page 693 of the textbook is worked in detail next. Use it as a study tool to master the material before you take the "real" test. If you want more practice, there are two other tests given at the end of this manual, multiple choice and free response. Practice with the type of test your instructor gives, so that you have a better idea of your strengths and weaknesses with various concepts.

1. (a) I = Prt = ($800)(0.28)(2) = **$448**
   (b) I = Prt = ($800)(0.28)(1/4) = **$56**
   (c) $800 + $448 = **$1248**

2. (a) (0.06)($360) = **$21.60**   (b) $360 + $21.60 = **$381.60**

3. (a) (0.20)($390) = **$78**   (b) $390 - $78 = **$312**

4. (a) Interest rate is 4% per period and there are 4 periods. The table gives 1.1699 for these numbers, so the accumulated amount is $100 × 1.1699 = **$116.99**; the interest is **$16.99**.
   (b) Interest rate is 2% per period and there are 8 periods. The table gives 1.1717 for these numbers, so the accumulated amount is $100 × 1.1717 = **$117.17**; the interest is **$17.17**.

5. (a) 0.05 × $185.76 = $9.29, so the minimum payment is **$10.**
   (b) Finance charge = 0.015 × $175.76 = **$2.64**.

6. (a) $179.64 - $50 = $129.64, so the finance charge is 0.015 × $129.64 = **$1.94**
   (b) New balance = $129.64 + $1.94 + $23.50 = **$155.08**.

7. (a) $6500 - $1500 = $5000 (unpaid balance)
       Interest = (0.12)($5000)(4) = **$2400**
   (b) Total to be paid off is $5000 + $2400 = $7400, so the monthly payment is $\frac{\$7400}{48}$ = **$154.17**.

8. (a) 12 × $18.10 = $217.20 (total amount to be paid)
       $217.20 - $200 = $17.20 (interest charged)
       $\frac{\$17.20}{2}$ = $8.60 (interest charge per $100). The closest table entry is $8.59 corresponding to an APR of $15\frac{1}{2}$%.
   (b) Here, r = 7 and n = 12, so the interest refund is $\frac{7 \times 8}{12 \times 13}$ × $17.20 = **$6.17**.
   (c) (7 × $18.10) - $6.17 = **$120.53** (amount needed to pay off the loan.

9. (a) (0.75)($50,000) = **$37,500** (amount of loan)
   (b) $50,000 - $37,500 = **$12,500** (down payment)
   (c) (0.03)($25,000) + (0.05)($25,000) = **$2000** (minimum down payment for FHA loan)
   (d) $50,000 - $2000 = **$48,000** (maximum FHA loan)

10. The table shows a payment of $12.00 per $1000 loan at 12% for 15 years, so the monthly mortgage payment (for principal and interest) is 37.5 × $12.00 = **$450.**

APPENDIX
THE METRIC SYSTEM

EXERCISE A.1

STUDY TIPS Start by memorizing the multiples, subdivisions and prefixes of the basic units in the metric system. (See Table A.1) Remember that these go in multiples and submultiples of 10, so that changing from one metric unit to another involves only a shift in the decimal point. For example, if you remember
Hecto Deka Unit Deci Centi
you will know that going from Hecto to Deci, three units right, involves a shift of the decimal point three units right. Thus, 382 hectoliters is equivalent to 382,000 deciliters. This idea works not only for liters but other metric measurements. The units for length, volume and weight are given in Tables A.2, A.3 and A.4 respectively. As far as temperature is concerned have you seen the temperature on one of those outdoor thermometers? Sometimes the temperature is given in degrees **Celsius**. This section also shows you how to convert temperatures from Celsius to **Fahrenheit** and vice versa but to do this you have to memorize the formulas.

1. (a) **1000** (kilo means thousand)
   (b) **0.001** (milli means one-thousandth)
   (c) **100** (centi means one-hundredth)
   (d) **1000** (kilo means thousand)

3. Answer **(c)**   5. Answer **(a)**   7. Answer **(a)**

9. (a) **8000** (km = 1000 m)   (b) **400** (m = 100 cm)
   (c) **34.09** (cm = 0.01 m)   (d) **4.94** (mm = 0.1 cm)

11. **2.10** (cm = 0.01 m)   13. **1500** (km = 1000 m)

15. (a) 300 × 52.5 = 15,750 cm = **157.5 m**
    (b) 50 × 52.5 = 2625 cm = **26.25 m**
    (c) 30 × 52.5 = 1575 cm = **15.75 m**

17. A meter is 100 cm = 10 × 10 cm, so a cubic meter = $(10 \times 10)^3$ cm^3 = $1000 \times (10^3$ cm$^3) =$ **1000 liters**. Note: 10^3 cm^3 = 1000 cm^3 = 1 liter.

19. 1000 ml = 1 liter, so the answer is **3.5**.

21. The volume is $50 \times 20 \times 10$ cm^3 = 10,000 cm^3 = **10 liters.**

23. Since 5/20 = 1/4, 1/4 of a liter or **250 ml** are needed.

25. (a) **14,000** (1 kg = 1000 g)   (b) **4800** (1 kg = 1000 g)
    (c) 2.8 g = **0.0028 kg** (1 g = 0.001 kg)
    (d) 3.9 g = **3900 mg** (1 g = 1000 mg)

27. 1 liter = 1000 cm^3, so 1 liter weighs **1000 g** or **1 kg.**

29. Answer **(b)** (See problem 27.)

31. $\frac{5}{9}(59 - 32) = 15$, so fill the blank with **15**.

33. $\frac{5}{9}(86 - 32) = 30$, so fill the blank with **30**.

35. $\frac{5}{9}(-22 - 32) = -30$, so fill the blank with **-30**.

37. $\frac{9}{5}(10) + 32 = 50$, so fill the blank with **50**.

39. $\frac{9}{5}(-10) + 32 = 14$, so fill the blank with **14.**

41. $\frac{5}{9}(500 - 32) = 260$. Thus, 500° F = **260° C**.

43. $\frac{9}{5}(41) + 32 = 105.8$. Thus, 41° C = **105.8° F**.

45. $\frac{5}{9}(98.6 - 32) = 37$, so 98.6° F = **37° C.**

47. (a) $\frac{5}{9}(41 - 32) = 5$, so use **5° C**. (b) $\frac{5}{9}(212 - 32) = 100$, so use **100° C.**

SECTION A.1 Metric Units of Measurement 329

49. $\frac{9}{5}(-78) + 32 = -108.4$, so $-78°$ C = **-108.4° F**

51. $2.75 \times 31 + 71.48 = 156.73$, so the person was about **157 cm** tall.

53. (i) and (d) (ii) and (c); (iii) and (e); (iv) and (a); (v) and (b)

APPENDIX A.2

STUDY TIPS Most people think that changing to the metric system means that you will have to be converting units from one system to the other. But you buy sodas and wine by the liter and you do not convert it to quarts! Conversions should be done only to give you a good idea of the relative size of the units involved. Table A5 will give you the conversion factors you need. Keep in mind that the inch is defined to be 2.54 cm exactly, that is, 1 in. = 2.54 cm. All the other items in table are approximate values, so it is a good idea to know how to do computations with approximate numbers.

1. $8 \times 2.54 =$ **20.32**
3. $12 \times 0.394 =$ **4.73**
5. $51 \times 0.914 =$ **46.6**
7. $3.7 \times 1.09 =$ **4.03**
9. $4 \times 1.61 =$ **6.44**
11. $3.7 \times 0.621 =$ **2.30**
13. $6 \times 0.454 =$ **2.72**
15. $5 \times 2.20 =$ **11.0**
17. $5 \times 0.946 =$ **4.73**
19. $8.1 \times 1.06 =$ **8.59**
21. $(75 \times 0.394) \div 12 =$ **2.46**
23. $(2.54)^3 =$ **16.39**
25. $2 \times 36 \times 2.54 =$ **183**
27. $78 \times 0.394 \div 36 =$ **0.854**
29. $40 \times 1.61 =$ **64 km/hr**
31. $63 \times 0.914 =$ **58**
33. $8848 \times 1.09 \times 3 =$ **28,900**
35. $1234 \times 0.454 =$ **560**
37. $30 \times 1.61 =$ **48**
39. $52 \times 2.20 =$ **114**
41. $2(52.3 + 96.84) =$ **298.3**
43. $14 \times 2000 \div 2200 =$ **12.7**
45. $(40 \times 1.61) \div (4 \times 0.946) =$ **17.0**

47. 1900 m = 1.9 km = 1.9 × 0.621 mi = **1.18 mi**

49. 316 ft = 3792 in., so 316 ft $5\frac{3}{4}$ in. = 3797.75 in., or (3797.75 × 2.54) cm = 9646.285 cm = **96.5 m**

51. 40 × 2.54 = 101.6, 26 × 2.54 = 66.04, 38 × 2.54 = 96.52. To the nearest cm, the measurements would be **102-66-97.**

53. 15 × 1.61 ≈ **24**

55. 40 pt = 20 qt ≈ 20 × 0.946 ≈ **19 liters**

57. 1234 × 0.454 × 1000 ≈ **560,000**

PART II **ADDITIONAL PRACTICE TESTS AND ANSWERS**

Chapter Tests   (2) for each Chapter   333

One fill-in-the-blank test (Form A) and one multiple choice test (Form B) for each chapter.

> Note: To insure complete mastery, the order of the questions in these tests do not follow the same order as the Practice Tests in the book.

Answers   Answers to the above tests follow the TEST BANK.

TEST A   **CHAPTER 1, SETS AND PROBLEM SOLVING**

1. Let n(A) = 20 and n(B) = 10.   Find n(A ∪ B) if:
   a. A ∩ B = ∅
   b. n(A ∩ B) = 5

2. Let n(A) = 24, n(B) = 12, and n(A ∪ B) = 32. Find:
   a. n(A ∩ B)
   b. n(𝒰) if n(A' ∩ B') = 6

3. In the figure, the circular regions represent the sets A and B, and the rectangular region represents 𝒰. What set is represented by the shaded region in the diagram?

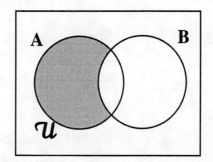

4. The figure shows the number of elements in the sets represented by the various regions in the diagram. Find the number of elements in the set A'

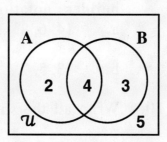

5. Upon checking 100 students, it is found that 40 are taking mathematics, 60 are taking biology, 50 are taking chemistry, 25 are taking mathematics and biology, 20 are taking mathematics and chemistry, 22 are taking biology and chemistry, and 10 are taking all three of these subjects.

_____   a. How many are taking mathematics but neither of the other two subjects?
_____   b. How many are taking chemistry but not biology?
_____   c. How many are taking none of the three subjects?

_____ 6. Show that the cardinal number of the sequence 2, 4, 8, . . ., 2^n, . . . is the same as the cardinal number of the sequence 1, 2, 3, . . ., n, ...

_____ 7. Identify the pattern and find the next three terms: 2, 7, 15, 26, 40, 57, . . .

_____ 8. Use set notation and list the elements of the set: $\{x \mid x$ is a positive even integer less than $10\}$.

9. Write the following sets verbally and using set builder notation
_____   a. $\{4, 6, 8, ... \}$

_____   b. $\{3, 5, 7, ... \}$

_____ 10. Write all the subsets of the set $\{1, 2, 3\}$.

11. Identify each of the following as the union of sets A and B, the intersection of sets A and B, or a difference of sets A and B.
_____   a. $\{x \mid x \notin A$ and $x \in B\}$
_____   b. $\{x \mid x \in A$ and $x \in B\}$
_____   c. $\{x \mid x \in A$ and/or $x \in B\}$

12. Let $\mathcal{U} = \{a, b, c, d, e\}$, $A = \{a, c, e\}$ and $B = \{b, c, e\}$
   Find:
   a. B'
   b. A ∪ B
   c. (A ∩ B)'
   d. $\mathcal{U}$ - (A ∩ B)'
   e. A ∩ B'

13. Let $\mathcal{U}$, A and B be as in Problem 12, and let C = {b, d, e}. Find:
   a. (A ∪ B) ∩ C
   b. (A' ∪ B) ∩ C'

14. Use a pair of Venn Diagrams to show that (A ∪ B)' = A' ∩ B'.

15. Draw a Venn diagram to illustrate the set A' ∩ B' ∩ C'.

16. Find the numbered regions in the figure that represent the following sets:

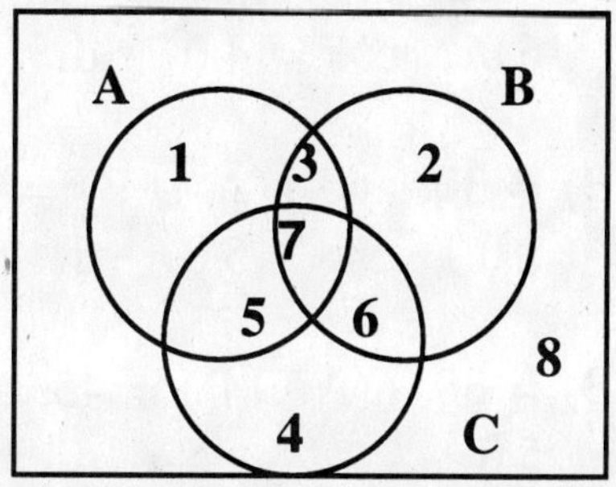

   a. A ∩ B' ∩ C
   b. (A ∩ B) ∪ C
   c. B - A
   d. (B ∪ C) ∩ (A ∪ C)

TEST B   CHAPTER 1, SETS AND PROBLEM SOLVING

1. In the figure, the numbered regions that identify the set A ∩ B' ∩ C are:

   a. 5   b. 1, 2,
   c. 1, 2, 3, 5, 6, 7   d. 1, 2
   e. 1, 2, 3, 8

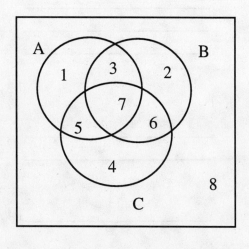

2. In the figure, the numbered regions that identify the set (A ∩ B) ∪ C are:
   a. 1, 2, 3   b. 3, 4, 5, 6, 7
   c. 3, 7   d. 5, 6, 7
   e. None of these

3. In the figure, the numbered regions that identify the set (B ∩ A) ∪ (B ∩ C) are:
   a. 1, 2, 3   b. 2, 3, 5, 6, 7
   c. 3, 7   d. 3, 6, 7
   e. None of these

4. If n(A) = 20, n(B) = 30 and n(A ∩ B) = 10, then n(A ∪ B) =
   a. 50   b. 60   c. 40   d. 45   e. 55

5. If n(A) = 20, n(B) = 30, and A ∩ B = ∅, then n(A ∪ B) =
   a. 50   b. 60   c. 40   d. 45   e. 55

6. If n(A) = 20, n(B) = 30, and n(A ∪ B) = 45, then n(A ∩ B) =
   a. 10   b. 20   c. 50   d. 5   e. 15

7. If n(A) = 20, n(B) = 30, n(A ∪ B) = 45, and n(A' ∩ B') = 10, then n(𝒰) =
   a. 95   b. 50   c. 55   d. 60   e. 65

8. In the figure, the shaded region represents the set:
   a. A ∩ B'   b. A' ∩ B
   c. A'   d. A ∪ B'
   e. None of these

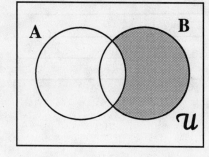

9. The figure shows the number of elements in the sets represented by the various regions in the figure. The number of elements in A' is
   a. 9   b. 2   c. 8
   d. 4   e. 7

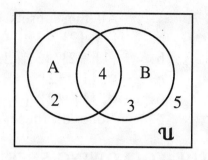

10. In the figure, n(A' ∩ B') =
    a. 9   b. 8   c. 7   d. 6   e. 5

11. Upon checking 100 students, it is found that 40 are taking mathematics, 60 are taking biology, 50 are taking chemistry, 25 are taking mathematics and biology, 20 are taking mathematics and chemistry, 22 are taking biology and chemistry, and 10 are taking all three of these subjects. The number of students taking none of the three subjects is:
    a. 0   b. 1   c. 3   d. 5   e. 7

12. Which of the following is correct?
    A' ∩ B' =
    a. (A ∩ B)'   b. (A ∪ B)'   c. A ∪ B
    d. A' ∪ B'   e. None of these.

13. Which of the following is correct?
    A ∩ (B ∪ C) =
    a. (A ∩ B) ∪ C   b. A ∪ (B ∩ C)   c. (A ∪ B) ∩ C
    d. (A ∩ B) ∪ (A ∩ C) e. None of these

14. For the sequence 3, 8, 16, 27, 41, . . . , the next three terms are:
    a. 51, 72, 85   b. 60, 75, 90   c. 58, 78, 101
    d. 58, 75, 93   e. None of these

15. The set of elements in {x | x is a positive odd integer less than 5} is:
    a. {1, 2, 3, 4}   b. {1, 3, 5}   c. {1, 3}
    d. {. . ., -1, 1, 3} e. None of these

16. The set {3, 5, 7, ...} is:
    a. The set of all odd integers.
    b. The set of all positive odd integers.
    c. The set of all odd natural numbers.
    d. The set of all positive odd integers greater or equal to 3.
    e. None of these

17. All the subsets of {1, 2, 3} are:
   a. {1}, {2}, {3}, {1, 2} {1, 3}, {2, 3}
   b. {1}, {2}, {3}, {1, 2} {1, 3}, {2, 3}, {1, 2, 3}
   c. ∅, {1}, {2}, {3}, {1, 2} {1, 3}, {2, 3}
   d. ∅, {1}, {2}, {3}, {1, 2} {1, 3}, {2, 3}, {1, 2, 3}
   e. None of these

THE FOLLOWING SETS ARE TO BE USED IN PROBLEMS 18, 19, 20. A AND B ARE SUBSETS OF A UNIVERSAL SET $\mathcal{U}$

   a. $\{x \mid x \in A \text{ and } x \notin B\}$ b. $\{x \mid x \notin A \text{ and } x \in B\}$
   c. $\{x \mid x \notin A \text{ and } x \notin B\}$ d. $\{x \mid x \in A \text{ and/or } x \in B\}$
   e. $\{x \mid x \in A \text{ and } x \in B\}$

18. Which one of these sets is the set $(A \cup B)'$?   a   b   c   d   e

19. Which one of these sets is the set $A \cap B'$?   a   b   c   d   e

20. If $\mathcal{U}$ = {a, b, c, d, e}, A = {a, c, e} and B = {a, d, e}, then $(A \cup B)'$ =
   a. {b, c, d}   b. {b}   c. {a, e}
   d. {a, c, d, e}   e. None of these

21. If $\mathcal{U}$, A and B are as in Problem 20 and C = {b, c, e}. Then $(A \cup B) \cap C'$ =
   a. $\mathcal{U}$ b. {a, d}   c. {a, b, d, e}
   d. ∅   e. None of these

22. A Venn diagram that illustrates the set $(A \cup C) \cap B'$ is:

a 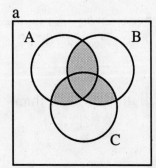   b   c 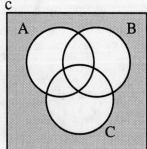   d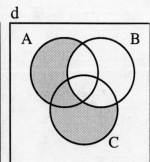

   e. None of these

TEST A   CHAPTER 2, LOGIC

_____ 1. Write the truth values for the → column in the following table.

p	q	p → (p ∨ q)
T	T	?
T	F	?
F	T	?
F	F	?

_____ 2. Show that the statement (~p ∨ r) ∧ (q ∨ r) is equivalent to (~p ∧ q) ∨ r.

_____ 3. Write the conditions under which the statement, "The day is warm or it is summer time," is a true statement.

_____ 4. Is the statement, "If 4 + 4 = 8, then 4 x 4 = 20," true or false?

The statement, **"If the weather is rainy, I stay indoors,"** will be used in problems 5-7.

_____ 5. Write the contrapositive of the given statement.

_____ 6. Write the converse of the given statement.

_____ 7. Write the inverse of the given statement.

_____ 8. If two statements **r** and **s** have the truth values shown in the table, which of the five relationships exist between **r** and **s**?

p	q	r	s		
T	T	T	T	a.	r → s
T	F	T	F	b.	s ⇔ r
F	T	F	F	c.	r ⇒ s
F	F	F	F	d.	s ⇒ r
				e.	None of these

_____ 9. Which of the following are tautologies?
    a. (p ∨ q) → (p ∧ q)   b. p ∧ ~p
    c. (p → q) ↔ (~p ∨ q)

_____10. Which of the following is a contradiction:
   a. $(p \wedge q) \vee (\sim p \wedge q)$ b. $(p \vee q) \rightarrow p$
   c. $q \vee \sim q$ d. $(p \wedge q) \rightarrow p$
   e. $\sim q \leftrightarrow q$

_____11. Use the given symbols to write the statement: "If it is the sixth month of the year (s) then it has thirty days (t)," in symbolic form.

_____12. Use the given symbols to write the statement: "Susie likes Paul (p), but Paul does not like Susie (~s)," in symbolic form.

_____13. Let h be "The test is hard," and p be "I shall pass it." Write in words the statement $\sim(h \vee p)$.

_____14. Let h and p be as in Problem 13. Write in words the statement $\sim h \vee p$.

_____15. Write in words the negation of the statement "I shall drive or I shall walk."

_____16. Write in words the negation of the statement "Some science fiction stories are interesting"

_____17. Write in words the negation of the statement, "All ripe peaches taste good."

The following truth table will be used in Problems 18 and 19.

p	q	$\sim p \wedge q$	$\sim(p \wedge q)$
T	T		
T	F		
F	T		
F	F		

_____18. Write the truth values for the statement $\sim p \wedge q$.

_____19. Write the truth values for the statement $\sim(p \wedge q)$.

_____20. Write the truth values for the ∧ column in the following table.

p	q	(p ∨ q) ∧ (~p ∨ ~q)
T	T	?
T	F	?
F	T	?
F	F	?

_____21. Is the argument   p ∨ q   valid or invalid?
                                               ~q
                                               ―――
                                               ∴ p

_____22. Select the argument that is **not** valid:
a. All fish can swim and all trout are fish.
Therefore, all trout can swim.

b. All cats have fur and all Siamese are cats
Therefore, all Siamese have fur.

c. All fish can swim and trout can swim.
Therefore, all trout are fish.

d. Every cat is an animal.
The Siamese is a cat.
Therefore, the Siamese is an animal.

e. Every book has covers.
The dictionary is a book.
Therefore, the dictionary has covers.

_____23. Use all premises to supply a valid conclusion for the argument:

All ball players read comics.
All kids play ball.
Patsy is a kid.

_____ 24. Write the following argument in symbolic form.

If the weather is bad (b), Josie will not go on a picnic (~g).
Josie will go on a picnic.
Therefore, the weather is not bad.

_____ 25. Draw a switching network corresponding to the statement ~p → q.

TEST B   CHAPTER 2, LOGIC

1. The correct entries for the → column of the following table are

p	q	p → (p ∨ q)
T	T	?
T	F	?
F	T	?
F	F	?

    a. FFTF
    b. FFFT
    c. TTTT
    d. FFFF
    e. TTTF

2. Which of the following statements is equivalent to (~p ∧ q) ∨ r?
    a. ~p ∧ (q ∨ r)   b. (~p ∨ r) ∧ (q ∨ r)
    c. ~(p ∧ ~q) ∨ r   d. (~p ∨ r) ∧ q   e. ~(p ∨ r) ∧ ~q

3. The statement, "The day is warm, or it is summer time" is true <u>only</u> under the following condition:
    a. The day is warm.
    b. It is summer time.
    c. The day is warm and it is summer time.
    d. The day is warm, or it is summer time, or both.
    e. None of these

4. The statement, "If 4 + 4 = 9, then 4 × 4 = 20," is
    a. True   b. False   c. A conjunction
    d. A disjunction   e. None of these

5. Given the statement, "If the weather is rainy, I stay indoors." Its contrapositive is
    a. If I stay indoors, then the weather is rainy.
    b. If the weather is not rainy, then I do not stay indoors.
    c. If I do not stay indoors, then the weather is not rainy.
    d. The weather is not rainy and I do not stay indoors.
    e. None of these.

6. Which of the statements in Problem 5 is the converse of the original given statement?   a   b   c   d   e

7. Which of the statements in Problem 5 is the inverse of the original given statement?   a   b   c   d   e

8. If two statements **r** and **s** have the truth values shown in the table, then

p	q	r	s
T	T	T	T
T	F	T	F
F	T	F	F
F	F	F	F

    a. $r \Leftrightarrow s$
    b. $s \Rightarrow r$
    c. $r \Rightarrow s$
    d. $r \rightarrow s$
    e. None of these

9. Which of the following are tautologies?
    1. $(p \vee q) \rightarrow (p \wedge q)$ 2. $p \wedge \sim p$ 3. $(p \rightarrow q) \leftrightarrow (\sim p \vee q)$
    a. 1 only   b. 2 only   c. 3 only
    d. 1 and 2 only   e. 1 and 3 only

10. Which of the following is a contradiction:
    a. $(p \wedge q) \vee (\sim p \wedge q)$ b. $(p \vee q) \rightarrow p$ c. $q \vee \sim q$
    d. $(p \wedge q) \rightarrow p$ e. $q \leftrightarrow \sim q$

11. The argument $p \vee q$ is:
                      $\sim q$
                      ─────
                      ∴ p

    a. Valid   b. Invalid   c. True   d. False   e. None of these

12. Select the argument that is **not** valid:

    a. All fish can swim and all trout are fish.
        Therefore, all trout can swim.

    b. All fish can swim and trout can swim.
        Therefore, all trout are fish.

    c. All cats have fur and all Siamese are cats
        Therefore, all Siamese have fur.

    d. Every cat is an animal.
        The Siamese is a cat.
        Therefore, the Siamese is an animal.

    e. Every book has covers.
        The dictionary is a book.
        Therefore, the dictionary has covers.

13. Which of the following (if any) is a valid conclusion using all the premises for the argument:

    All ball players read comics.
    All kids play ball.
    Patsy is a kid.
    Therefore:

    a. Patsy can play ball.   b. Patsy cannot play ball.
    c. Patsy reads comics   d. Patsy is a good ball player.
    e. No valid conclusion is given.

14. Read the following argument. Then select its symbolic form.

    If the weather is bad (p), Josie will not go on a picnic (~q).
    Josie will go on a picnic.
    Therefore, the weather is not bad.

    a. $p \to \sim q$ b. $p \to q$ c. $p \to q$ d. $p \to q$ e. $p \to q$
     q $\sim p$ q $\sim q$ q
     ─────   ─────   ─────   ──────   ─────
     ∴ ~p   ∴ ~q   ∴ p   ∴ ~q   ∴ ~p

15. The statement $\sim p \to q$ corresponds to the switching network

    a. A—P—Q'—B   b. A—P'—Q—B   c. A [P / Q] B

    d. A [P' / Q] B   e. None of these

16. Using the given symbols, the statement, "If it is the sixth month of the year (s), then it has 30 days (t)," is written:
    a. $s \wedge t$ b. $s \leftrightarrow t$ c. $t \to s$
    d. $s \to t$ e. None of these

17. Using the given symbols, the statement, "Susie likes Paul (p), but Paul does not like Susie (~s)," is written:
    a. $s \to p$ b. $p \to s$ c. $s \wedge \sim p$
    d. $p \wedge \sim s$ e. $s \vee \sim p$

18. Let h be "The test is hard," and p be "I shall pass it." The statement ∼(h ∧ p) is:
    a. The test is not hard and I shall not pass it.
    b. It is not the case that the test is hard and I shall pass it.
    c. The test is not hard, and I shall pass it.
    d. The test is not hard, but I shall not pass it.
    e. None of these

19. Let h and p be as in Problem 18. The statement ∼h ∧ p is which one of the statements in Problem 18?
    a   b   c   d   e

20. The negation of the statement, "I shall drive or I shall walk," is:
    a. I shall not drive and I shall not walk.
    b. I shall not drive or I shall not walk.
    c. I shall drive and I shall not walk.
    d. I shall not drive, but I shall walk.
    e. None of these

21. The negation of the statement, "Some science fiction stories are interesting," is:
    a. Some science fiction stories are not interesting.
    b. Not all science fiction stories are interesting.
    c. If it is not a science fiction story, then it is not interesting.
    d. If it is not a science fiction story, then it is interesting.
    e. No science fiction story is interesting.

22. The negation of the statement, "All ripe peaches taste good," is:
    a. Some ripe peaches do not taste good.
    b. No ripe peaches taste good.
    c. If it is a ripe peach, then it does not taste good.
    d. All ripe peaches do not taste good.
    e. If it is not a ripe peach, then it does not taste good.

The following truth table will be used in Problems 23 and 24.

p	q	1	2
T	T	F	F
T	F	F	T
F	T	T	T
F	F	F	T

23. Column **1** gives the truth values of
    a. p ∨ ~q   b. ~(p ∨ q)   c. ~p ∧ q
    d. ~(p ∧ q)   e. p ∧ ~q

24. Column **2** gives the truth values of
    a. p ∨ ~q   b. ~(p ∨ q)   c. ~p ∧ q
    d. ~(p ∧ q)   e. p ∧ ~q

25. The correct entries for the ∧ column of the following table are

p	q	(p ∨ q) ∧ (~p ∨ ~q)
T	T	?
T	F	?
F	T	?
F	F	?

   a. FTTF
   b. TFFT
   c. FTTT
   d. TFFF
   e. FTFF

TEST A   **CHAPTER 3, NUMERATION SYSTEMS**

_____ 1. Write in decimal notation:
$5 \times 10^3 + 2 \times 10^2 + 9 \times 10$

2. Write in decimal notation:
_____ a. 1101_2
_____ b. BA_{16}
_____ c. 324_5
_____ d. 751_8

3. Change the number 31 to the stated base:
_____ a. base five
_____ b. base two
_____ c. base sixteen
_____ d. base eight

4. Compute in base two:
_____ a. $1101_2 + 111_2$
_____ b. $1101_2 - 111_2$

5. Compute in base two:
_____ a. $1011_2 \times 101_2$

_____ b. $101_2 \overline{\smash{)}100011_2}$

6. Write in Egyptian numerals:
_____ a. 63
_____ b. 861

7. Write in Babylonian numerals:
_____ a. 63
_____ b. 861

8. Write in Roman numerals:
_____ a. 46
_____ b. 17,000

_____ 9. Write in decimal notation: ▼ ＜ ▼▼▼▼

_____ 10. Write in decimal notation: ∩∩∩ ||||

11. Write in decimal notation:
    a. XLVIII
    b. $\overline{\text{LXVII}}$

12. Perform the multiplication 22 × 24 using
    a. the Egyptian method of successive duplication.
    b. the Egyptian method of mediation and duplation.

13. Fill in the blanks:
    a. $x^a \cdot x^b =$ _____
    b. $\dfrac{x^m}{x^n} =$ _____, $m > n$, $x \neq 0$
    c. $x^0 =$ _____, $x \neq 0$

14. Perform the indicated operations and leave the answers in exponential form:
    a. $2^6 \cdot 2^7$
    b. $\dfrac{2^8}{2^3}$

15. Write in expanded notation:
    a. 3508
    b. 862

TEST B **CHAPTER 3, NUMERATION SYSTEMS**

1. $\dfrac{2^8}{2^3} =$

   a. $\dfrac{8}{3}$ b. 2^5 c. 1^5
   d. $2^{8/3}$ e. None of these

2. The number 3508 written in expanded notation is:
   a. $3 \times 10^3 + 5 \times 10^2 + 8$ b. $35 \times 10^2 + 8$
   c. $3 \times 10^3 + 508$ d. $3 \times 10^3 + 50 \times 10 + 8$
   e. None of these

3. Written in decimal notation, $5 \times 10^3 + 2 \times 10^2 + 9 \times 10 =$
   a. 5290   b. 5029   c. 5209
   d. 529   e. None of these

4. What is the answer to the following multiplication problem? 1011_2
                                                                                                                     $(\times)$ 101_2

   a. 1001101_2 b. 1010101_2 c. 110111_2
   d. 112221 e. None of these

5. What is the answer to the following subtraction problem? 1011_2
                                                                                                                      $(-)$ 101_2

   a. 9910   b. 9910_2 c. 1110_2
   d. 110   e. 110_2

6. In Egyptian numerals, 861 is

   a. ∩∩∩ 9999 | b. 999 ∩∩∩∩ |
       ∩∩∩ 9999   999 ∩∩∩∩

   c. 9999 ∩∩∩ | d. ∩∩∩ 9999   e. None of
       9999 ∩∩∩   ∩∩∩ 999 | these.

7. In Babylonian numerals, 65 is
   a. ⟨⟨⟨⟨⟨⟨ ▼▼▼▼▼ b. ▼ ▼▼▼▼▼
   c. ▼▼▼▼▼▼ ▼▼▼▼▼ d. ▼ ▼▼▼▼▼
   e. None of these

8. In Roman numerals, 1986 is
   a. I IX VIII VI   b. MCMLXXXVI   c. CMMXXXLVI
   d. IXX VIII VI   e. None of these

9. In decimal notation, the Babylonian numeral ▼ ⟨ ▼▼▼▼ is:
   a. 24   b. 74   c. 65
   d. 614   e. None of these

10. In decimal notation, the Egyptian numeral ∩∩∩ |||| is
    a. 34   b. 43   c. 64
    d. 1114   e. None of these

11. In decimal notation, the Roman numeral XLVIII is
    a. 1058   b. 10,508   c. 43
    d. 48   e. 68

12.
    | 1 | 24 |
    | 2 | 48 |
    | 4 | 96 |
    | 8 | 192 |
    | 16 | 384 |
    | | ? |

    The multiplication 22 × 24 by successive duplication is shown. What numbers should be marked with \ so that the corresponding numbers in the right hand column add up to the correct answer?
    a. 1, 2, 4 only   b. 2, 4, 16 only   c. All of them
    d. 1 and 16 only   e. None of these

13.
```
13    24
 6    48
 3    96
 1   192
```
The multiplication of 24 by 13 by mediation and duplation is shown. Which numbers in the right hand column should be circled and added to obtain the product?

    a.   All of them   b.   24 and 192 only
    c.   24, 48, and 192 only   d.   48 and 92 only
    e.   24, 96, and 192 only

14.   Which of the following is (are) **INCORRECT**?

    1. $x^m \cdot x^n = x^{mn}$ 2. $\dfrac{x^a}{x^b} = x^{a-b}$ for $a > b$ 3. $x^0 = 0$ for $x \neq 0$

    a.   2 only   b.   3 only   c.   1 and 3 only
    d.   All three   e.   They are all correct.

15. $2^6 \cdot 2^7 =$
    a. 4^{13} b. 2^{13} c. 2^{42}
    d. 4^{42} e.   None of these

TEST A CHAPTER 4, NUMBER THEORY AND THE REAL NUMBERS

1. Find a rational number with a denominator of 28 and equal to $\frac{3}{4}$.

2. Find the reciprocal of:
   a. $\frac{5}{7}$
   b. $-\frac{4}{5}$
   c. $3\frac{7}{16}$
   d. -14

3. Perform the indicated operations:
   a. $(-\frac{4}{9}) \times \frac{11}{18}$
   b. $(-\frac{4}{9}) \div (-\frac{11}{18})$

4. Write as decimals:
   a. $\frac{5}{8}$
   b. $\frac{5}{6}$

5. Write in simplest form as the quotient of two integers:
   a. $0.\overline{63}$
   b. $3.2555\ldots$

6. Write:
   a. 50.0601 in expanded form.
   b. $4 \times 10^3 + 5 \times 10^2 + 2 \times 10^{-1} + 5 \times 10^{-3}$ in decimal form.

_____ 7. Do the following calculation and write the answer in scientific notation:
$(9 \times 10^5) \times (6 \times 10^{-9})$

_____ 8. Which of the following numbers is (are) irrational?
$\sqrt{36}, \sqrt{42}, \sqrt{144},$ 2.131313 . . ., 0.2020020002

_____ 9. a. Find a rational number between $0.\overline{2}$ and 0.25.

_____   b. Find an irrational number between $0.\overline{2}$ and 0.25.

_____ 10. The diameter of a circle is 4 inches. To the nearest tenth of an inch what is the circumference of this circle? (Use $\pi \approx 3.14$.)

                11. Tell whether the underlined item is used as a cardinal number, an ordinal number, or for identification.

_____   a. Joe came <u>third</u> in the election.

_____   b. Susan's lottery ticket won <u>five</u> dollars.

_____   c. Bill's auto license number was 272-864.

                12. What properties of the system of natural numbers are used in the following equations?

_____   a. $(7 + 6) + 3 = (7 + 3) + 6$

_____   b. $9(10 + 3) = 90 + 27$

_____   c. $(2 \times 39) \times 5 = (2 \times 5) \times 39$

_____ 13. Write 286 as a product of primes.

_____ 14. Write the prime numbers between 40 and 50.

_____ 15. Is 123 prime or composite?

16. Consider the numbers 6345, 718, 849, 1650
    Identify the numbers divisible by:
_____ a. 2

_____ b. 3

_____ c. 5

_____ 17. Find the GCF of 135, and 225 and reduce the fraction $\frac{135}{225}$ to lowest terms.

_____ 18. Find the LCM of 90, 135 and 225 and perform the indicated operations:
$$\frac{1}{90} + \frac{1}{225} - \frac{1}{135}$$

_____ 19. What fractional part of Julian's budget goes for entertainment if $\frac{3}{5}$ of his budget is for school expenses, $\frac{1}{5}$ is for gasoline, $\frac{1}{20}$ is for savings and the remainder is for entertainment?

_____ 20. Evaluate: $6 \times 12 \div 4 \times 10^4 - 2(-8 + 4) \times 10^3$

21. Perform the indicated operations:
_____ a. 4.35 + 5.7
_____ b. 8.32 - 4.73
_____ c. 0.39 × 5.6
_____ d. 16.74 ÷ 5.4

_____ 22. The three sides of a triangle are measured as 18.4, 6.68, and 19.66 cm, respectively. What is the perimeter of this triangle? (Give the answer to the correct number of places.)

_____ 23. The dimensions of a rectangular court are measured as 83.6 by 50.5 feet. What is the area of this court? (Give the answer to the correct number of places.)

_____ 24. The circumference of a ball is measured as 11.5 inches. What is the diameter of this ball? (Use $\pi \approx 3.14$ and give the answer to the correct number of places.)

25. Write as a decimal:
_____ a. 23%
_____ b. 5.75%
_____ c. 0.34%

26. Write as a percent:
_____ a. 0.54
_____ b. 2.43
_____ c. $\frac{3}{5}$
_____ d. $\frac{5}{9}$ (Answer to one decimal place.)

_____ 27. A 2-liter bottle of soda sells for 69 cents and costs the store 49 cents. Find the percent of profit, correct to two decimal places.

_____ 28. 50 of the employees of Company A make between $25,000 and $50,000 annually. If this represents 20% of the company's employees, how many employees does Company A have?

29. Simplify as much as possible:
_____ a. $\sqrt{338}$
_____ b. $\sqrt{73}$
_____ c. $\frac{4}{\sqrt{48}}$
_____ d. $\sqrt{\frac{128}{49}}$

30. Perform the indicated operations and simplify:

    a. $\sqrt{15} \times \sqrt{5}$

    b. $\dfrac{\sqrt{72}}{\sqrt{8}}$

    c. $\sqrt{54} - \sqrt{24}$

    d. $\sqrt{48} + \sqrt{18} - \sqrt{12}$

31. For a right triangle with sides of length a, b and hypotenuse c, $c^2 = a^2 + b^2$. If the sides of a right triangle are 6 and 9 inches long, what is the length of the hypotenuse? Simplify your answer.

32. Classify the given numbers by making a check mark in the appropriate row:

Number Set	$\sqrt{8}$	-0.19	$\sqrt{25}$	0.666...	$-2\frac{1}{2}$
Natural					
Integers					
Rational					
Irrational					
Real					

33. Classify as an arithmetic or a geometric sequence:
    a. 3, 6, 12, 24, . . .
    b. 4, 7, 10, 13, . . .

34. For the sequence 6, 9, 12, 15, . . ., find:
    a. The first term
    b. The common difference
    c. The sum of the first ten terms
    d. The sum of the first n terms

35. For the sequence 4, -2, 1, $-\frac{1}{2}$, ..., find:

_____ a. a_1
_____ b. r
_____ c. The sum of the first five terms.
_____ d. The sum of the first n terms

36. Use the sum of an infinite geometric sequence to write the following repeating decimals as fractions in lowest terms:

_____ a. 0.888 ...
_____ b. 1.121212 ...
_____ c. 2.777 ...

TEST B   CHAPTER 4, NUMBER THEORY AND THE REAL NUMBERS

1. A rational number between $0.\overline{2}$ and 0.25 is
   a. 0.222 . . .   b. 0.232332333 . . .   c. 0.24567 . . .
   d. 0.22333 . . .   e. 0.24999 . . .

2. The diameter of a circle is 4 inches. To the nearest tenth of an inch, what is the circumference of this circle? (Use $\pi \approx 3.14$.)
   a. 12.6 in.   b. 12.56 in.   c. 12.5 in.
   d. 12 in.   e. None of these

3. $0.39 \times 5.6 =$
   a. 2.2   b. 2.20   c. 2.19
   d. 2.18   e. 2.184

4. The three sides of a triangle are measured as 18.4, 6.68, and 19.66 centimeters long, respectively. What is the perimeter of the triangle?
   a. 45 cm   b. 44.8 cm   c. 44.74 cm
   d. 44.70 cm   e. 44 .7 cm

5. The dimensions of a rectangular court are measured to be 83.6 by 50.5 feet. What is the area of this court?
   a. 4200 sq ft   b. 4222 sq ft   c. 4221.8 sq ft
   d. 4220 sq ft   e. 4221.80 sq ft

6. The circumference of a ball is measured as 11.5 in. What is the diameter of this ball? (Use $\pi \approx 3.14$.)
   a. 3.662 in.   b. 3.7 in.   c. 3.66 in.
   d. 3.6624 in.   e. None of these

7. Written as a decimal, 5.75% is
   a. 0.575   b. 0.00575   c. 0.0575
   d. 5.75   e. None of these

8. Written as a percent, correct to one decimal place, $\frac{5}{9}$ is
   a. 0.6%   b. 5.6%   c. 56%
   d. 55.6%   e. None of these

9. A 2-liter bottle of soda sells for 69 cents and costs the store 49 cents. Find the percent of profit, correct to two decimal places.
   a. 28%   b. 40.82%   c. 28.99%
   d. 40.80%   e. None of these

10. 50 of the employees of Company A make between $25,000 and $50,000 annually. If this represents 20% of all the employees, how many employees does Company A have?
    a. 200   b. 250   c. 500   d. 400
    e. None of these

11. In simplified form, $\sqrt{648}$ =
    a. $2\sqrt{162}$ b. $9\sqrt{8}$ c. 25.5
    d. $18\sqrt{2}$.   e. None of these

12. In simplified form, $\sqrt{\dfrac{128}{49}}$ =
    a. $\dfrac{16}{\sqrt{98}}$ b. $\dfrac{\sqrt{128}}{7}$ c. $4\sqrt{\dfrac{8}{49}}$
    d. $\dfrac{16}{7\sqrt{2}}$ e. $\dfrac{8\sqrt{2}}{7}$

13. In simplified form, $\sqrt{15} \times \sqrt{5}$ =
    a. $\sqrt{75}$ b. $3\sqrt{5}$ c. $5\sqrt{3}$
    d. $\sqrt{20}$ e. None of these

14. $\sqrt{48} + \sqrt{18} - \sqrt{12}$ =
    a. $\sqrt{66} - \sqrt{12}$ b. $3\sqrt{2} + 2\sqrt{3}$ c. $3\sqrt{6}$
    d. $\sqrt{48} + \sqrt{6}$ e. None of these

15. Which one of the following is an arithmetic sequence?
    a. 2, 3, 5, 8, ...   b. 4, 8, 16, 32, ...   c. 2, 5, 8, 11, ...
    d. 1, 0, 1, 0, ...   e. 0, 2, 3, 7, ...

16. The sum of the first ten terms of the sequence 2, 5, 8, 11, ... is
    a. 185   b. 155   c. 220
    d. 370   e. None of these

17. For the sequence $1, -\dfrac{1}{2}, \dfrac{1}{4}, -\dfrac{1}{8}, \ldots$, which of the following is correct? (Note: S_6 means the sum of the first six terms.)
    a. $r = -\dfrac{1}{2}, S_6 = -\dfrac{21}{32}$ b. $r = \dfrac{1}{2}, S_6 = \dfrac{63}{32}$ c. $r = -\dfrac{1}{2}, S_6 = \dfrac{21}{32}$
    d. $S_6 = -\dfrac{3}{2}$.   e. None of these

18. The repeating decimal 1.888 . . . =
    a. $1\frac{888}{1000}$ b. $\frac{17}{9}$ c. $\frac{236}{125}$
    d. 1.89   e. None of these

19. The underlined word in the sentence "This is the fifth test." is used as:
    a. a cardinal number   b. an ordinal number
    c. identification   d. a natural number
    e. None of these

20. What addition properties of the system of natural numbers are used in the equation: 7 + 6 + 3 = 7 + 3 + 6 = (7 + 3) + 6 ?
    a. Associativity only
    b. Commutativity only
    c. The distributive property only
    d. The associative and the distributive properties only
    e. Commutativity and associativity only

21. Written as a product of primes, 286 =
    a. 2•143   b. 2•11•13   c. 1•2•11•13
    d. 22•13   e. None of these

22. All the prime numbers between 50 and 60 are:
    a. 51, 53 and 57   b. 53 and 59   c. 51, 57 and 59
    d. 53 and 57   e. None of these

23. The number 123 is:
    a. a prime number   b. a composite number
    c. an even number   d. neither prime nor composite
    e. None of these

24. The number 45 is divisible by:
    a. 5 only   b. 1, 3, and 5 only   c. 1, 3 and 15 only
    d. 1, 3, 5, 9, and 15 only   e. 1, 3, 5, 9, 15 and 45 only

25. The GCF of 135 and 225 is
    a. 3   b. 5   c. 15   d. 45   e. 9

26. When written in simplest reduced form $\frac{135}{225}$ =
    a. $\frac{9}{15}$ b. $\frac{15}{25}$ c. $\frac{3}{5}$ d. $\frac{45}{75}$ e. None of these

27. The LCM of 12, 24 and 30 is
   a. 2   b. 3   c. 6   d. 30   e. 120

28. When written in simplest <u>reduced</u> form, $\frac{5}{12}+\frac{7}{24}-\frac{7}{30}=$
   a. $\frac{57}{120}$ b. $\frac{1}{3}$ c. 0   d. $\frac{19}{40}$ e. $\frac{38}{80}$

29. A pattern for a wedding outfit calls for $17\frac{1}{3}$ yards of lace trim for the dress and $6\frac{1}{2}$ yards of lace trim for the veil. How many yards of lace trim does the pattern call for?
   a. $17\frac{5}{6}$ yd   b. $10\frac{5}{6}$ yd   c. $23\frac{5}{6}$ yd
   d. $18\frac{2}{3}$ yd   e. None of these

30. -4 - (-12) =
   a. 6   b. -16   c. 8   d. -8   e. None of these

31. 48 ÷ 8 × 3 + 19 =
   a. 126   b. 21   c. 37   d. 1.12   e. None of these

32. The reciprocal of $-\frac{5}{7}$ is:
   a. $\frac{7}{5}$ b. $-\frac{7}{5}$ c. $\frac{5}{7}$ d. $-(-\frac{5}{7})$ e. None of these

33. $(-\frac{4}{9}) \div (\frac{11}{18}) =$
   a. $-\frac{44}{162}$ b. $-\frac{162}{44}$ c. $\frac{8}{11}$
   d. $-\frac{8}{11}$ e. None of these

34. Written as a decimal, $\frac{5}{6}=$
   a. 0.833   b. $0.\overline{83}$ c. $0.8\overline{3}$
   d. $0.8\frac{1}{3}$ e. None of these

35. Written as the quotient of two integers $0.\overline{63}$ =
    a. $\dfrac{63}{100}$
    b. $\dfrac{7}{11}$
    c. $\dfrac{21}{32}$
    d. 0.636363 . . .
    e. None of these

36. Written in standard decimal notation, 3.86×10^6 =
    a. 3,860,000
    b. 386,000
    c. $386 \times 100,000$
    d. $386 \times 10,000$
    e. None of these

37. In scientific notation, $(9 \times 10^5) \times (6 \times 10^{-9})$ =
    a. 54×10^{-4}
    b. 54×10^{-45}
    c. 5.4×10^{-3}
    d. 1.5×10^{-4}
    e. None of these

38. Which one of the following numbers is irrational?
    a. $\sqrt{36}$
    b. $\sqrt{42}$
    c. $\sqrt{144}$
    d. 2.131313 . . .
    e. 0.2020020002

TEST A   CHAPTER 5, EQUATIONS, INEQUALITIES, PROBLEM SOLVING

_____ 1. Factor $x^2 - 5x + 6$.

_____ 2. Factor $x^2 - 4x - 5$.

3. Solve:

_____   a. $(x + 2)(x - 3) = 0$

_____   b. $x(x - 3)(x + 4) = 0$

_____ 4. Solve by factoring: $x^2 + 8x + 12 = 0$.

_____ 5. Solve by factoring: $x^2 - 2x - 15 = 0$.

_____ 6. Solve by the quadratic formula: $3x^2 + 2x - 5 = 0$

_____ 7. Solve by the quadratic formula: $4x^2 - 4x - 5 = 0$

_____ 8. Solve: $4x^2 - 81 = 0$

_____ 9. Solve: $6x^2 - 49 = 0$

_____ 10. The hypotenuse of a right triangle is 4 cm longer than one leg and 2 cm longer than the other leg. What are the dimensions of this triangle?

_____ 11. Suppose you rent a car for one day at the rate of $25 per day plus 20 cents per mile. How many miles could you drive for a rental charge of $90?

_____ 12. It is known that corresponding sides of two similar rectangles are proportional. Joey drew a 5 in. by 8 in. rectangle and wants to draw a similar rectangle with the shorter side of length 11 in. How long must the longer side be?

_____ 13. The cost of gasoline per hour for running a certain car is directly proportional to the square of the speed. If the cost is $2 per hr for a speed of 40 mph, what is the cost per hour for a speed 60 mph?

_____ 14. At constant temperature, the volume of an enclosed gas varies inversely as the pressure. If the volume is 12 in.^{3} when the pressure is 24 lb per in.^{2}, what is the volume if the pressure is increased to 36 lb per in.^{2}?

15. If the replacement set is the set of integers, find the solution set of
_____ a. x + 4 = 7
_____ b. x - 4 = -9

16. If the replacement set is the set of integers, find the solution set of
_____ a. x + 9 > 11
_____ b. x + 4 ≤ -4 + 3x

17. Solve:
_____ a. 2 + x = 5 - 8x
_____ b. 2x - 2 ≤ 8 - 3x

18. Graph the solution set of
_____ a. x + 2 > 0
_____ b. -x + 3 > 5 - 2x

_____ 19. Graph the solution set of x - 1 ≥ 3 and x ≥ 2.

_____ 20. Graph the solution set of x - 3 ≤ 0 and x + 1 ≥ -2.

_____ 21. Graph the solution set of x ≥ 1 or x + 2 ≤ 1.

_____ 22. Graph the solution set of x + 1 < 0 or x - 1 < -3.

_____ 23. Graph the solution set of |x| = 4.

_____ 24. Graph the solution set of |x| ≥ 3.

_____ 25. Graph the solution set of |x - 1| < 1.

_____ 26. Graph the solution set of |x + 2| < 2.

_____ 27. Graph the solution set of |x| > 3.

_____ 28. Graph the solution set of |x - 2| > 2.

TEST B CHAPTER 5, EQUATIONS, INEQUALITIES, PROBLEM SOLVING

1. When factored, $x^2 - 6x + 5 =$
   a. $(x + 5)(x - 1)$ b. $(x - 5)(x + 1)$ c. $(x - 1)(x - 5)$
   d. $(x + 1)(x + 5)$ 5. None of these

2. The factored form of $x^2 - 2x - 8$ is
   a. $(x - 2)(x + 4)$ b. $(x + 2)(x - 4)$ c. $(x - 2)(x - 4)$
   d. $(x - 8)(x + 1)$ e. None of these

3. The solution set of $x(x - 3)(x + 4) = 0$ is
   a. $\{3, 4\}$ b. $(-3, 4)$ c. $\{-4, 3\}$
   d. $\{-4, 0, 3\}$ e. $\{-3, 0, 4\}$

4. The solution set of $x^2 - 13x + 12 = 0$ is
   a. $x = 1, x = 12$ b. $x = -1, x = -12$ c. $\{1, 12\}$
   d. $\{-12, -1\}$ e. None of these

5. The solutions of $x^2 + 3x - 10 = 0$ are
   a. $x = -5, x = 2$ b. $x = -2, x = 5$ c. $x = -10, x = 1$
   d. $x = -1, x = 10$ e. None of these

6. The solution set of $5x^2 + 3x - 2 = 0$ is
   a. $\left\{\dfrac{-3 \pm \sqrt{29}}{10}\right\}$ b. $\left\{-\dfrac{2}{5}, 1\right\}$ c. $\{-5, 2\}$
   d. $\left\{-1, \dfrac{2}{5}\right\}$ e. None of these

7. The solutions of $5x^2 + 8x - 4 = 0$ are
   a. $x = \dfrac{2}{5}, x = -2$ b. $x = -\dfrac{2}{5}, x = 2$ c. $x = \dfrac{1}{5}, x = 4$
   d. $x = \dfrac{2}{5}, x = 2$ e. None of these

8. The solution set of $4x^2 - 81 = 0$ is
   a. $\left\{\dfrac{9}{2}\right\}$ b. $\left\{\dfrac{9}{4}, \dfrac{9}{4}\right\}$ c. $\left\{-\dfrac{9}{2}, \dfrac{9}{2}\right\}$
   d. $\left\{-\dfrac{9}{2}\right\}$ e. None of these

9. The solution set of $6x^2 - 49 = 0$ is
   a. $\{-\frac{6}{7}, \frac{6}{7}\}$
   b. $\{\frac{7}{6}\}$
   c. $\{-\frac{7\sqrt{6}}{6}, \frac{7\sqrt{6}}{6}\}$
   d. $\{-\frac{7}{6}\}$
   e. None of these

10. The hypotenuse of a right triangle is 4 cm longer than one leg and 2 cm longer than the other leg. Thus, the length of the hypotenuse is
    a. 12 cm b. 10 cm c. 6 cm d. 9 cm
    e. None of these

11. Suppose you rent a car for one day at the rate of $25 per day plus 20 cents per mile. How many miles could you drive for a rental charge of $90?
    a. 300 b. 350 c. 400
    d. 325 e. None of these

12. It is known that corresponding sides of similar rectangles are proportional. A given rectangle is 5 in. by 8 in. and the shorter side of a similar rectangle is 11 in. long. The length of the other side of the second rectangle is
    a. 16.5 in. b. 17 in. c. $17\frac{3}{5}$ in. d. 18.6 in.
    e. None of these

13. The cost of gasoline per hour for running a certain car is directly proportional to the square of the speed. If the cost is $2 per hr for a speed of 40 mph, what is the cost per hr for a speed of 60 mph?
    a. $4.00 b. $3.50 c. $4.25 d. $4.50
    e. None of these

14. At constant temperature, the volume of an enclosed gas varies inversely as the pressure. If the volume is 12 cubic inches when the pressure is 24 pounds per square inch, what is the volume when the pressure is increased to 36 pounds per square inch?
    a. 6 in.^{3} b. 8 in.^{3} c. 9 in.^{3} d. 7 in.^{3}
    e. None of these

15. If the replacement set is the set of integers, the solution set of $x + 4 = 7$ is
    a. -3 b. 3 c. $\{-3\}$
    d. $\{3\}$ e. None of these

16. If the replacement set is the set of integers, the solution set of
    x - 4 = -9 is
    a. 5   b. {5}   c. -5
    d. {-5}   e. None of these

17. If the replacement set is the set of integers, the solution set of
    x + 9 > 11 is
    a. {2, 3, 4, ...}   b. 2, 3, 4, ...   c. {3, 4, 5, ...}
    d. 3, 4, 5, ...   e. None of these

18. If the replacement set is the set of integers, the solution set of
    x + 4 ≤ -4 + 3x is
    a. {3, 4, 5, ...}   b. {4, 5, 6, ...}   c. {x | x ≤ 4}
    d. {x | x ≥ 4}   e. None of these

19. The solution of 2 + x = 5 - 8x is
    a. x = 3   b. x = -1/3   c. x = 1/3
    d. x = -3   e. None of these

20. The solution set of 2x - 2 ≤ 8 - 3x is
    a. x ≤ 2   b. {x | x ≥ 2}   c. x ≥ 2
    d. {x | x ≤ 2}   e. None of these

21. The graph of the solution set of x + 4 ≥ 1 - 2x is
    a.   b.
    c.   d.
    e. None of these

22. The graph of the solution set of the compound statement
    x - 1 ≥ 3 **and** x > -4 is
    a.   b.
    c.   d.
    e. None of these

23. The graph of the solution set of the compound statement
    x - 2 ≤ 0 **and** x + 1 > 2 is
    a. 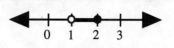 b.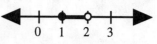
    c. 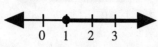 d.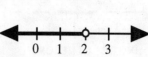
    e. None of these

24. The graph of the solution set of the compound statement
    x ≥ 1 **or** x + 2 ≤ 1 is
    a. 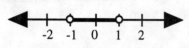 b.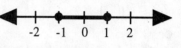
    c. 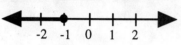 d.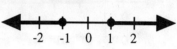
    e. None of these

25. The graph of the solution set of the compound statement
    x + 1 ≤ 0 **or** x - 1 < -3 is
    a. 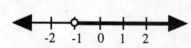 b.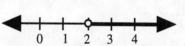
    c. 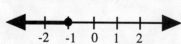 d.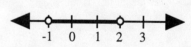
    e. None of these

26. The graph of the solution set of |x| < 1 is
    a. 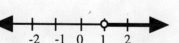 b.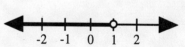
    c. 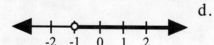 d.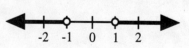
    e. None of these

27. The graph of the solution set of $|x + 1| > 1$ is

a.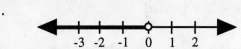
b.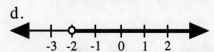

c.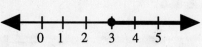
d.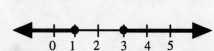

e. None of these

28. The graph of the solution set of $|x - 2| \leq 1$ is

a.
b.

c.
d.

e. None of these

TEST A   CHAPTER 6, FUNCTIONS AND GRAPHS

_____ 1. Find the domain and range of the relation R = {(4, -1), (2, -2), (1, 1)}.

_____ 2. Find the domain and range of the relation R = {(x, y) | y = 3x}.

_____ 3. Find the domain and range of the relation R = {(x, y) | y ≤ 2x, x and y positive integers less than 6}.

_____ 4. Which of the following relations is (are) functions?
a. $\{(x, y) | y^2 = 2x + 1\}$
b. $\{(x, y) | y = 2x^2 + 1\}$
c. {(2, 3), (3, 3), (4, 3)}

5. A function is defined by $f(x) = 2x - x^2$. Find:
_____ a. f(0)
_____ b. f(1)
_____ c. f(-2)

_____ 6. The daily cost for renting a car ($20 per day plus $0.15 per mile) is given by C(m) = 20 + 0.15m, where m is the number of miles driven. If a person paid $53.75 for one day's rental, how many miles did the person drive?

7. Graph the relation R = {(x, y) | y = 2x, x an integer between -2 and 2, inclusive}.

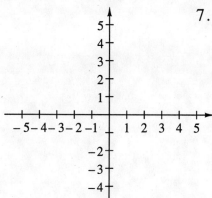

8. Graph the relation $Q = \{(x,y) \mid x + y \leq 2,$ x and y nonnegative integers$\}$.

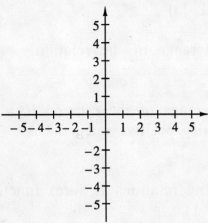

9. Graph the function defined by $g(x) = x^2 - 2$, x an integer and $-2 \leq x \leq 2$.

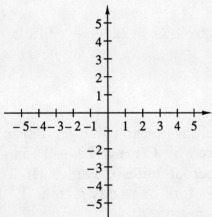

10. Graph the function defined by $f(x) = 2 - 2x$

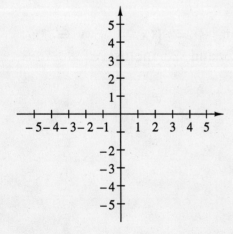

11. Graph the equation 2x - 3y = - 6.

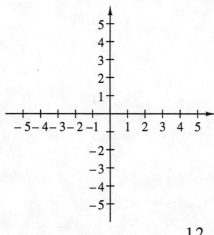

12. Find the distance between the two points:
    a. (1, 0), (3, -4)
    b. (7, -2), (7, -12)

13. Find the slope of the line that goes through the two points (-2, 1) and (-4, -5).

14. Find the general equation of the line in Problem 13.

15. a. Find the slope-intercept form of the equation of the line that goes through the point (-2, 4) and has slope -3.
    b. Find the slope-intercept form of the line 3x + 4y = 8. What is the slope and what is the y-intercept?

16. Determine whether or not the two given lines are parallel. If they are not parallel, find the coordinates of the point of intersection.
    a. 2x - y = 7, 3y = 6x - 15
    b. y = 4 - 2x, 6x + 2y = 9

17. Find the general equation of the line that passes through the point (3, 4) and is parallel to the line 2x + y = -4.

18. Find the point of intersection of the lines x + y = 6 and 2x - y = 0.

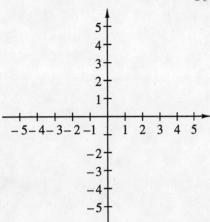

19. Graph the solution set of the inequality $2y - 3x \leq 6$.

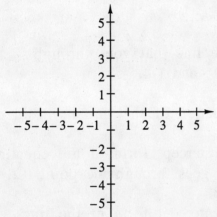

20. Graph the solution set of the system of inequalities: $3x + y \geq 6$ and $x + y \geq 2$

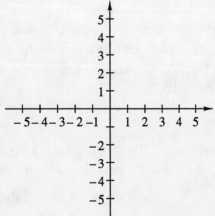

21. Graph the solution set of the system of inequalities: $2x + 3y \leq 6$, $x \geq y$, $y \geq 0$

_____ 22. Solve the following system if possible. If not possible, explain why
$$y = 3x - 3$$
$$9x - 3y = 6$$

_____ 23. Find the maximum value of $C = 3x + 2y$
subject to the constraints:
$x + 2y \geq 6$, $0 \leq x \leq 2$, and $0 \leq y \leq 4$

_____ 24. Find the minimum value of $P = x - 2y$
subject to the constraints:
$x - y \leq 2$, $x + y \leq 4$, $x \geq 0$, $0 \leq y \leq 2$

_____ 25. Two machines produce the same item. Machine A can produce 10 items per hour and machine B can produce 12 items per hour. At least 420 of the items must be produced each 40-hour week, but the machines cannot be operated at the same time. If it costs $30 per hour to operate A and $40 per hour to operate B, determine how many hours per week to operate each machine to meet the production requirement at minimum machine cost.

TEST B   CHAPTER 6, FUNCTIONS AND GRAPHS

1. The domain of the relation R = {(1, 1), (2, -2), (4, -1)} is
   a. {-2, -1, 1}   b. {1, 2, 4}   c. {-2, -1, 1, 2, 4}
   d. {-2, -1, 1, 1, 2, 4}   e. None of these

2. The range of the relation R = {(x, y) | y = 3x} is
   a. The positive real numbers   b. The positive integers
   c. The integers   d. The real numbers
   e. None of these

3. The range of the relation {(x,y) | y ≤ 2x, x and y positive integers less than 6} is
   a. {1, 2, 3, 4, 5}   b. {1, 2, 3, 4}   c. {1, 2, 3}
   d. {1, 2}   e. {1}

4. Which of the following relations are functions?
   a. $\{(x, y) | y^2 = 2x + 1\}$ b. $\{(x, y) | y = 2x^2 + 1\}$
   c. {(2, 3), (3, 3), (4, 3)}
   a. a only   b. b only   c. b and c only
   d. a and b only   e. None of these

5. If a function is defined by $f(x) = 2x - x^2$, then f(2) equals
   a. 6   b. 2   c. 8   d. $4x - x^2$ e. 0

6. The daily cost of renting a car is C(m) = 20 + 0.15m dollars, where m is the number of miles driven. If a person paid $53.75 for one day's rental, the number of miles the person drove is
   a. 175   b. 472   c. 205   d. 225
   e. 200

7. The graph of R = {(x, y) | y = -2x, x an integer between -2 and 2, inclusive} is
   a.   b.   c.   d.
   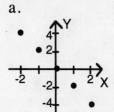 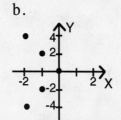 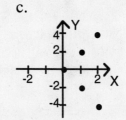 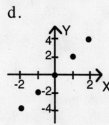

   e. None of these

8. The graph of the relation Q = { (x,y) | x + y ≤ 2, x and y nonnegative integers} is

a. 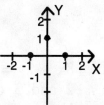   b. 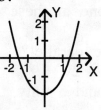   c. 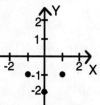   d.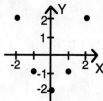

e.   None of these

9. The graph of the function defined by g(x) = x² - 2, x an integer and -2 ≤ x ≤ 2 is

a. 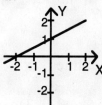   b. 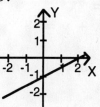   c. 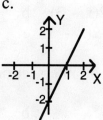   d.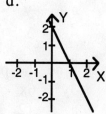

e.   None of these

10. The graph of f(x) = 2 - 2x is

a. 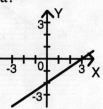   b. 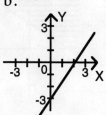   c. 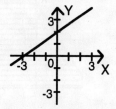   d.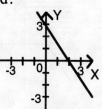

e.   None of these

11. The graph of the equation 2x - 3y = -6 is

a.   b.   c.   d.

e.   None of these

12. The distance between (1, 0) and (3, -4) is
    a. $3\sqrt{2}$ b. 8   c. $\sqrt{-6}$
    d. $2\sqrt{5}$ e. None of these

13. The slope of the line through (-2, 1) and (-4, -5) is
    a. 1/3   b. -1/3   c. 3
    d. -3   e. None of these

14. The general equation of the line through (1, 2) and (-5, 4) is
    a. -x - 3y = 7   b. x + 3y = 7   c. x + 3y = -7
    d. x - 3y = 7   e. None of these

15. The slope and the y-intercept of the line 3x - 4y = -12 are, respectively,
    a. 4/3, -3   b. 3/4, -3   c. 3/4, 3
    d. 4/3, 3   e. None of these

16. Which of the following lines are parallel?
    1. y = 4 - 4x   2. 6x - 2y = 9   3. 8x + 2y = 9
    a. 1 and 2 only   b. 1 and 3 only   c. 2 and 3 only
    d. All three are parallel.   e. None of these

17. The general equation of the line passing through the point (3, 4) and parallel to the line 2x - y = -4 is:
    a. y - 4 = 2(x - 3)   b. y = 2x   c. 2x - y = 2
    d. y - 4 = -2(x - 3)   e. -2x + y = -8

18. Find the point of intersection (if there is one) of the lines 2x - y = 7 and 3y = 6x - 15
    a. (2, 3)   b. (-2, 3)   c. (-2, -3)
    d. (2, -3)   e. There is none.

19. The graph of the solution set of $2y - 3x \leq 6$ is
    a.   b.   c.   d.

    e. None of these

20. The graph of the solution set of the system of inequalities
    $3x + y \geq 6$ and $x + y \geq 2$ is:
    a.   b.   c.   d.

    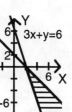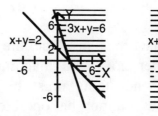

    e.   None of these

21. The graph of the solution set of the system of inequalities
    $3x + 2y \geq 6$, $x \geq y$,   and $y \geq 0$ is
    a.   b.   c.   d.

    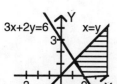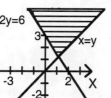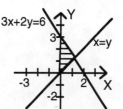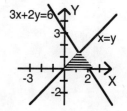

    e.   None of these

22. Which system of equations has no solution:
    a. $x + 2y = 9$ b. $x + 2y = 9$
         $x + 2y = 7$ $4x + 8y = 36$

    c. $x + 2y = 9$ d. $x - 2y = 9$
         $x - 2y = 7$ $x + 2y = 9$
    e.   All of the systems have solutions

23. The maximum value of $C = 3x + 2y$ subject to the constraints
    $x + 2y \geq 6$, $0 \leq x \leq 2$, and $0 \leq y \leq 4$ is
    a.   12   b.   14   c.   10
    d.   4   e.   None of these

24. The minimum value of P = x - 2y subject to the constraints
x - y ≤ 2, x + y ≤ 4, x ≥ 0, and 0 ≤ y ≤ 2   is
   a.   0
   b.   2
   c.   -2
   d.   -4
   e.   None of these

25. Two machines produce the same items. Machine A can produce 10 items per hour and machine B can produce 12 items per hour. At least 420 of the items must be produced each 40-hour week, but the machines cannot be operated at the same time. If it costs $30 per hour to operate A and $40 per hour to operate B, find the number of hours per week machines A and B, respectively, should be operated to minimize the cost.
   a.   10 and 30
   b.   30 and 10
   c.   40 and 0
   d.   0 and 35
   e.   None of these

TEST A CHAPTER 7, GEOMETRY

_____1. A rectangular plot of ground is to be enclosed with 180 yd of fencing. If the plot is twice as long as it is wide, what are its dimensions?

_____2. A 4 cm by 6 cm rectangle has all four of its vertices on a circle. Find the circumference of the circle. (Leave your answer in terms of π.)

_____3. In Problem 3, find the area of the region that is inside the circle and outside the rectangle. (Leave your answer in terms of π.)

_____4. A circular cylinder has a circumference of 33 in. Use $\frac{22}{7}$ as the approximate value of π and find the radius of this cylinder.

_____5. The width of a rectangle is 10 feet and its diagonal is 26 ft long. Find the length of this rectangle.

_____6. A window is in the shape of a rectangle surmounted by a semicircle. The width of the window is 6 ft, which is also the diameter of the circle, and the height of the rectangular part is 5 ft. Find the total area of this window.

_____7. A circle of diameter 6 in. has its center at the center of a square of side 6 in. Find the area of the region that is inside the square and outside the circle.

_____8. The dimensions of a rectangular box are 2 ft by 4 ft by 5 ft. Find the total surface area of this box.

_____9. A solid consists of a cone and a hemisphere mounted base to base. If the radius of the common base is 2 in. and the volume of the cone is equal to the volume of the hemisphere, what is the height of the cone?

_____10. A sphere and a circular cylinder have the same radius, 8 cm. If the two solids have equal total surface areas, what is the height of the cylinder?

_____11. The base of a pyramid is a triangle whose base is 6 ft and whose height is 3 ft. If the pyramid is of height 5 ft, what is the volume of this pyramid?

_____12. Which two of the following statements are incorrect?
  1. A network with three odd vertices is not traversable.
  2. A network with exactly two odd vertices is traversable.
  3. A network with four odd vertices is traversable.
  4. A network with four even and no odd vertices is traversable.

_____13. You have two buttons, one with two holes and the other with four holes. Are these buttons topologically equivalent? Why?

_____14. State the genus of each of the buttons in Problem 13.

_____15. In the equation $y = 2x(1-x)$, take $x = 0.3$ and find y_1. Then let $x = y_1$ and find y_2. After one more step what is the value of y correct to 3 decimal places?

_____16. Start with a square that is one inch on a side and replace the middle quarter of each side by adding on a square whose sides are parallel to the sides of the original square. What is the total area of the resulting figure?

_____ 17. X, Y, and Z are three points, in order from left to right on a line AB. What does $\overrightarrow{XZ} \cap \overrightarrow{YX}$ describe?

_____ 18. The edges of the two bases of a rectangular prism form how many pairs of parallel lines?

_____ 19. Through how many degrees does the hour hand of a clock turn in going from 1 o'clock to 6 o'clock?

_____ 20. If $\angle A$ and $\angle B$ are supplementary and $m\angle A = 3m\angle B$, find the measure of $\angle A$.

_____ 21. In a triangle ABC, $m\angle A = 38°$ and $m\angle B = 48°$. Find $m\angle C$.

_____ 22. In a triangle ABC, $m\angle C = 4m\angle A = 4m\angle B$. Find $m\angle C$.

_____ 23. In the word SUNSHADE, which letters are simple but not closed broken lines?

_____ 24. Triangles ABC and XYZ are similar, with $m\angle A = m\angle X$ and $m\angle B = m\angle Y$. If AB, BC, and AC are 2 in., 3 in., and 4 in. long, respectively, and XY is 3 in. long, find the length of YZ.

_____ 25. What is the measure of one of the interior angles of a regular polygon of nine sides?

TEST B   CHAPTER 7, GEOMETRY

1. A rectangular plot of ground is to be enclosed with 180 yd of fencing. If the plot is twice as long as it is wide, what are its dimensions?
   a. 30 by 60 yd   b. 20 by 40 yd   c. 35 by 70 yd
   d. 25 by 50 yd   e. None of these

2. A 4 cm by 6 cm rectangle has all four of its vertices on a circle. Find the circumference of the circle. (Leave your answer in terms of π.)
   a. 7π cm   b. $6\sqrt{2}\,\pi$ cm   c. 8π cm
   d. $2\sqrt{13}\,\pi$ cm   e. None of these

3. In Problem 2, find the area of the region that is inside the circle and outside the rectangle. (Leave your answer in terms of π.)
   a. $(9\pi - 24)$ cm²   b. $(10\pi - 24)$ cm²   c. $(13\pi - 24)$ cm²
   d. $(7\pi - 24)$ cm²   e. None of these

4. A circular cylinder has a circumference of 33 in. Use $\frac{22}{7}$ as the approximate value of π and find the radius of this cylinder.
   a. 10 in.   b. 14 in.   c. $5\frac{1}{4}$ in.
   d. 9 in.   e. None of these

5. The width of a rectangle is 10 feet and the diagonal is 26 feet long. Find the length of the rectangle.
   a. 28 ft   b. 24 ft   c. 30 ft
   d. 32 ft   e. None of these

6. A window is in the shape of a rectangle surmounted by a semicircle. The width of the window is 6 ft, which is also the diameter of the circle, and the height of the rectangular part is 5 ft. Find the total area of this window.
   a. $(30 + 9\pi)$ ft²   b. $(30 + 6\pi)$ ft²   c. $(30 + 4.5\pi)$ ft²
   d. $(30 + 4\pi)$ ft²   e. None of these

7. A circle of diameter 6 in. has its center at the center of a square of side 6 in. Find the area of the region that is inside the square and outside the circle.
   a. $(36\pi - 36)$ in.²   b. $(36 - 6\pi)$ in.²   c. $(36 - 9\pi)$ in.²
   d. $(9\pi - 36)$ in.²   e. None of these

8. The dimensions of a rectangular box are 2 ft by 4 ft by 5 ft. The total surface area of this box is
    a. 38 ft^2 b. 40 ft^2 c. 60 ft^2
    d. 76 ft^2 e. None of these

9. A solid consists of a cone and a hemisphere mounted base to base. If the radius of the common base is 2 in. and the volume of the cone is equal to the volume of the hemisphere, what is the height of the cone?
    a. 2 in.   b. 4 in.   c. 6 in.
    d. 8 in.   e. 10 in.

10. A sphere and a circular cylinder have the same radius, 8 cm. If the two solids have equal total surface areas, what is the height of the cylinder?
    a. 2 cm   b. 4 cm   c. 6 cm
    d. 8 cm   e. None of these

11. The base of a pyramid is a triangle whose base is 6 ft and whose height is 3 ft. If the pyramid is 5 ft high, what is its volume?
    a. 10 ft^3 b. 15 ft^3 c. 20 ft^3
    d. 25 ft^3 e. 30 ft^3

12. Which two of the following statements are incorrect?
    1. A network with three odd vertices is not traversable.
    2. A network with two odd vertices is traversable.
    3. A network with a four odd vertices is traversable.
    4. A network with four even and no odd vertices is traversable.
    a. 1 and 2   b. 2 and 3   c. 1 and 3
    d. 1 and 4   e. 3 and 4

13. Two buttons with an even number of holes
    1. are always topologically equivalent.
    2. are never topologically equivalent.
    3. could be topologically equivalent.
    4. are always of the same genus.
    5. could be of the same genus.
    Which two of these are correct?
    a. 1 and 4   b. 2 and 5   c. 3 and 4   d. 3 and 5
    e. 2 and 4

14. If a button has an even number of holes, then its genus is always
    a. 2 b. 4 c. equal to the number of holes d. 1
    e. None of these

15. In the equation y = 2x(1 - x), take x = 0.4 and find y_1. Then let x = y_1 and find y_2. Next let x = y_2 and find y_3. To three decimal places, y_3 =
    a. 0.420 b. 0.480 c. 0.490 d. 0.500 e. None of these

16. Start with a square that is one inch on a side and replace the middle quarter of each side by adding on a square whose sides are parallel to the sides of the original square. How many square inches is the total area of the resulting figure?
    a. $1\frac{1}{2}$ b. 2 c. $1\frac{3}{4}$ d. $1\frac{1}{4}$ e. None of these

17. X, Y, and Z are three points, in order from left to right on a line AB. What does $\overline{XZ} \cap \overline{YZ}$ describe?
    a. Point Y b. $\overline{XY}$ c. $\overline{YZ}$ d. $\overline{XZ}$
    e. None of these

18. The edges of the two bases of a rectangular prism form how many pairs of parallel lines?
    a. 4 b. 8 c. 19
    d. 12 e. None of these

19. Through how many degrees does the hour hand of a clock turn in going from 1 o'clock to 6 o'clock?
    a. 100 b. 120 c. 150
    d. 110 e. None of these

20. If $\angle A$ and $\angle B$ are supplementary and $m\angle A = 3m\angle B$, find the measure of $\angle A$.
    a. 120° b. 135° c. 150°
    d. 160° e. None of these

21. In a triangle ABC, $m\angle A = 38°$ and $m\angle B = 48°$. Find $m\angle C$.
    a. 90° b. 92° c. 94°
    d. 66° e. 96°

22. In a triangle ABC, m∠C = 4m∠A = 4m∠B. Find m∠C.
    a. 80°   b. 00°   c. 160°
    d. 100°   e. 120°

23. In the word SUNSHADE, which letters are simple but not closed broken lines?
    a. N, H, A and E   b. N and A   c. N and E
    d. H and E   e. Only N

24. Triangles ABC and XYZ are similar, with m∠A = m∠X and m∠B = m∠Y. If AB, BC, and AC are 2 in., 3 in., and 4 in. long, respectively, and XY is 3 in. long, find the length of YZ.
    a. 5 in.   b. 6 in.   c. 6.5 in.
    d. 4.5 in.   e. None of these

25. The measure of one of the interior angles of a regular polygon of nine sides is
    a. 100°   b. 110°   c. 120°
    d. 140°   e. 150°

TEST A   CHAPTER 8,   MATHEMATICAL SYSTEMS AND MATRICES

IN PROBLEMS (1-5), FIND THE ANSWER IN CLOCK ARITHMETIC.

_____ 1. $5 + 10 =$

_____ 2. $4 - 10 =$

_____ 3. $4 \times 7 =$

_____ 4. $6 \times 9 =$

_____ 5. $\dfrac{2}{5} =$

IN PROBLEMS (6-9), {0,1,2,3,4} IS THE REPLACEMENT SET FOR n.

_____ 6.   Find the value of n if $4 \times 3 \equiv n \pmod 5$.

_____ 7.   Find the value of n if $2 + 3 \equiv n \pmod 5$.

_____ 8.   Find the value of n if $2 - 4 \equiv n \pmod 5$.

_____ 9.   Find the value of n if $\dfrac{4}{3} \equiv n \pmod 5$.

IN PROBLEMS (10-11), FIND ALL POSSIBLE REPLACEMENTS FOR n FOR WHICH THE GIVEN CONGRUENCE IS TRUE.

_____ 10.   a. $5 + n \equiv 1 \pmod 3$
_____   b. $6 - n \equiv 4 \pmod 5$

_____ 11.   a. $5n \equiv 1 \pmod 7$
_____   b. $\dfrac{n}{4} \equiv 1 \pmod 3$

THE FOLLOWING TABLE, WHICH DEFINES AN OPERATOR ∗ ON THE SET S = {@, $. %}, IS USED IN PROBLEMS (12-17).

∗	#	@	%
#	%	#	@
@	#	@	%
%	@	%	#

_____ 12. Is the set S closed under operation ∗? Explain.

_____ 13. Is the operation ∗ commutative? Explain.

_____ 14. What is the identity element for the operation ∗?

_____ 15. For the operation ∗, find the inverse of #.

_____ 16. For the operation ∗, find the inverse of @.

_____ 17. For the operation ∗, find the inverse of %.

IN PROBLEMS (18-19), SUPPOSE THAT a S b MEANS TO SELECT THE SECOND OF THE TWO NUMBERS a AND b, AND a L b MEANS TO SELECT THE LARGER OF THE TWO NUMBERS (IF THE NUMBERS ARE EQUAL, SELECT THE NUMBER).

_____ 18. Is S distributive over L? Explain.

_____ 19. Is L distributive over S? Explain.

THE SET A = {%, @} AND THE OPERATIONS * AND # AS DEFINED IN THE FOLLOWING TABLES WILL BE USED IN PROBLEMS (20-23).

*	%	@
%	%	@
@	@	%

#	%	@
%	%	%
@	%	@

_____ 20. Find the result of the given operations:
a. (% * @) # %

_____ b. (% # @) * (% * @).

_____ 21. Is the set A together with the operation * a group?

_____ 22. Is the set A together with the operation # a group?

_____ 23. Does the set A together with the two operations * and # form a field?

_____ 24. Which of the following payoff matrices is (are) strictly determined?

(i) $\begin{bmatrix} -2 & 2 \\ 3 & -3 \end{bmatrix}$ (ii) $\begin{bmatrix} 1 & 2 \\ 0 & 1 \end{bmatrix}$ (iii) $\begin{bmatrix} 1 & -3 & 3 \\ -4 & 2 & -2 \\ 2 & 4 & 2 \end{bmatrix}$

_____ 25. The following payoff matrix is strictly determined. What is the row player's optimum strategy and what is the value of the game?

$\begin{bmatrix} 3 & 2 \\ 2 & -1 \end{bmatrix}$

_____ 26. The following payoff matrix is not strictly determined. What fraction of the time should the row player play each row and what is the corresponding payoff value?

$\begin{bmatrix} 1 & 2 \\ 2 & -1 \end{bmatrix}$

_____ 27. Given the three matrices below, find x and y so that 3A - B = C.

$$A = \begin{bmatrix} 2 & x \\ 4 & y \end{bmatrix}, B = \begin{bmatrix} 1 & -1 \\ 4 & 3 \end{bmatrix}, C = \begin{bmatrix} 5 & -2 \\ 8 & 9 \end{bmatrix}$$

_____ 28. Calculate AB and BA for the matrices

$$A = \begin{bmatrix} 3 & -4 \\ 1 & -1 \end{bmatrix}, B = \begin{bmatrix} -1 & 4 \\ 1 & 3 \end{bmatrix}$$

_____ 29. In order to build three types of a rolling toy, the Roller Toy Company makes the following demand chart:

TYPE	FRAMES	WHEELS	CHAINS
I	1	2	1
II	1	4	2
III	1	3	2

Suppose that Roller Toy gets an order for 30 type I, 40 type II, and 50 type III of these toys. Use matrices to find the number of frames, wheels, and chains needed to fill this order.

_____ 30. Roller Toy finds that frames cost $3 each, wheels $1 each, and chains $2 each. Use matrices to find the total cost of these items for each type of toy in Problem 29.

FOR PROBLEMS (31-34), LET

$$A = \begin{bmatrix} 1 & 3 & -1 \\ 0 & 2 & 1 \\ -1 & 1 & 2 \end{bmatrix} \text{ and } B = \begin{bmatrix} -1 & 1 & 1 \\ 2 & 2 & 0 \\ 2 & 3 & -1 \end{bmatrix}$$

_____ 31. Find 3A - 2B.

_____ 32. Find A - B.

_____ 33. Find $(A - B)^2$.

_____ 34. Find AB and BA.

_____ 35. Use matrices to solve the system:

$x + 2y = 4$
$y + 2z = -1$
$z + 3x = 5$

_____ 36. Upon cashing her paycheck for $475, Polly found that she received all five-, ten-, and twenty-dollar bills. If there were 40 bills in all, and as many twenties as fives, how many of each bill did she get?

_____ 37. The augmented matrix of a system of three linear equations in three unknowns x, y, z is reduced to the following form. Find the solution of this system.

$$\begin{bmatrix} 1 & 2 & 3 & | & 8 \\ 0 & 3 & 1 & | & 9 \\ 0 & 0 & 2 & | & 6 \end{bmatrix}$$

_____ 38. Suppose that the matrix in Problem 37 is changed to read as follows. Find the solutions of this new system.

$$\begin{bmatrix} 1 & 2 & 3 & | & 8 \\ 0 & 3 & 1 & | & 9 \\ 0 & 0 & 0 & | & 0 \end{bmatrix}$$

_____ 39 Suppose that the matrix in Problem 37 is changed to read as follows. What can be said about this system of equations?

$$\begin{bmatrix} 1 & 2 & 3 & | & 8 \\ 0 & 3 & 1 & | & 9 \\ 0 & 0 & 0 & | & 6 \end{bmatrix}$$

TEST B   CHAPTER 8,   MATHEMATICAL SYSTEMS AND MATRICES

1. In clock arithmetic, 5 + 10 =
   a. 1   b. 2   c. 3   d. 4   e. 16

2. In clock arithmetic, 4 - 10 =
   a. 1   b. 3   c. 4   d. 5   e. 6

3. In clock arithmetic, 4 × 7 =
   a. 2   b. 4   c. 6   d. 8   e. 28

4. In clock arithmetic, $\frac{2}{5}$ =
   a. 12   b. 5   c. 15   d. 10   e. None of these

5. A value of n that satisfies 2 + 3 ≡ n (mod 5) is
   a. 0   b. 1   c. 2   d. 3   e. 4

6. A value of n that satisfies 4 × 3 ≡ n (mod 5) is
   a. 1   b. 2   c. 3   d. 4   e. 5

7. A value of n that satisfies $\frac{4}{3}$ ≡ n (mod 5) is
   a. 1   b. 2   c. 3   d. 4   e. None of these

8. If 5 + n ≡ 1 (mod 3), then all possible values of n are given by
   a. n = 2   b. n = 3 + 3k, k an integer   c. n = -1
   d. n = 2 + 3k, k an integer   e. None of these

9. If $\frac{n}{2}$ ≡ 2 (mod 3), then all possible values of n are given by
   a. n = 4   b. n = 2   c. n = 1
   d. n = 1 + 3k, k an integer   e. n = 2 + 3k, k an integer

THE FOLLOWING TABLE, WHICH DEFINES AN OPERATION * ON THE SET {@, %, $}, IS USED IN PROBLEMS (10-11).

*	@	%	$
@	$	@	%
%	@	%	$
$	%	$	@

10. Which of the following are correct?
    (1) The set S is closed under the operation *.
    (2) The operation * is commutative.
    (3) The identity element for * is %.
    a. All three b. (1) and (2) only c. (1) and (3) only
    d. (2) and (3) only e. None is correct

11. Which of the following is (are) correct?
    (1) The inverse of @ is %.
    (2) The inverse of $ is @.
    (3) The inverse of % is %.
    a. (1) only b. (2) only c. (3) only
    d. (1) and (2) only e. (2) and (3) only

IN PROBLEMS (12-14), a, b, AND c REPRESENT REAL NUMBERS. LET a S b MEAN TO SELECT THE SECOND OF THE TWO NUMBERS a AND b, AND LET a L b MEAN TO SELECT THE LARGER OF THE TWO NUMBERS (IF THE NUMBERS ARE EQUAL, SELECT THE NUMBER).

12. Which of the following is (are) correct?
    (1) The set of real numbers is closed under the operation S.
    (2) The operation S has the commutative property a S b = b S a.
    (3) The operation S has the associative property
        a S (b S c) = (a S b) S c.
    a. (1) only b. (2) only c. (3) only
    d. (1) and (3) only e. (2) and (3) only

13. Which of the following are correct?
    (1) The set of real numbers is closed under the operation L.
    (2) The operation L has the commutative property a L b = b L a.
    (3) The operation L has the associative property
        a L (b L c) = (a L b) L c.
    a. (1) only b. (1) and (2) only
    c. (2) and (3) only d. (1) and (3) only
    e. All are correct.

14. Which of the following is (are) correct?
    (1) S is distributive over L, that is, a S (b L c) = (a S b) L (a S c).
    (2) L is distributive over S, that is, a L (b S c) = (a L b) S (a L c).
    (3) For all real numbers a, b, c, a L (b S c) = a S (b L c).
    a. (1) and (2) only
    b. (1) and (3) only
    c. (2) and (3) only
    d. (3) only
    e. All three are correct.

THE SET A = {%, @} AND THE OPERATIONS * AND # AS DEFINED IN THE FOLLOWING TABLES WILL BE USED IN PROBLEMS (15-21).

*	%	@		#	%	@
%	%	@		%	%	%
@	@	%		@	%	@

15. (% * @) # % =
    a. @ # @
    b. @ * %
    c. @
    d. %
    e. None of these

16. (% # @) * (% * @) =
    a. @ # @
    b. @ * %
    c. @
    d. %
    e. None of these

17. Which of the following is (are) correct?
    (1) The set A is closed under *.
    (2) * has the associative property.
    (3) The identity element for * is %.
    a. (1) and (2) only
    b. (1) and (3) only
    c. All three
    d. (1) only
    e. (3) only

18. Which of the following is (are) correct?
    (1) % and @ are their own inverses under *.
    (2) * does not have the commutative property.
    (3) The set A together with the operation * is a commutative (abelian) group.
    a. (1) and (2) only
    b. (1) and (3) only
    c. (2) and (3) only
    d. All three
    e. (1) only

19. Which of the following is (are) correct?
    (1) The set A is closed under #.
    (2) # has the associative property.
    (3) The identity element for # is @.
    a. (1) and (2) only   b. (1) and (3) only
    c. All three   d. (1) only   e. (3) only

20. Which of the following is (are) correct?
    (1) @ is its own inverse, and % has no inverse under #.
    (2) # has the commutative property.
    (3) The set A together with the operation # is a commutative (abelian) group.

    a. (1) and (2) only   b. (1) and (3) only
    c. (2) and (3) only   d. (1) only   e. (2) only

21. Which of the following is (are) correct?
    (1) The set A together with the operation * is a group but not a commutative group.
    (2) The set A together with the operation # is <u>not</u> a group.
    (3) The set A together with the operations * and # is a field.

    a. (1) and (2) only   b. (2) and (3) only
    c. (1) and (3) only   d. All three   e. (2) only

22. Which of the following payoff matrices is (are) strictly determined?

    (i) $\begin{bmatrix} -2 & 2 \\ 3 & -3 \end{bmatrix}$ (ii) $\begin{bmatrix} 1 & 2 \\ 0 & 1 \end{bmatrix}$ (iii) $\begin{bmatrix} 1 & -3 & 3 \\ -4 & 2 & -2 \\ 2 & 4 & 2 \end{bmatrix}$

    a. (i) and (ii) only   b. (i) only   c. (ii) only
    d. (iii) only   e. (ii) and (iii) only

23. The following payoff matrix is strictly determined. What is the row player's optimum strategy and what is the value of the game?

    $$\begin{bmatrix} 3 & 2 \\ 2 & -1 \end{bmatrix}$$

    a. Play Row 1, value 3   b. Play Row 2, value 2
    c. Play Row 1, value 2   d. Play Row 2, value -1
    e. None of these

24. The following payoff matrix is not strictly determined. What fraction of the time should the row player play each row and what is the corresponding payoff value?

$$\begin{bmatrix} 1 & 2 \\ 2 & -1 \end{bmatrix}$$

   a. Play each row one-half the time, payoff value 2.
   b. Play Row 1 two-thirds of the time and Row 2 one-third of the time, payoff value 2.
   c. Play Row 1 all the time, payoff value 2.
   d. Play Row 1 three-fourths of the time and Row 2 one-fourth of the time, payoff value $1\frac{1}{4}$.
   e. None of these

25. Given the three matrices below, find x and y so that 3A - B = C.

$$A = \begin{bmatrix} 2 & x \\ 4 & y \end{bmatrix}, B = \begin{bmatrix} 1 & -1 \\ 4 & 3 \end{bmatrix}, C = \begin{bmatrix} 5 & -2 \\ 8 & 9 \end{bmatrix}$$

   a. x = -1, y = -4   b. x = 4, y = -1   c. x = -1, y = 4
   c. x = -4, y = -1   e. None of these

26. Calculate AB for the matrices $A = \begin{bmatrix} 3 & -4 \\ 1 & -1 \end{bmatrix}, B = \begin{bmatrix} -1 & 4 \\ 1 & 3 \end{bmatrix}$.

   a. $\begin{bmatrix} -3 & -16 \\ 1 & -3 \end{bmatrix}$ b. $\begin{bmatrix} 1 & 0 \\ 0 & 1 \end{bmatrix}$ c. $\begin{bmatrix} -7 & 0 \\ -2 & 1 \end{bmatrix}$ d. $\begin{bmatrix} 2 & 0 \\ 0 & 2 \end{bmatrix}$

   e. None of these

27. In order to build three types of a rolling toy, the Roller Toy Company makes the following demand chart:

TYPE	FRAMES	WHEELS	CHAINS
I	1	2	1
II	1	4	2
III	1	3	2

Suppose that Roller Toy gets an order for 30 type I, 40 type II, and 50 type III of these toys.

If $D = \begin{bmatrix} 1 & 2 & 1 \\ 1 & 4 & 2 \\ 1 & 3 & 2 \end{bmatrix}$, $A = \begin{bmatrix} 30 & 40 & 50 \end{bmatrix}$, $B = \begin{bmatrix} 30 \\ 40 \\ 50 \end{bmatrix}$,

then the matrix M giving the numbers of frames, wheels, and chains needed for this order is the product
a.   DA   b.   AD   c.   DB
d.   BD   e.   None of these

28. Suppose frames cost $3 each, wheels $1 each, and chains $2 each. Let D be as in Problem 27, let $C = \begin{bmatrix} 3 & 1 & 2 \end{bmatrix}$, and let

$K = \begin{bmatrix} 3 \\ 1 \\ 2 \end{bmatrix}$. The matrix that gives the total cost of these items for each type of toy is the product
a.   DC   b.   KD   c.   CD
d.   DK   e.   None of these

The following two matrices will be used in Problems (29-32).

$$A = \begin{bmatrix} 1 & 3 & -1 \\ 0 & 2 & 1 \\ -1 & 1 & 2 \end{bmatrix} \quad \text{and} \quad B = \begin{bmatrix} -1 & 1 & 1 \\ 2 & 2 & 0 \\ 2 & 3 & -1 \end{bmatrix}$$

29. $3A - 2B =$
   a. $\begin{bmatrix} 5 & 7 & -5 \\ -4 & 0 & 3 \\ 8 & 3 & -7 \end{bmatrix}$
   b. $\begin{bmatrix} 5 & 7 & -5 \\ -4 & 2 & 3 \\ -7 & -3 & 8 \end{bmatrix}$
   c. $\begin{bmatrix} 5 & -7 & 5 \\ -4 & -2 & -3 \\ -7 & 3 & -8 \end{bmatrix}$

   d. $3A - 2B$ is not defined   e. None of these

30. $A - B =$
   a. $\begin{bmatrix} -2 & 2 & 2 \\ -2 & 1 & 0 \\ -3 & -2 & 3 \end{bmatrix}$
   b. $\begin{bmatrix} 2 & 2 & -2 \\ -2 & 0 & 1 \\ -3 & -2 & 3 \end{bmatrix}$
   c. $\begin{bmatrix} 2 & 2 & -2 \\ 2 & 0 & 1 \\ 3 & -2 & -3 \end{bmatrix}$

   d. $A - B$ is not defined   e. None of these

31. $(A - B)^2 =$
   a. $\begin{bmatrix} 6 & 8 & 8 \\ 7 & 6 & -7 \\ -11 & -2 & 13 \end{bmatrix}$
   b. $\begin{bmatrix} -6 & 8 & -8 \\ -7 & -6 & -7 \\ -11 & 2 & 13 \end{bmatrix}$
   c. $\begin{bmatrix} 6 & 8 & -8 \\ -7 & -6 & 7 \\ -11 & -12 & 13 \end{bmatrix}$

   d. $(A - B)^2$ is not defined   e. None of these

32. $BA =$
   a. $\begin{bmatrix} -2 & 0 & 4 \\ 2 & 10 & 0 \\ 3 & 11 & -1 \end{bmatrix}$
   b. $\begin{bmatrix} -2 & 0 & -4 \\ 2 & 2 & 0 \\ 3 & 11 & -1 \end{bmatrix}$
   c. $\begin{bmatrix} -2 & -1 & -4 \\ 2 & 10 & 0 \\ 3 & 11 & 1 \end{bmatrix}$

   d. BA is not defined   e. None of these

33. The solution by matrices of the system
$$x + 2y = 4$$
$$y + 2z = -1$$
$$z + 3x = 5$$
   gives:
   a. $x = -2, y = 1, z = 1$ b. $x = 2, y = 1, z = -1$ c. $x = 2, y = -1, z = 1$
   d. $x = 1, y = 2, z = 3$ e. $x = -2, y = -2, z = 1$

34. Upon cashing her paycheck for $475, Polly found that she received all five-, ten-, and twenty-dollar bills. If there were 40 bills in all, and as many twenties as fives, how many ten-dollar bills did she get?
    a. 5   b. 10   c. 15   d. 8   e. None of these

35. The augmented matrix of a system of three linear equations in three unknowns x, y, z is reduced to the form at the right. The solution of this system is
$$\begin{bmatrix} 1 & 2 & 3 & | & 8 \\ 0 & 3 & 1 & | & 9 \\ 0 & 0 & 2 & | & 6 \end{bmatrix}$$

    a.   x = -5, y = -2, z = 3   b.   x = -5, y = 2, z = 4
    c.   x = -5, y = 2, z = 3   d.   x = -5, y = -2, z = 4
    e.   None of these

36. The augmented matrix of a system of three linear equations in three unknowns x, y, z is reduced to the form at the right. If z = 3k, where k is any real real number, the solution of the system is:
$$\begin{bmatrix} 1 & 2 & 3 & | & 8 \\ 0 & 3 & 1 & | & 9 \\ 0 & 0 & 0 & | & 0 \end{bmatrix}$$

    a.   x = 2 - 7k, y = 3 + k, z = 3k   b.   x = 2 + 7k, y = 3 - k, z = 3k
    c.   x = 2 - 7k, y = 3 - k, z = 3k   d.   The system has no solution.
    e.   None of these

37. The augmented matrix of a system of three linear equations in three unknowns x, y, z is reduced to the form at the right. This system of equations has
$$\begin{bmatrix} 1 & 2 & 3 & | & 5 \\ 0 & 3 & 1 & | & 8 \\ 0 & 0 & 0 & | & 4 \end{bmatrix}$$

    a.   the solution z = 0.   b.   infinitely many solutions.
    c.   the solution z = k, k any real number.
    d.   the solution z = 4.   e.   no solution.

TEST A   **CHAPTER 9, COUNTING TECHNIQUES**

1. Compute:
   a.   C(5, 3)   b.   C(6, 5)

   _____

   _____

2. Compute:
   a.   C(8,8)   b. $\dfrac{C(5,3)}{C(5,2)}$

   _____

   _____

3. How many different subsets of 3 letters does the set {a, b, c, d, e, f} have?

   _____

4. Two cards are drawn in succession and without replacement from a standard deck of 52 cards. How many different sets of two cards are possible?

   _____

5. Three spades are picked from the 13 spades in a standard deck of 52 cards. How many different sets of three spades are possible?

   _____

6. Johnny has 5 coins, a penny, a nickel, two dimes, and a quarter. How many different sums of money can Johnny form by using 2 of the coins?

   _____

7. On a certain night, a TV station schedules 7 half-hour programs including the 11 o'clock news. Suppose you want to watch 4 of these programs:
   a. How many choices do you have?
   b. If one of the programs you want to watch is the 11 o'clock news, how many choices do you have?

   _____

   _____

8. How many distinct arrangements (in a row) can be made with the letters in the word BOXING?

   _____

9. How many distinct arrangements (in a row) can be made with the letters in the word SPINELESS?

_____10. The A-1 Company needs 4 skilled employees. 1 to be a foreman, and 3 to be helpers. If the company has 6 competent applicants, in how many different ways can these employees be selected?

_____11. On a recent market day, 2242 different stocks were traded on the New York Stock Exchange. Of these 2242 stocks, 917 advanced, 787 declined, and 538 were unchanged in price. Suppose at the end of the day, you marked a, d, or n after each stock traded according as that stock advanced, declined, or did not change. How many distinct arrangements of all the a's, d's, and n's are possible? (Leave your answer in indicated form. Do not try to simplify.)

_____12. A student wants a sandwich and a drink for lunch. At the snackbar, three kinds of sandwiches are available: chicken (C), ham (H) and tuna (T). The available drinks are ginger ale (G), milk (M) and root beer (R). Make a tree diagram to show all the possible lunches for this student.

_____13. A restaurant offers a choice of 2 soups, 6 entrees, and 3 desserts. How many different meals consisting of a soup, an entree, and a dessert are possible?

14. Two dice are tossed.
_____ a. How many different outcomes are possible?
_____ b. In how many ways could you get a sum of 5?

_____15. Two cards are drawn in succession and without replacement from a standard deck of 52 cards. In how many ways could these be a red king and a spade in that order?

_____ 16. An airline has 3 flights from city A to city B and 6 flights from city B to city C. In how many ways could you fly from city A to city C, using this airline?

17. Compute:
_____ a. 9!   b. $\dfrac{9!}{7!}$
_____

18. Compute:
_____ a. 3! x 5!   b. 3! + 5!
_____

19. Compute:
_____ a. P(4, 4)   b. P(5, 5)
_____

20. Compute:
_____ a. P(8, 2)   b. P(6, 4)
_____

_____ 21. In how many different ways can six people be arranged in a row for a group picture?

_____ 22. Two married couples are posing for a group picture. They are to be seated in a row of four chairs, with the two wives together in the middle. In how many ways can this be done?

_____ 23. Sally has 6 rabbits, 2 black, 2 white, and 2 black and white. In how many ways can Sally select 2 of her rabbits and include at least 1 black rabbit?

_____ 24. How many counting numbers less than 46 are divisible by 3 or by 5?

_____ 25. How many different sums of money can be made from a set of coins consisting of a penny, a nickel, a dime, a quarter, and a half-dollar if exactly four coins are to be used?

TEST B   CHAPTER 9, COUNTING TECHNIQUES

1. The number C(5, 3) =
   a. 10   b. 15   c. 20   d. P(5, 3)   e. None of these

2. The number $\dfrac{C(5,3)}{C(5,2)}$ =
   a. 3/2   b. C(5, 1)   c. 5   d. 1   e. None of these

3. How many different subsets of 3 letters each does the set {a, b, c, d, e, f} have?
   a. 6   b. 10   c. 20   d. 30   e. None of these

4. Two cards are drawn in succession and without replacement from a standard deck of 52 cards. How many different sets of two cards are possible?
   a. P(52, 2)   b. (2!)(52)(51)   c. 52 × 51
   d. $\dfrac{52 \times 51}{2!}$ e. None of these

5. Three spades are picked from the 13 spades in a standard deck of 52 cards. How many different sets of three spades are possible?
   a. 286   b. P(13, 3)   c. (3!)(13)   d. 13 × 12 × 11
   e. None of these

6. Johnny has 5 coins, a penny, a nickel, two dimes, and a quarter. How many different sums of money can he form by using 2 of the coins?
   a. C(5, 2)   b. 5   c. 7   d. 3   e. None of these

7. On a certain night, a TV station schedules 7 half-hour programs. If you want to watch 4 of these programs, how many choices do you have?
   a. 7!   b. P(7, 4)   c. 35   d. 28.   e. None of these

8. How many distinct arrangements (in a row) can be made with the letters in the word BOXING?
   a. 6   b. C(6, 6)   c. 120   d. 720   e. None of these

9. How many distinct arrangements (in a row) can be made with the letters in the word SPINELESS?
   a. C(9, 9)   b. P(9, 9)   c. $\dfrac{9!}{2!3!}$ d. 6!
   e. None of these

10. The A-1 Company needs 4 skilled employees, 1 to be a foreman, and 3 to be helpers. If the company has 6 competent applicants, in how many different ways can these employees be selected?
    a. 60   b. C(6, 4)   c. P(6, 4)   d. 30
    e. None of these

11. On a recent market day, 2242 different stocks were traded on the New York Stock Exchange. Of these 2242 stocks, 917 advanced, 787 declined and 538 were unchanged in price. Suppose at the end of the day, you marked <u>a</u>, <u>d</u>, or <u>n</u> after each stock traded according as that stock advanced, declined, or did not change. How many distinct arrangements of all the a's, d's, and n's are possible? (Leave your answer in indicated form. Do not try to simplify.)
    a. P(2242, 917)   b. (917)(787)(538)   c. C(2242, 917)
    d. $\dfrac{2242!}{917!787!538!}$ e. None of these

12. A student wants a sandwich and a drink for lunch. At the snackbar, three kinds of sandwiches are available: chicken, ham and tuna. The available drinks are ginger ale, milk and root beer. How many different choices for a sandwich and a drink does this student have?
    a. 3   b. 6   c. 9   d. 12   e. None of these

13. A restaurant offers a choice of 2 soups, 6 entrees, and 3 desserts. How many different meals consisting of a soup, an entree, and a dessert are possible?
    a. 11   b. 12   c. 24   d. 36   e. None of these

14. Two dice are tossed. In how many ways could you get a sum of 5?
    a. 2   b. 3   c. 5   d. 4   e. None of these

15. Two cards are drawn in succession and without replacement from a standard deck of 52 cards. In how many ways could these be a red king and a spade in that order?
    a. 26   b. 13   c. (2!)(13!)   d. 52
    e. None of these

16. An airline has 3 flights from city A to city B and 6 flights from city B to city C. In how many ways could you fly from city A to city C, using this airline?
    a. P(9, 2)   b. C(9, 2)   c. 18   d. 9
    e. None of these

17. The number 3! × 4! =

    a. 12!   b. 144   c. 72   d. 36   e. None of these

18. The number $\frac{9!}{7!}$ =

    a. 2!   b. 18   c. 36   d. 72   e. None of these

19. The number P(5, 5) =

    a. $\frac{5!}{4!}$ b. 5!   c. 1   d. 5   e. None of these

20. The number P(8, 2) =

    a. $\frac{8!}{2!}$ b. 2 × 8!   c. 72   d. 56   e. None of these

21. In how many ways can six people be arranged in a row for a group picture?
    a. 6^6 b. C(6, 6)   c. 720   d. 360   e. None of these

22. Two married couples are posing for a group picture. They are to be seated in a row of four chairs, with the two wives together in the middle. In how many ways can this be done?
    a. 4   b. 8   c. 12   d. 24   e. None of these

23. Sally has 6 rabbits, 2 black, 2 white, and 2 black and white. In how many ways can Sally select 2 of her rabbits and include at least 1 black rabbit?
    a. 6   b. 8   c. 12   d. 10   e. None of these

24. How many counting numbers less than 46 are divisible by 3 or by 5?
    a. 24   b. 21   c. 15   d. 9   e. None of these

25. How many different sums of money can be made from a set of coins consisting of a penny, a nickel, a dime, a quarter, and a half-dollar if exactly four coins are to be used?
    a. 5   b. P(5, 4)   c. 120   d. 20   e. None of these

TEST A   CHAPTER 10, PROBABILITY

_____ 1. Two fair dice are rolled. Find the probability that the sum turning up is 9, given that the first die turns up an even number.

2. Two fair dice are rolled. Find the probability that:
_____ a. They show a sum of 10.
_____ b. The first die turns up an odd number.
_____ c. Are these two events independent? Explain?

3. A certain prescription drug produces side effects in 3% of the patients. Three patients that have taken this drug are selected at random. Find the probability that:
_____ a. All three had side effects.
_____ b. None of the three had side effects.

_____ 4. Rosie has to take a Math course and an English course, both of which are available at 9 a.m., 10 a.m., and 11 a.m. If Rosie picks a schedule at random, what is the probability that she will have Math at 9 a.m. and English at 11 a.m.?

_____ 5. The probability that a cassette tape is defective is 0.03. If two tapes are selected at random, what is the probability that both are good?

6. A card is selected at random from a standard deck of 52 cards. Find the odds in favor of the card being:
_____ a. A red face card (Jack, Queen, King).
_____ b. Not a red face card.

7. The probability of an event occurring is 3/5. Find the odds:
_____ a. In favor of the event occurring.
_____ b. Against the event occurring.

8. The odds in favor of an event occurring are 3 to 5. Find:
   a. The odds against the event occurring?
   b. The probability that the event will not occur?

9. A single fair die is rolled twice. If exactly one six turns up, you receive $5 and, if two sixes turn up, you receive $10; otherwise, you get nothing. What is a fair price to pay for playing this game?

10. The probabilities of being an "instant winner" of $25 or $50 in a certain lottery are $\frac{1}{500}$ and $\frac{1}{1000}$, respectively. What is the mathematical expectation of being an "instant winner" of $25 or $50?

11. A single fair die is rolled. Find the probability of obtaining:
    a. A number different from both 1 and 2.
    b. A number greater than or equal to 4.

12. A box contains 2 red balls, marked R_1, and R_2, and 3 white balls, marked W_1, W_2, and W_3.
    a. Two balls are drawn in succession without replacement. Find the number of elements in the sample space for this experiment. (We are interested in which balls are drawn and the order in which they are drawn.)
    b. Do Part (a) if the balls are drawn in succession with replacement.

13. A box contains 7 balls numbered from 1 to 7. If a ball is taken at random from the box, find the probability that it is:
    a. An even-numbered ball.
    b. Ball number 3.
    c. Not ball number 3.

14. Two cards are drawn at random and without replacement from a standard deck of 52 cards. Find the probability that:
    a. Both cards are red.
    b. Neither card is a Jack, Queen, or King.

15. A card is drawn at random from a standard deck of 52 cards and is then replaced. A second card is then drawn. Find the probability that:
    a. Both cards are red.
    b. Neither card is a King or a Queen

16. A fair coin is tossed 3 times. What is the probability of obtaining at least one head?

17. An urn contains 3 white, 3 black, and 2 red balls. Find the probability of obtaining in a single random draw:
    a. A white ball or a red ball.
    b. A ball that is <u>not</u> white.

18. A student estimates that the probability of his passing Chemistry or English is 0.8, the probability of his passing Chemistry is 0.7, but his probability of passing both is 0.5. What should be his estimate of the probability of his passing English?

19. Two cards are drawn in succession and without replacement from a standard deck of 52 cards. What is the probability that they are both red face cards (Jack, Queen, King)?

20. Two fair dice are rolled. Find the probability that the sum turning up is 10, given that the first die turns up 6.

TEST B   CHAPTER 10,   PROBABILITY

1. A fair die was rolled twice and an even number turned up the first time. What is the probability that the sum of the numbers turning up was 9?
   a.   0   b.   1/9   c.   1/6   d.   1/18   e.   1/10

2. A certain prescription drug produces side effects in 3% of the patients. Three patients who have taken this drug are selected at random. The probability that all three of these patients had side effects is
   a.   0.09   b.   0.009   c.   0.0009   d.   0.000027
   e.   0.00000027

3. In Problem 2, the probability that none of the three patients had side effects is
   a.   0.97   b. $(0.97)^2$ c. $(0.97)^3$ d. $(0.97)^4$
   e.   None of these

4. Susie has to take a Math course and an English course, both of which are available at 9 a.m., 10 a.m., and 11 a.m. If Susie picks a schedule at random, the probability that she will have Math at 9 a.m. and English at 11 a.m. is
   a.   1/6   b.   1/3   c.   1/4   d.   3/4
   e.   2/3

5. The probability that a cassette tape is defective is 0.03. If two tapes are selected at random, what is the probability that both are good?
   a.   0.9994   b.   0.9609   c.   0.9409   d. $(0.03)^2$
   e.   None of these

6. A card is selected at random from a standard deck of 52 cards. The odds in favor of the card being a red face card (Jack, Queen, King) are
   a.   23 to 3   b.   3 to 23   c.   3 to 26   d.   6 to 52
   e.   None of these

7. The probability that a certain event will occur is 3/5. The odds that the event will not occur are
   a.   3 to 5   b.   5 to 3   c.   2 to 5   d.   2 to 3
   e.   None of these

8. The odds in favor of an event occurring are 3 to 5. The probability that the event will <u>not</u> occur is
   a. 3/5   b. 3/8   c. 5/3   d. 2/5
   e. 5/8

9. A single fair die is rolled twice. If exactly one 6 comes up, you receive $5, and if two 6's come up, you receive $10; otherwise, you get nothing. A fair price to pay for playing this game is
   a. $1.67   b. $4   c. $5   d. $6
   e. $7.50

10. The probabilities of being an "instant winner" of $25 or $50 in a certain lottery are $\frac{1}{500}$ and $\frac{1}{1000}$, respectively. The mathematical expectation of being an "instant winner" of $25 or $50 is
    a. $1   b. $3.50   c. 20 cents   d. 10 cents
    e. None of these

11. A single fair die is rolled. The probability of obtaining a number different from both 1 and 2 is
    a. 1/6   b. 1/3   c. 2/3   d. 5/6
    e. None of these

12. A single fair die is rolled. The probability of obtaining a number greater than or equal to 4 is
    a. 1/3   b. 1/2   c. 2/3   d. 5/6
    e. None of these

13. A box contains two red balls marked R_1, and R_2, and three white balls marked W_1, W_2 and W_3. Two balls are drawn in succession and without replacement. Suppose that we are interested in which balls and in what order they are drawn. The number of elements in the sample space for this experiment is
    a. 5   b. 10   c. 15   d. 20
    e. None of these

14. Two cards are drawn at random and without replacement from a standard deck of 52 cards. The probability that both cards are red is
    a. 13/51   b. 26/51   c. 25/102   d. 1/4
    e. None of these

15. If two cards are drawn as in Problem 4, the probability that neither card is a King, Queen or Jack is
    a. 10/17   b. 10/13   c. 1/40   d. $\dfrac{C(12,2)}{C(52,2)}$
    e. None of these

16. A fair coin is tossed 3 times. The probability of getting at least one head is
    a. 1/2   b. 3/4   c. 7/8   d. 15/16
    e. None of these

17. An urn contains three white balls, three black balls, and two red balls. The probability of obtaining a white or a red ball in a single draw is
    a. 5/8   b. 3/8   c. 1/2   d. 1
    e. None of these

18. For the urn of Problem 17, the probability of obtaining a ball that is <u>not</u> white is
    a. 5/8   b. 3/8   c. 1/2   d. 1
    e. None of these

19. Tommy estimates that the probability of his passing Chemistry or English is 0.8, the probability of his passing Chemistry is 0.7, but his probability of passing both is 0.5. What should be his estimate of the probability of his passing English?
    a. 0.3   b. 0.4   c. 0.5   d. 0.6
    e. 0.7

20. Two cards are drawn in succession and without replacement from a standard deck of 52 cards. The probability that they are both red face cards (Jack, Queen, King) is
    a. $\dfrac{3}{26}$ b. $\dfrac{6}{13}$ c. $\dfrac{12\times11}{52\times51}$ d. $\left(\dfrac{3}{13}\right)^2$
    e. None of these

TEST A   CHAPTER 11, STATISTICS

1. A college dean wants to find out which courses students enjoy. The dean decides to conduct a survey of a sample of 25 students from the Physics Department. Do these 25 students correspond to a simple random sample of the student body? Explain your answer.

THE FOLLOWING SCORES WERE MADE ON A SCHOLASTIC APTITUDE TEST BY A GROUP OF 25 HIGH SCHOOL SENIORS. THIS SET OF SCORES IS TO BE USED IN PROBLEMS (1-3).

$$
\begin{array}{ccccc}
87 & 66 & 87 & 81 & 96 \\
65 & 90 & 85 & 86 & 92 \\
93 & 79 & 94 & 74 & 86 \\
97 & 64 & 93 & 75 & 88 \\
77 & 85 & 63 & 72 & 73 \\
\end{array}
$$

_____ 2. Group the scores into intervals of $60 \leq s < 65$, $65 \leq s < 70$, and so on. Then make a frequency distribution with this grouping.

_____ 3a. Make a histogram for the frequency distribution in Problem 1.

_____ 3b. Make a frequency polygon for the preceding distribution.

4. During a certain week in the winter, the following minimum temperatures were recorded in an eastern city: 20, 28, 24, 28, 31, 39, 40 (all in degrees F).

_____ a. Find the mean of these temperatures.
_____ b. Find the mode.
_____ c. Find the median low temperature for the week.

_____ 5. a. Find the range of the temperatures in Problem 4.

_____ b. Find the standard deviation of these temperatures.

6. A fair coin is tossed 100 times and the number of heads is recorded. If this experiment is repeated many times, the number of heads will form an approximately normal distribution with a mean of 50 and a standard deviation of 5.
   a. Within what limits should we expect the number of heads in 100 tosses to lie?
   b. What is the probability that heads will occur fewer than 45 times in 100 tosses?

7. A normal distribution consists of 1000 scores with a mean of 100 and a standard deviation of 10.
   a. About how many of the scores are above 120?
   b. About how many of the scores are below 90?
   c. About how many of the scores are between 80 and 90?

8. A testing program shows that the breaking points of fishing lines made from a certain plastic fiber are normally distributed with a mean of 12 lb and a standard deviation of 1.2 lb. Find the probability that one of these lines selected at random has a breaking point of
   a. more than 12 lb.
   b. less than 9.6 lb.

9. The scores on a multiple-choice test taken by 2000 students were normally distributed with a mean of 65 and a standard deviation of 5. Find the z-score corresponding to the test score
   a. 70
   b. 73

_____10. Lizzie scored 87 in an English test and 85 in a history test. The mean score in the English test was 83 with a standard deviation of 8, and the mean score in the history test was 82 with a standard deviation of 6. If the scores were normally distributed, which of Lizzie's scores was the better?

_____11. With the data given in Problem 9, find the probability that a randomly selected student will have a score between 65 and 70. (Use Table II in the back of the book.)

_____12. Here are the yield rates of 5 of the 30 Dow Jones industrial stocks during the second week of August, 1992 Make a vertical bar graph of these yield rates.

STOCK	YIELD RATE
AT & T	3.0%
IBM	4.9%
GMotr	4.3%
Merck	1.9%
Texaco	5.0%

_____13. Here are the prices of IBM common stock at the end of each week for a six week period in 1992. Make a line graph of these data.

WEEK	PRICE
1	87.25
2	86
3	88
4	83.5
5	82.25
6	78.5

_____14. A restaurant manager asked their patrons what features they liked best. Here is the listing. Make a horizontal bar graph of these data.

Self-service bar	65%
Varied menu	45%
All-you-can-eat specials	40%
Sandwich bar	30%

_____ 15. A typical family budget is as follows:
Monthly Family Budget
Savings $ 250
Housing 450
Clothing 250
Food 750
Other 300
$2000
Make a circle graph for this budget

_____ 16. A bar graph with bars of equal width shows the 1992 sales and the predicted 1995 sales of the ABC Company. If the bar for the 1992 sales is 1.5 in. long and for the projected 1995 sales is 1.8 in. long, what is the percent increase projected for the 1995 sales over the 1992 sales?

_____ 17. The testing department of Circle Tire Company checks a random sample of 200 of a certain tire that the company makes and finds a defective tread on 4 of these tires. In a batch of 10,000 of these tires, how many are expected to have defective treads?

_____ 18. In a large county, 30,000 high school students took a reading comprehension test, and 2400 of these students got a rating of "excellent." In a random sample of 100 of these students, how many should be expected to have gotten an excellent rating on this test?

_____ 19. By graphing the five points in the following table, and drawing the best line you can "between" these points, estimate the value of y for x = 9.

x	1	2	3	4	5
y	5.0	7.0	9.5	10.5	12.0

20. What kind of correlation would you expect for the indicated ordered pairs?

_____ a. (person's weight, person's salary)
_____ b. (speed of your driving, your gas mileage)
_____ c. (number of hr of running practice, number of sec in which the runner can do the 100 m dash)
_____ d. (the number of hours you practice doing problems on the computer, the number of problems you can do in one hour)

TEST B   CHAPTER 11,   STATISTICS

During a certain week in the winter, the following minimum temperatures were recorded in an eastern city: 20, 28, 24, 28, 31, 39, 40 (all in degrees F). These data are to be used in Problems (1-5).

1. The mean of these temperatures is
   a. 30   b. 31   c. 32   d. 33
   e. None of these

2. The mode of these temperatures is
   a. 30   b. 40   c. 39   d. 28
   e. There is none.

3. The median of these temperatures is
   a. 30   b. 28   c. 26   d. 29
   e. None of these

4. The range of these temperatures is
   a. 5 degrees   b. 10 degrees   c. 15 degrees
   d. 20 degrees   e. 25 degrees

5. The standard deviation of these temperatures is
   a. $\sqrt{12}$ b. $\sqrt{24}$ c. $\sqrt{48}$ d. $\sqrt{163/3}$
   e. None of these

6. A fair coin is tossed 100 times and the number of heads is recorded. If this experiment is repeated many times, the number of heads will form an approximately normal distribution with a mean of 50 and a standard deviation of 5. Within what limits should the number of heads in 100 tosses be expected to lie?
   a. 45 and 55   b. 40 and 60   c. 35 and 65   d. 30 and 70
   e. None of these

7. For the coin tossing experiment of Problem 6, what is the probability that heads will occur fewer than 45 times in 100 tosses?
   a. 0.025   b. 0.135   c. 0.16   d. 0.5
   e. None of these

THE FOLLOWING INFORMATION IS TO BE USED IN PROBLEMS (8-10): A CERTAIN NORMAL DISTRIBUTION CONSISTS OF 1000 SCORES WITH A MEAN OF 100 AND A STANDARD DEVIATION OF 10.

8. About how many of the scores are above 120?
   a. 80   b. 25   c. 160   d. 135
   e. None of these

9. About how many of the scores are below 90?
   a. 80   b. 25   c. 160   d. 135
   e. None of these

10. About how many of the scores are between 80 and 90?
    a. 80   b. 125   c. 160   d. 135
    e. None of these

11. A testing program shows that the breaking points of fishing lines made from a certain plastic fiber are normally distributed with a mean of 12 lb and a standard deviation of 1.2 lb. The probability that one of these lines selected at random has a breaking point of more than 12 lb is:
    a. 0.1   b. 0.2   c. 0.3   d. 0.4
    e. 0.5

12. For the fishing lines of Problem 11, the probability that one of these lines selected at random has a breaking point of less than 9.6 lb is:
    a. 0.5   b. 0.025   c. 0.135   d. 0.34
    e. None of these

13. The scores on a multiple-choice test taken by 2000 students were normally distributed with a mean of 65 and a standard deviation of 5. The z-score corresponding to a test score of 73 is:
    a. 1   b. 1.6   c. 2   d. 2.5
    e. None of these

14. The following line graph shows the price of IBM common stock for a six week period in 1992. Estimate the total percent decrease in the price for the six weeks.
    a. 20%   b. 25%   c. 15%   d. 5%   e. 10%

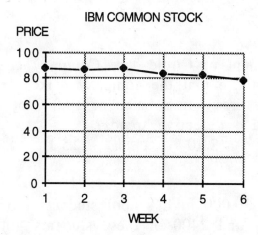

15. A restaurant manager asked their patrons what features they liked best and obtained the following listing:

    | Self-service bar | 65% |
    | Varied menu | 45% |
    | All-you-can-eat specials | 40% |
    | Sandwich bar | 30% |

    Suppose a bar graph with bars of equal width is to be made for this listing. If a bar 4 in. long represents 100%, what would be the length of the bar for the "All-you-can-eat specials" ?
    a.   1.2 in.   b.   1.4 in.   c.   1.6 in.   d.   1.8 in.
    e.   None of these

16. A typical family budget is as follows:

    Monthly Family Budget
    | Savings | $ 250 |
    | Housing | 450 |
    | Clothing | 250 |
    | Food | 750 |
    | Other | 300 |
    | | $2000 |

    If a circle graph is drawn for this budget, how many degrees should be in the angle corresponding to Food?
    a.   75   b.   105   c.   135   d.   150
    e.   None of these

17. A bar graph with bars of equal width shows the 1992 sales and the predicted 1995 sales of the ABC Company. If the bar for the 1992 sales is 1.5 in. long and for the projected 1995 sales is 1.8 in. long, what is the percent increase projected for the 1995 sales over the 1992 sales?
   a.   30%   b.   35%   c.   25%   d.   20%
   e.   None of these

18. The testing department of Circle Tire Company checks a random sample of 200 of a certain tire that the company makes and finds a defective tread on 4 of these tires. In a batch of 10,000 of these tires, how many are expected to have defective treads?
   a.   150   b.   200   c.   250   d.   300
   e.   350

19. In a large county, 30,000 high school students took a reading comprehension test, and 2400 of these students got a rating of "excellent." In a random sample of 100 of these students, how many should be expected to have gotten an excellent rating on this test?
   a.   1   b.   3   c.   5   d.   6
   e.   8

20. Which of the following ordered pairs would be expected to show a positive correlation?

   i.   (Person's weight, Person's salary)
   ii.   (Speed of your driving, Your gasoline mileage)
   iii. (Number of hours of running practice, Number of seconds in which runner can do the 100 m dash)
   iv.   (Number of hours you practice doing problems on the computer, Number of problems you can do in 1 hour)

   a.   i and ii   b.   ii and iii   c.   iii and iv   d.   i and iv
   e.   iv only

TEST A   CHAPTER 12, YOUR MONEY AND YOUR MATH

1. The Easy Loan Company charges 28% simple interest (annual) for a 2-year, $600 loan. Find:
   a. The total interest on this loan.
   b. The interest for three months.
   c. The total amount to be paid to the loan company at the end of two years.

2. A state has a 6% sales tax. Find:
   a. The sales tax on a microwave oven priced at $320.
   b. The total cost of this oven.

3. In a sale, a store offers a 20% discount on a freezer chest that is normally priced at $380.
   a. How much is the discount?
   b. What is the sale price of the freezer?

4. Here is a portion of a compound interest table to use in this problem.

	Amount (in $) to which $1 will accumulate in n periods under compound interest				
n	2%	4%	6%	8%	10%
1	1.0200	1.0400	1.0600	1.0800	1.1000
2	1.0404	1.0609	1.1236	1.1664	1.2100
3	1.0612	1.1249	1.1910	1.2597	1.3310
4	1.0824	1.1699	1.2625	1.3605	1.4641
5	1.1041	1.2167	1.3382	1.4693	1.6105
6	1.1262	1.2653	1.4185	1.5869	1.7716
7	1.1487	1.3159	1.5036	1.7138	1.9487
8	1.1717	1.3686	1.5938	1.8509	2.1436

Find the accumulated amount and the interest earned for:
   a. $100 at 8% compounded semiannually for 4 years.
   b. $100 at 8% compounded quarterly for $1\frac{1}{2}$ years.

5. A credit card holder is obligated to pay his balance in full if it is less than $10. Otherwise, the minimum payment is $10 or 5% of the balance, whichever is more. Suppose that a customer received a statement listing the balance as $292.75.
a. Find the minimum payment due.
b. The finance charge is 1.5% per month. What will be the amount of this charge on the next monthly statement if the customer makes only the minimum payment?

6. Brenda Brown received a statement showing that she owed a balance of $210 to a department store where she had a revolving charge account. Brenda made a payment of $50 and charged an additional $35. If the store charges 1.5% on the unpaid balance, find:
a. The finance charge on the next monthly statement.
b. The new balance.

7. A car costing $8500 can be bought with $1500 down and 10% add-on interest to be paid in 48 equal installments.
a. What is the total interest charge?
b. What is the monthly payment?

Use the following table to solve problem 8.

True Annual Interest Rate for a 12-Payment Plan

	14%	$14\frac{1}{2}$%	15%	$15\frac{1}{2}$%	16%
Finance Charge per $100 of the amount financed	7.74	8.03	8.31	8.59	8.88

8. John Bishop borrows $300 and agrees to pay $27.20 per month for twelve months.
a. What is the APR for this transaction?
b. If John decided to pay off the balance of the loan after 5 months (with 7 payments remaining), use the Rule of 78 to find the amount of the interest refund.
c. Find the amount needed to pay off the loan.

9. The Nakos family wants to buy a $70,000 house.
a. If a bank is willing to loan them 75% of the price of the house, what would be the amount of the loan?
b. What would be the down payment for this house?
c. If they decide to obtain an FHA loan instead, what would be the minimum down payment? (Recall that FHA requires a down payment of 3% of the first $25,000 and 5% of the balance up to the maximum loan amount of $67,500.)
d. What would be the maximum FHA loan they could get?

10. Refer to Problem 9. Suppose the Nakos family contracted for a 15-year mortgage at 12% with the bank that loaned them 75% of the price of the house. What is the monthly payment for principal and interest? (Use the following table.)

| | Monthly Payment ($) for each $1000 Borrowed | | |
RATE	10 years	15 years	20 years
11%	13. 78	11.37	10.32
12%	14.35	12.00	11.01
13%	14.93	12.65	11.72

TEST B   CHAPTER 12,   YOUR MONEY AND YOUR MATH

1. A loan company charges 28% simple annual interest for a 2-year, $600 loan. The total interest on this loan is
   a. $168   b. $336   c. $936
   d. $1008   e.   None of these

2. The total amount to be paid on a 2-year, $600 loan at 28% simple annual interest is
   a. $168   b. $336   c. $936
   d. $1008   e.   None of these

3. What is the sales tax on a microwave oven priced at $320 if the sales tax rate is 6%?
   a. $18   b. $192   c. $19.20
   d. $339.20   e.   None of these

4. The total cost of the microwave oven in Problem 3 is
   a. $406   b. $300.80   c. $360
   d. $339.20   e.   None of these

5. A store offers a 20% discount on a freezer that is normally priced at $360. How much is the discount on this freezer?
   a. $72   b. $80   c. $304
   d. $280   e.   None of these

6. The sale price of the freezer in Problem 5 is:
   a. $72   b. $80   c. $288
   d. $280   e.   None of these

7. Here is a portion of a compound interest table to use in this problem.

   Amount (in $) to which $1 will accumulate in n periods under compound interest

n	2%	4%	6%	8%	10%
1	1.0200	1.0400	1.0600	1.0800	1.1000
2	1.0404	1.0609	1.1236	1.1664	1.2100
3	1.0612	1.1249	1.1910	1.2597	1.3310
4	1.0824	1.1699	1.2625	1.3605	1.4641
5	1.1041	1.2167	1.3382	1.4693	1.6105
6	1.1262	1.2653	1.4185	1.5869	1.7716
7	1.1487	1.3159	1.5036	1.7138	1.9487
8	1.1717	1.3686	1.5938	1.8509	2.1436

If $100 is invested at 8% compounded semiannually for 4 years, the accumulated amount is:
- a. $116.99
- b. $136.86
- c. $117.17
- d. $126.53
- e. None of these

8. A customer received a statement with a $292.75 balance. She was obligated to pay her balance in full if it was less than $10. Otherwise, the minimum payment was $10 or 5% of the balance, whichever is more. The minimum payment due was
   - a. $10
   - b. $292.75
   - c. $14.64
   - d. $282.75
   - e. None of these

9. If, in Problem 8, the finance charge is 1.5% per month on the unpaid balance, what will this charge be on the next statement if the customer makes only the minimum payment?
   - a. $4.17
   - b. $278
   - c. $290
   - d. $4.00
   - e. None of these

10. Brenda Brown received a statement with a $230 balance. She made a $50 payment and charged an additional $35. If the finance charge is 1.5% per month on the unpaid balance, the new balance on the next monthly statement will be
    - a. $2.70
    - b. $232.70
    - c. $195
    - d. $204.17
    - e. None of these

11. A car priced at $8500 can be bought with $1500 down and 10% add-on interest to be paid in 48 equal monthly installments. The total interest charge would be
    - a. $850
    - b. $1700
    - c. $700
    - d. $70
    - e. $2800

12. The monthly payment for the car of Problem 11 would be
    - a. $160.42
    - b. $215.83
    - c. $204.17
    - d. $194.79
    - e. None of these

13. John Bishop borrowed $300 and agreed to pay $27.20 per month for 12 months. After 5 months, John decided to pay off the balance (7 payments remaining). According to the Rule of 78, his interest refund would be
    - a. $15.40
    - b. $190.40
    - c. $65.28
    - d. $9.70
    - e. $9.48

14. In Problem 13, the amount needed to pay off the loan after 5 months is
    a. $190.40   b. $180.92   c. $175
    d. $126.45   e. None of these

15. The Turner family wants to buy a $70,000 house. If the bank will loan them 75% of the price of the house, what will the down payment be?
    a. $52,500   b. $4500   c. $17,500
    d. $15,000   e. None of these

16. What would be the down payment on an FHA loan on the house of Problem 15? (Recall that FHA requires a 3% down payment on the first $25,000 and 5% of the balance up to a $67,500 maximum.)
    a. $2000   b. $3000   c. $2500
    d. $7500   e. None of these

17. The maximum FHA loan that could be obtained to buy the house in Problems 15-16 is
    a. $60,000   b. $70,000   c. $67,000
    d. $62,500   e. None of these

The following table is to be used in Problems 18-20.

| **Monthly Payment ($) for Each $1000 Borrowed** | | | |
Rate	10 years	15 years	20 years
11%	13.78	11.37	10.32
12%	14.35	12.00	11.01
13%	14.93	12.65	11.72

18. A family contracted for a 15-year mortgage at 12% on a 75% loan to buy a $70,000 house. Their monthly payment for principal and interest was
    a. $720   b. $540   c. $495.45
    d. $630   e. None of these

19. The total interest charge on the house of Problem 18 amounts to
    a. $6300   b. $8400   c. $60,900
    d. $52,200   e. None of these

20. Suppose the term of the mortgage of Problem 18 was extended to 20 years. What would the difference in the monthly payment be?
    a. $51.97   b. $0.93   c. $93
    d. $11.88   e. None of these

PART III

ANSWERS TO ALL ADDITIONAL PRACTICE TESTS

ANSWERS TO STUDENT'S PRACTICE TESTS

CHAPTER 1, TEST A

1. (a) 30   (b) 25   2. (a) 4   (b) 38

3. A - B or A ∩ B'   4. 8

5. (a) 5   (b) 28   (c) 7

6.
$$\begin{array}{ccccc} 2 & 4 & 8 & \ldots & 2^n \ldots \\ \updownarrow & \updownarrow & \updownarrow & & \updownarrow \\ 1 & 2 & 3 & \ldots & n \ldots \end{array}$$
This one-to-one correspondence shows that the two sets have the same cardinal number.

7. 77, 100, 126.   Add **5** to the first term 2, to get 7
Add **8** to the second term, to get 15
Add **11** to the third term, to get 26
and so on.

8. {2, 4, 6, 8}

9. (a) The set of positive even integers, starting with 4.
   $\{x \mid x = 2n,\ n = 2, 3, 4, \ldots\}$
   (b) The set of positive odd integers, starting with 3.
   $\{x \mid x = 2n + 1,\ n = 1, 2, 3, \ldots\}$

10. ∅, {1}, {2}, {3}, {1, 2}, {1, 3}, {2, 3}, {1, 2, 3}

11. (a) B - A   (b) A ∩ B   (c) A ∪ B

12. (a) {a, d}   (b) {a, b, c, e}   (c) {a, b, d}   (d) {c, e}   (e) {a}

13. (a) {b, e}   (b) {c}

14.   15.

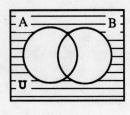
(A ∪ B)'

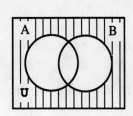
A' ∩ B'

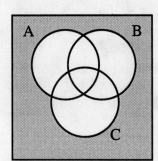

16. (a) Region 5   (b) Regions 3, 4, 5, 6, 7
    (c) Regions 2, 6   (d) Regions 3, 4, 5, 6, 7

CHAPTER 1, TEST B

1. (a)   2. (b)   3. (d)   4. (c)   5. (a)   6. (d)

7. (c)   8. (b)   9. (c)   10. (e)   11. (e)   12. (b)

13. (d)   14. (c)   15. (c)   16. (d)   17. (d)   18. (c)

19. (a)   20. (b)   21. (b)   22. (d)

CHAPTER 2, TEST A

1. TTTT

2. Both have the truth values TFTFTTTF for the order p, q, r in the truth tables.

3. The statement is true if the day is warm, or if it is summer, or if the day is warm and it is summer.

4. False, because the premise is true and the conclusion is false.

5. If I do not stay indoors, then the weather is not rainy.

6. If I stay indoors, then the weather is rainy.

7. If the weather is not rainy, then I do not stay indoors.

8. $s \Rightarrow r$ 9. $(p \rightarrow q) \leftrightarrow (\sim p \vee q)$ 10. $q \leftrightarrow \sim q$

11. $s \rightarrow t$ 12. $p \wedge \sim s$ 13. It is not the case that the test is hard or I shall pass it.

14. The test is not hard or I shall pass it.

15. I shall not drive and I shall not walk.

16. No science fiction stories are interesting.

17. Some ripe peaches do not taste good.

18. FFTF 19. FTTT 20. FTTF

21. Valid 22. (c) is not valid.

23. Patsy reads comics.

24. $b \to {\sim}g$
    $\underline{g}$
    $\therefore {\sim}b$

25.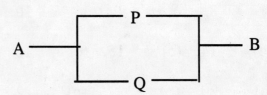

CHAPTER 2, TEST B

1. (c) 2. (b) 3. (d) 4. (a) 5. (c) 6. (a)

7. (b) 8. (b) 9. (c) 10. (e) 11. (a) 12. (b)

13. (c) 14. (a) 15. (c) 16. (d) 17. (d) 18. (b)

19. (c) 20. (a) 21. (e) 22. (a) 23. (c) 24. (d)

25. (a)

CHAPTER 3, TEST A

1. 5290 2. (a) 13 (b) 186 (c) 89 (d) 489

3. (a) 111_5 (b) 11111_2 (c) $1F_{16}$ (d) 37_8

4. (a) 10100_2 (b) 110_2 5. (a) 110111_2 (b) 111_2

6. (a) ∩∩∩III
       ∩∩∩
   (b) 9999∩∩∩I
       9999∩∩

7. (a) ▼ ▼▼▼ (b) < ▼▼▼▼ < < ▼

8. (a) XLVI (b) $\overline{\text{XVII}}$ 9. 74

10. 34 11. (a) 48 (b) 67,000

12. (a)
```
      2 2    2 4
     \11     4 8
     \ 5     9 6
       2    1 9 2
     \ 1    3 8 4
            ─────
             528
```
(b)
```
       1       2 4
       2      ┌──┐
              │48│
              └──┘
       4      ┌──┐
              │96│
              └──┘
       8     1 9 2
              ┌────┐
      16      │3 8 4│
              └────┘
      ───────────
      22      528
```

13. (a) x^{a+b} (b) x^{m-n} (c) 1   14. (a) 2^{13} (b) 2^5

15. (a) $3 \times 10^3 + 5 \times 10^2 + 8$ (b) $8 \times 10^2 + 6 \times 10 + 2$

CHAPTER 3, TEST B

1. (b)   2. (a)   3. (a)   4. (c)   5. (e)

6. (c)   7. (d)   8. (b)   9. (b)   10. (a)

11. (d)   12. (b)   13. (e)   14. (c)   15. (b)

CHAPTER 4, TEST A

1. $\frac{21}{28}$

2. (a) $\frac{7}{5}$ (b) $-\frac{5}{4}$ (c) $\frac{16}{55}$ (d) $-\frac{1}{14}$

3. (a) $-\frac{22}{81}$ (b) $\frac{8}{11}$ 4. (a) 0.625   (b) 0.8333 . . . or $0.8\overline{3}$

5. (a) $\frac{7}{11}$ (b) $\frac{293}{90}$

6. (a) $5 \times 10 + 6 \times 10^{-2} + 1 \times 10^{-4}$ (b) 4500.205

7. 5.4×10^{-3} 8. $\sqrt{42}$

9. (a) 0.24 (Other answers are possible.)
   (b) 0.241121231234 . . . (Other answers are possible.)

10. 12.6 in.

11. (a) Ordinal (b) Cardinal (c) Identification

12. (a) The associative and commutative properties of addition.
    (b) The distributive property of multiplication over addition.
    (c) The associative and commutative properties of multiplication.

13. $286 = 2 \times 11 \times 13$
14. 41, 43 and 47

15. Composite, $123 = 3 \times 41$

16. (a) 718 and 1650
    (b) 6345, 849, and 1650
    (c) 6345 and 1650

17. GCF = 45; $\dfrac{3}{5}$
18. LCM = 1350; $\dfrac{11}{1350}$

19. $\dfrac{3}{20}$
20. 188,000

21. (a) 10.05 (b) 3.59 (c) 2.184 (d) 3.1

22. 44.7 cm
23. 4220 ft^2
24. 3.66 in.

25. (a) 0.23 (b) 0.0575 (c) 0.0034

26. (a) 54% (b) 243% (c) 60% (d) 55.6%

27. 40.82%
28. 250

29. (a) $13\sqrt{2}$ (b) $\sqrt{73}$ is in simplest form. (c) $\dfrac{\sqrt{3}}{3}$ (d) $\dfrac{8\sqrt{2}}{7}$

30. (a) $5\sqrt{3}$ (b) 3 (c) $\sqrt{6}$ (d) $2\sqrt{3} + 3\sqrt{2}$

31. $3\sqrt{13}$ in.

32. For $\sqrt{8}$, check: **Irrational, Real**
    For -0.19, check: **Rational, Real**
    For $\sqrt{25}$, check: **Natural, Integers, Rational, Real**
    For 0.666 . . ., check: **Rational, Real**
    For $-2\frac{1}{2}$, check: **Rational, Real**

33. (a)   A geometric sequence   (b)   An arithmetic sequence

34. (a) 6   (b) 3   (c) $S_{10} = 195$ (d) $S_n = \dfrac{3n(n+3)}{2}$

35. (a) 4   (b) $-\dfrac{1}{2}$ (c) $\dfrac{11}{4}$ (d) $\dfrac{2^n - (-1)^n}{3(2^{n-3})}$

36. (a) $\dfrac{8}{9}$ (b) $\dfrac{37}{33}$ (c) $\dfrac{25}{9}$

CHAPTER 4, TEST B

1. (d)   2. (a)   3. (e)   4. (e)   5. (d)

6. (c)   7. (c)   8. (d)   9. (b)   10. (b)

11. (d)   12. (e)   13. (c)   14. (b)   15. (c)

16. (b)   17. (c)   18. (b)   19. (b)   20. (e)

21. (b)   22. (b)   23. (b)   24. (e)   25. (d)

26. (c)   27. (e)   28. (d)   29. (c)   30. (c)

31. (c)   32. (b)   33. (d)   34. (c)   35. (b)

36. (a)   37. (c)   38. (b)

CHAPTER 5, TEST A

1. $(x - 2)(x - 3)$

2. $(x + 1)(x - 5)$

3. (a) $x = -2$ or $x = 3$ (b) $x = -4$ or $x = 0$ or $x = 3$

4. $x = -6$ or $x = -2$

5. $x = -3$ or $x = 5$

6. $x = -\dfrac{5}{3}$ or $x = 1$

7. $x = \dfrac{1 \pm \sqrt{6}}{2}$

8. $x = \pm \dfrac{9}{2}$

9. $x = \pm \dfrac{7\sqrt{6}}{6}$

10. 6 cm, 8 cm, 10 cm

11. 325

12. $17\dfrac{3}{5}$ in.

13. $4.50

14. 8 in.^{3}

15. (a) {3}   (b) {-5}

16. (a) {3, 4, 5, . . .}   (b) {4, 5, 6, . . .}

17. (a) $\dfrac{1}{3}$ (b) {x | x ≤ 2}

18. (a) [number line: open circle at -2, shaded right]
    (b) [number line: open circle at 2, shaded left]

19. [number line: closed circle at 4, shaded left]

20. [number line: closed circles at -3 and 3, shaded between]

21. [number line: closed circles at -1 and 1, shaded between]

22. [number line: open circle at -1, shaded right]

23. [number line: closed circles at -4 and 4, shaded between]

24. [number line: closed circles at -3 and 3, shaded outside]

25. [number line: open circles at 0 and 2, shaded between]

26. [number line: open circles at -2 and 2, shaded between]

27. [number line: open circles at -3 and 3, shaded outside]

28. [number line: open circles at 0 and 4, shaded outside]

CHAPTER 5, TEST B

1. (c)   2. (b)   3. (d)   4. (c)   5. (a)

6. (d)   7. (a)   8. (c)   9. (c)   10. (b)

11. (d)   12. (c)   13. (d)   14. (b)   15. (d)

16. (d)   17. (c)   18. (b)   19. (c)   20. (d)

21. (c)   22. (d)   23. (a)   24. (d)   25. (c)

26. (e)   27 (a)   28. (a)

CHAPTER 6, TEST A

1. Domain: {1, 2, 4}
   Range: {-2, -1, 1}

2. Domain: {x | x is a real number}
   Range: {y | y is a real number}

3. Domain: {1, 2, 3, 4, 5,}
   Range: {1, 2, 3, 4, 5}

4. (b) and (c) only

5. (a) 0 (b) 1 (c) -8

6. 225

7. 8. 9.

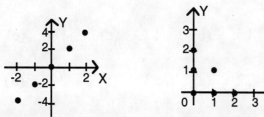
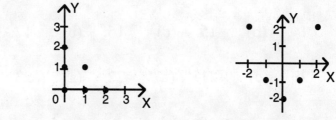

10. 11. 12. (a) $2\sqrt{5}$ b. 10

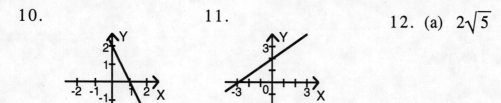

13. slope = 3

14. $3x - y = -7$

15. (a) $y = -3x - 2$ (b) $y = -\frac{3}{4}x + 2$; Slope: $-\frac{3}{4}$, y-intercept: 2

16. (a) Lines are parallel   17. $2x + y = 10$

    (b) Intersect at $(\frac{1}{2}, 3)$ 18. $(2, 4)$

19. 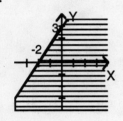   20. 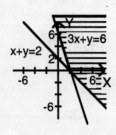   21.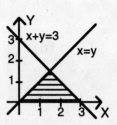

22. No solution. The lines are parallel.

23. 14   24. -4 25. Operate A 30 hr, B 10 hr

CHAPTER 6, TEST B

1. (b)   2. (d)   3. (a)   4. (c)   5. (e)   6. (d)

7. (a)   8. (c)   9. (d)   10. (d)   11. (c)   12. (d)

13. (c)   14. (b)   15. (c)   16. (b)   17. (c)   18. (e)

19. (b)   20. (c)   21. (a)   22. (a)   23. (b)   24. (d)

25. (b)

CHAPTER 7, TEST A

1. 30 yd by 60 yd   2. $2\sqrt{13}\,\pi$ cm   3. $(13\pi - 24)$ cm^2

4. $5\frac{1}{4}$ in.   5. 24 ft   6. $(30 + 4.5\pi)$ ft^2

7. $(36 - 9\pi)$ in.^{2} 8. 76 ft^2 9. 4 in.

10. 8 cm   11. 15 ft³   12. 1 and 3

13. No. You can make only two cuts in the two-hole button and four cuts in the four-hole one without cutting them into two pieces.

14. The two-hole one is of genus 2; the four-hole one is of genus 4.

15. 0.500   16. $1\frac{1}{4}$ sq. in.   17. ●—● YX

18. 12   19. 150   20. 135°

21. 94°   22. 120°   23. N only

24. 4.5 in.   25. 140°

CHAPTER 7, TEST B

1. (a)   2. (d)   3. (c)   4. (c)   5. (b)   6. (c)

7. (c)   8. (d)   9. (b)   10. (d) 11. (b) 12. (c)

13. (d) 14. (c) 15. (d) 16. (d) 17. (c) 18. (d)

19. (c) 20. (b) 21. (c) 22. (e) 23. (e) 24. (d)

25. (d)

CHAPTER 8, TEST A

1. 3   2. 6   3. 4   4. 6   5. 10

6. 2   7. 0   8. 3   9. 3

10. (a) n = 2 + 3k, k any integer   (b) n = 2 + 5k, k any integer

11. (a) n = 3 + 7k, k any integer   (b) n = 1 + 3k, k any integer

12. Yes. All the table entries are elements of S.

13. Yes. The table is symmetric to the diagonal from upper left to lower right.

14. @   15. %   16. @   17. #

18. Yes. a L (b S c) = a L c and (a L b) S (a L c) = a L c. Thus, a L (b S c) = (a L b) S (a L c).

19. Yes. a S (b L c) = b L c and (a S b) L (a S c) = b L c. Thus, a S (b L c) = (a S b) L (a S c).

20. (a) (% * @) # % = @ # % = %
    (b) (% # @) * (% * @) = % * @ = @

21. Yes.   22. No. % has no inverse.   23. Yes.

24. (ii) and (iii)   25. Play Row 1. Value = 2

26. Play Row 1 three-fourths of the time and Row 2 one-fourth of the time. Payoff value is $1\frac{1}{4}$.

27. $x = -1$, $y = 4$ 28. $AB = \begin{bmatrix} -7 & 0 \\ -2 & 1 \end{bmatrix}$ $BA = \begin{bmatrix} 1 & 0 \\ 6 & -7 \end{bmatrix}$

29. AD = [120   370   210] So 120 frames, 370 wheels, and 210 chains are needed.

30. $DK = \begin{bmatrix} 7 \\ 11 \\ 10 \end{bmatrix}$ so the total cost of parts for the Type I is $7, for the Type II is $11 and for the Type III is $10.

31. $\begin{bmatrix} 5 & 7 & -5 \\ -4 & 2 & 3 \\ -7 & -3 & 8 \end{bmatrix}$ 32. $\begin{bmatrix} 2 & 2 & -2 \\ -2 & 0 & 1 \\ -3 & -2 & 3 \end{bmatrix}$ 33. $\begin{bmatrix} 6 & 8 & -8 \\ -7 & -6 & 7 \\ -11 & -12 & 13 \end{bmatrix}$

34. $AB = \begin{bmatrix} 3 & 4 & 2 \\ 6 & 7 & -1 \\ 7 & 7 & -3 \end{bmatrix}$ $BA = \begin{bmatrix} -2 & 0 & 4 \\ 2 & 10 & 0 \\ 3 & 11 & -1 \end{bmatrix}$ 35. x = 2, y = 1, z = -1

36. 15 fives, 10 tens and 15 twenties.   37. x = -5, y = 2, z = 3

38. Let z = 3k, where k is any real number. Then solve the first two equations to get y = 3 - k and x = 2 - 7k. Thus, the solution is x = 2 - 7k, y = 3 - k, z = 3k, k any real number.

39. The system has no solution.

CHAPTER 8, TEST B

1. (c)   2. (e)   3. (b)   4. (d)   5. (a)   6. (b)

7. (c)   8. (d)   9. (d)   10. (a)   11. (e)   12. (d)

13. (e)   14. (a)   15. (d)   16. (c)   17. (c)   18. (b)

19. (c)   20. (e)   21. (b)   22. (e)   23. (c)   24. (d)

25. (c)   26. (c)   27. (b)   28. (d)   29. (b)   30. (b)

31. (c)   32. (a)   33. (b)   34. (b)   35. (c)   36. (c)

37. (e)

CHAPTER 9, TEST A

1. (a) 10   (b) 6   2. (a) 1   (b) 1

3. 20   4. 1326   5. 286   6. 7

7. (a) 35   (b) 20   8. 720   9. 30,240

10. 60   11. $\dfrac{2242!}{917!\ 787!\ 538!}$

12.

```
    ┌─ C ┬─ G    (Chicken and ginger ale)
    │    ├─ M    (Chicken and milk)
    │    └─ R    (Chicken and root beer)
    │
────┼─ H ┬─ G    (Ham and ginger ale)
    │    ├─ M    (Ham and milk)
    │    └─ R    (Ham and root beer)
    │
    └─ T ┬─ G    (Tuna and ginger ale)
         ├─ M    (Tuna and milk)
         └─ R    (Tuna and root beer)
```

13. 3 6

14. (a) 36 (b) 4

15. 26

16. 18

17. (a) 362,880   (b) 72   18. (a) 720   (b) 126

19. (a) 24   (b) 120   20. (a) 56   (b) 360

21. 720   22. 4   23. 10   24. 21

25. 5

CHAPTER 9, TEST B

1. (a)   2. (d)   3. (c)   4. (d)   5. (a)   6. (c)

7. (c)   8. (d)   9. (c)   10. (a)   11. (d)   12. (c)

13. (d)   14. (d)   15. (a)   16. (c)   17. (b)   18. (d)

19. (b)   20. (d)   21. (c)   22. (a)   23. (d)   24. (b)

25. (a)

CHAPTER 10, TEST A

1. 1/18

2. (a) 1/12 (b) 1/2
    (c) P(A) = 1/12, P(B) = 1/2, P(A ∩ B) = 1/36. P(A ∩ B) ≠ P(A) × P(B), so the events are not independent.

3. (a) $(0.03)^3 = 0.000027$ (b) $(0.97)^3 \approx 0.91$
4. 1/6   5. $(0.97)^2 = 0.9409$

6. (a) 3 to 23 (b) 23 to 3   7. (a) 3 to 2 (b) 2 to 3

8. (a) 5 to 3 (b) 5/8   9. $1.67

10. 10 cents   11. (a) 2/3 (b) 1/2

12. (a) 20 (b) 25   13. (a) 3/7 (b) 1/7 (c) 6/7

14. (a) 25/102 (b) 10/17   15. (a) 1/4 (b) 121/169

16. 7/8   17. (a) 5/8 (b) 5/8

18. 0.6   19. $\dfrac{5}{442} \approx 0.011$ 20. 1/6

CHAPTER 10, TEST B

1. (d)   2. (d)   3. (c)   4. (a)   5. (c)   6. (b)

7. (d)   8. (e)   9. (a)   10. (d)   11. (c)   12. (b)

13. (d)   14. (c)   15. (a)   16. (c)   17. (a)   18. (a)

19. (d)   20. (e)

CHAPTER 11, TEST A

1. No. Only the students in the Physics Department have a chance of being selected.

2.
SCORE	TALLY	FREQ
$60 \leq S < 65$	II	2
$65 \leq S < 70$	II	2
$70 \leq S < 75$	III	3
$75 \leq S < 80$	III	3
$80 \leq S < 85$	I	1
$85 \leq S < 90$	IIII II	7
$90 \leq S < 95$	IIII	5
$95 \leq S < 100$	II	2

3a and 3b.

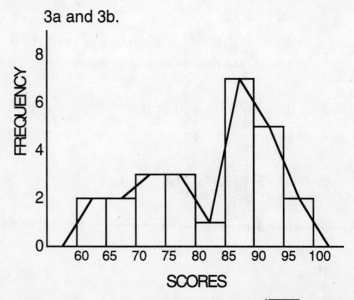

4. (a) $30^\circ F$ (b) $28^\circ F$ (c) $28^\circ F$ 5. (a) $20^\circ F$ (b) $\sqrt{\dfrac{163}{3}} \approx 7.37$

6. (a) 35 and 65   (b) 0.16   7. (a) 25   (b) 160   (c) 135

8. (a) 0.5   (b) 0.025   9. (a) 1   (b) 1.6

10. Neither. They correspond to the same z-score   11. 0.341

12.

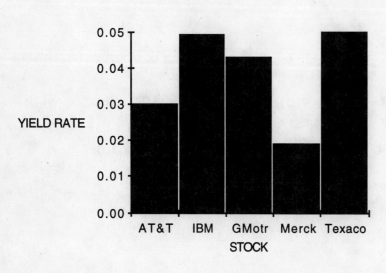

13.

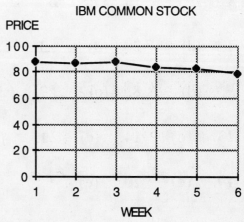

14.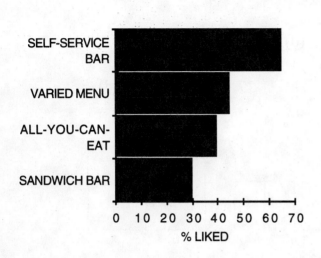

15.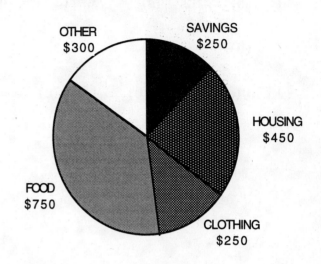

16. 20%   17. 200   18. 8   19. 19

20. (a) Zero   (b) Negative   (c) Negative   (d) Positive

CHAPTER 11, TEST B

1. (a)   2. (d)   3. (b)   4. (d)   5. (d)   6. (c)

7. (c)   8. (b)   9. (c)   10. (d)   11. (e)   12. (b)

13. (b)   14. (e)   15. (c)   16. (c)   17. (d)   18. (b)

19. (e)   20. (e)

CHAPTER 12, TEST A

1. (a) $336   (b) $42   (c) $936   2. (a) $19.20   (b) $339.20

3. (a) $76   (b) $304   4. (a) $136.86; $36.86
   (b) $112.62; $12.62

5. (a) $14.64   (b) $4.17   6. (a) $2.40   (b) $197.40

7. (a) $2800   (b) $204.17   8. (a) 16%   (b) $9.48   (c) $180.92

9. (a) $52,500   (b) $17,500   (c) $3000   (d) $67,000   10. $630

CHAPTER 12, TEST B

1. (b) 2. (c) 3. (c) 4. (d) 5. (a) 6. (c)

7. (b) 8. (c) 9. (a) 10. (d) 11. (e) 12. (c)

13. (e) 14. (b) 15. (c) 16. (b) 17. (c) 18. (d)

19. (c) 20. (a)